Deep Neural Network Design for Radar Applications

Related titles on radar:

Advances in Bistatic Radar Willis and Griffiths
Airborne Early Warning System Concepts, 3rd Edition Long
Bistatic Radar, 2nd Edition Willis
Design of Multi-Frequency CW Radars Jankiraman
Digital Techniques for Wideband Receivers, 2nd Edition Tsui
Electronic Warfare Pocket Guide Adamy
Foliage Penetration Radar: Detection and characterisation of objects under trees Davis
Fundamentals of Ground Radar for ATC Engineers and Technicians Bouwman
Fundamentals of Systems Engineering and Defense Systems Applications Jeffrey
Introduction to Electronic Warfare Modeling and Simulation Adamy
Introduction to Electronic Defense Systems Neri
Introduction to Sensors for Ranging and Imaging Brooker
Microwave Passive Direction Finding Lipsky
Microwave Receivers with Electronic Warfare Applications Tsui
Phased-Array Radar Design: Application of radar fundamentals Jeffrey
Pocket Radar Guide: Key facts, equations, and data Curry
Principles of Modern Radar, Volume 1: Basic principles Richards, Scheer and Holm
Principles of Modern Radar, Volume 2: Advanced techniques Melvin and Scheer
Principles of Modern Radar, Volume 3: Applications Scheer and Melvin
Principles of Waveform Diversity and Design Wicks *et al.*
Pulse Doppler Radar Alabaster
Radar Cross Section Measurements Knott
Radar Cross Section, 2nd Edition Knott *et al.*
Radar Design Principles: Signal processing and the environment, 2nd Edition Nathanson *et al.*
Radar Detection DiFranco and Rubin
Radar Essentials: A concise handbook for radar design and performance Curry
Radar Foundations for Imaging and Advanced Concepts Sullivan
Radar Principles for the Non-Specialist, 3rd Edition Toomay and Hannan
Test and Evaluation of Aircraft Avionics and Weapons Systems McShea
Understanding Radar Systems Kingsley and Quegan
Understanding Synthetic Aperture Radar Images Oliver and Quegan
Radar and Electronic Warfare Principles for the Non-Specialist, 4th Edition Hannen
Inverse Synthetic Aperture Radar Imaging: Principles, algorithms and applications Chen and Martorella
Stimson's Introduction to Airborne Radar, 3rd Edition Baker, Griffiths and Adamy
Test and Evaluation of Avionics and Weapon Systems, 2nd Edition McShea
Angle-of-Arrival Estimation Using Radar Interferometry: Methods and applications Holder
Biologically-Inspired Radar and Sonar: Lessons from nature Balleri, Griffiths and Baker
The Impact of Cognition on Radar Technology Farina, De Maio and Haykin
Novel Radar Techniques and Applications, Volume 1: Real aperture array radar, imaging radar, and passive and multistatic radar Klemm, Nickel, Gierull, Lombardo, Griffiths and Koch
Novel Radar Techniques and Applications, Volume 2: Waveform diversity and cognitive radar, and target tracking and data fusion Klemm, Nickel, Gierull, Lombardo, Griffiths and Koch
Radar and Communication Spectrum Sharing Blunt and Perrins
Systems Engineering for Ethical Autonomous Systems Gillespie
Shadowing Function From Randomly Rough Surfaces: Derivation and applications Bourlier and Li
Photo for Radar Networks and Electronic Warfare Systems Bogoni, Laghezza and Ghelfi
Multidimensional Radar Imaging Martorella
Radar Waveform Design Based on Optimization Theory Cui, De Maio, Farina and Li
Micro-Doppler Radar and Its Applications Fioranelli, Griffiths, Ritchie and Balleri
Maritime Surveillance with Synthetic Aperture Radar Di Martino and Antonio Iodice
Electronic Scanned Array Design Williams
Advanced Sparsity-Driven Models and Methods for Radar Applications Gang Li

Deep Neural Network Design for Radar Applications

Edited by
Sevgi Zubeyde Gurbuz

The Institution of Engineering and Technology

Published by SciTech Publishing, an imprint of The Institution of Engineering and Technology, London, United Kingdom

The Institution of Engineering and Technology is registered as a Charity in England & Wales (no. 211014) and Scotland (no. SC038698).

First published 2020

The Institution of Engineering and Technology
Michael Faraday House
Six Hills Way, Stevenage
Herts, SG1 2AY, United Kingdom

www.theiet.org

British Library Cataloguing in Publication Data
A catalogue record for this product is available from the British Library

ISBN 978-1-78561-852-9 (hardback)
ISBN 978-1-78561-853-6 (PDF)

Typeset in India by MPS Limited

To my mother:
my best friend,
hero, and
guardian angel.

You had many accomplishments,
but your greatest was somehow
turning this wild girl into an engineer.

Forever live in my heart you will.

Contents

About the editor

Sevgi Zubeyde Gurbuz is an assistant professor of electrical and computer engineering at the University of Alabama, USA. She received the SPIE Defense and Commercial Sensing Rising Researcher Award in 2020. Her research interests are in radar signal processing and machine learning for applications ranging from human activity and gait analysis for remote health monitoring in biomedical engineering to American Sign Language and gesture recognition for human–computer interaction, and multimodal remote sensing for earth sciences.

Acknowledgements

We would like to acknowledge the contributions of the following technical reviewers, who carefully read chapters of interest and offered valuable suggestions:

Moeness G. Amin – *Villanova University, USA*
Christopher F. Barnes – *Georgia Institute of Technology, USA*
John Ball – *Mississippi State University, USA*
Baris Erol – *Siemens Corporate Technology, USA*
Francesco Fioranelli – *Delft University of Technology, The Netherlands*
Maria S. Greco – *University of Pisa, Italy*
Hugh Griffiths – *University College London, UK*
Ali Cafer Gurbuz – *Mississippi State University, USA*
Youngwook Kim – *California State University, Fresno, USA*
Mehmet Kurum – *Mississippi State University, USA*
Azarakhsh Jalalvand – *Ghent University, Belgium*
Benjamin Lewis – *U.S. Air Force Research Laboratory, USA*
Eric Mason – *U.S. Naval Research Laboratory, USA*
George Stantchev – *U.S. Naval Research Laboratory, USA*
David Tahmoush – *University of Kansas, USA*
Simon Wagner – *Fraunhofer Institute for High Frequency Physics and Radar Techniques FHR, Germany*
Feng Xu – *Fudan University, China*
Birsen Yazici – *Rensselaer Polytechnic Institute, USA*

Prologue: perspectives on deep learning of RF data

Sevgi Zubeyde Gurbuz[1] *and Eric S. Mason*[2]

Deep learning (DL) is a new area of machine learning (ML) that in recent years has surpassed state-of-the-art approaches in computer vision and natural language processing to achieve previously unimaginable advancements to challenging problems. Unlike conventional methods, deep neural networks (DNNs) trade off domain expertise in feature engineering and model training with automatic feature learning that relies on the availability of massive amounts of data. While on the one hand, the large number of network parameters enables greater flexibility and permits DNNs to model more complex relationships, training a DNN takes a significant amount of time. For example, a DNN may require as much as multiple weeks or months of training time, running on high performance computer clusters or application-specific integrated circuits. This is in contrast with traditional ML algorithms that would only require several hours at most on a laptop computer. On the other hand, the inference phase of DNNs is extremely rapid, such that it can be run in real time on low size, weight, and power hardware, such as embedded processors. The most significant consequence of DL, however, is that to a certain extent the DNN functions akin to a wizard conjuring behind a curtain: we are enthralled with the results achieved, but we have absolutely no idea of ***why*** the network performs so well!

In traditional ML, the process of feature engineering reflects the analyst's understanding of the underlying relationships in the problem. The features extracted often have some physical or mathematically tangible meaning, so that analyses of which features are more effective to classification sheds light on the problem itself. DL short circuits the analyst and provides an end-to-end solution. The flexibility of the network can enable the DNN to learn relationships that might otherwise have gone unnoticed by the analyst, but, at the end of the day, it may be difficult to ascribe any meaning or insights to the resulting network. This in turn has given rise to a new branch of artificial intelligence (AI), known as *explainable AI*, which attempts to formulate results in a way understandable by human experts.

Nevertheless, DL has been effective in image and speech processing, primarily because of the large amounts of training data available, as well as the inductive bias of designed architectures and training algorithms, motivated by the structure of the data.

[1]Department of Electrical and Computer Engineering, The University of Alabama, Tuscaloosa, AL, USA
[2]U.S. Naval Research Laboratory, Washington, DC, USA

For example, in the case of natural language processing, recurrent neural networks model long-term dependencies of time sequences, which leverages the inherent nature of speech as a sequence formed by a hierarchical combination of sounds, words, and sentences. In computer vision, the convolutional filtering layers that consist of convolutional neural networks (CNNs) provide a means of capturing spatial features, motivated by classical methods used in image processing. Moreover, convolution with a small filter provides local features and equivariance that enables the network to handle situations where the object shifts in space. The small size of the filter together with the rectified linear unit activation function help eliminate background, while pooling marginalizes nuisance variables to reveal invariant statistics that concisely describe the object of interest.

It has only been in the mid-2010s that we have seen a resurgence of interest in application of DL to radar datasets, initially for the classification of synthetic aperture radar (SAR) images and 2D time–frequency representations of micro-Doppler, and even more recently to other radar functions such as detection, tracking, and image reconstruction. Concurrent with these initial investigations by radar engineers, there has been an explosion of civilian applications of radar technologies to biomedical engineering, human–computer interaction (HCI) for the design of smart environments, Internet of Things, gesture recognition, and automotive sensing. ML experts, predominantly computer scientists, who initially learned their skills by working on images and knew very little about RF sensing, are applying DNNs to radar datasets quite naive of the purpose and structure of radar signals.

Why not? DL has been advocated as a technique that "*directly learns* relevant features and data representations ***without requiring explicit domain knowledge***** or human feature engineering." So what could possibly be the problem with computer scientists, who do not know radar, or radar engineers, who are relatively unfamiliar with DL, directly applying off-the-shelf DNN architectures originally designed for other domains?

Indeed, off-the-shelf DNNs have provided improved results on well-defined RF signal classification problems. And surely, when applying a new idea for the first time, it is perfectly natural to initially attempt already developed solutions. However heartening such initial results are, the key point that ***we can and must do better!*** A typical RF signal environment is plagued with a host of non-ideal effects that are dynamic in nature and often difficult or impossible to accurately model. This reality has driven research into adaptive signal processing, phased array antenna design and beamforming, space–time adaptive processing, knowledge-aided signal processing, and more recently, the integration of AI into the radar transceiver itself to construct a fully adaptive or cognitive radar. DL can be a driving force in the design of next-generation radar systems, but it is critical that DNNs be designed so that they can deal with environmental dynamics, such as interference, jamming, missing data, clutter and noise, and an increasingly complex and congested electromagnetic environment.

*Excerpted from the website of the First International Conference on Deep Learning Theory and Applications (DeLTA 2020, http://www.delta.scitevents.org/).

A key limiting factor in the design of any DNN for radar applications is that although DL depends on large amounts of data for training, typically the requisite amount of data is ***not available*** in RF applications. Radar datasets are usually costly to acquire, either due to the nature of the sensing problem, e.g. airborne platform required, or the human time and effort required. In civilian applications requiring human subject testing, such as biomedical and human–computer interaction (HCI) applications, Institutional Review Board approval is required. The time required to get approval and recruit participants as well as the cost to reimburse participation in studies all makes acquisition of large amounts of training data challenging. Moreover, in many cases it may not be possible to collect data under the scenario for which the DNN is to be applied: consider space-based, airborne, and battlefield sensing, for example.

Thus, to apply DL to radar problems, we must develop physics-aware solutions that provide an alternative to training on massive datasets by leveraging radar signal processing, as well as sensor and target models. Oftentimes, this means developing simulations or exploiting *a priori* knowledge to reduce complexity and improve accuracy and robustness. Computer scientists and radar engineers must work together to find unique solutions to challenging RF sensing scenarios.

With all that being said, we use the rest of this prologue to outline in detail, what the authors feel are some of the current problems in applying DL to radar and some ideas on research directions that are important to fully exploit the power of DL in radar.

P.1 The need for novel DNN architectures

While architectures designed with other applications in mind can be useful in radar applications, it is important to keep in mind that there is no guarantee that the desired results will translate. Not only must DNN architectures be separately reassessed for application to radar, but also new architectures that are uniquely suited to the phenomenology of RF signals must be developed. To illustrate this point, let us consider four cases, detailed as follows:

- **Case 1:** In the 2014 ImageNet Large Scale Visualization Challenge (ILSVC), the winning DNN was GoogLeNet, a network comprised of nine inception modules. The design of the inception module was an innovation that used multiple filters at different resolutions to extract coarse, medium, and fine features. Rather than specifying the size of convolutional filters, as done in CNNs, the inception module enables the network itself to decide which resolutions are optimal. Max pooling and 1×1 convolutions are used for dimensionality reduction. The multi-resolution design of GoogLeNet surpassed the performance of VGGnet, a 16-layer CNN, in ILSVC 2014. Yet, when applied to a 12-class radar micro-Doppler recognition problem, ***the exact opposite was found to be true***—VGGnet outperformed GoogLeNet[†]. What is different about the radar data that caused these networks to achieve different relative performance?

[†]M. S. Seyfioğlu and S. Z. Gürbüz, "Deep Neural Network Initialization Methods for Micro-Doppler Classification With Low Training Sample Support," in IEEE Geoscience and Remote Sensing Letters, vol. 14, no. 12, pp. 2462–2466, 2017.

- **Case 2:** In RF applications, typically the measured data includes not just the signal of interest, but also significant components of clutter, noise, and interference. Thus, characterization of DNN performance just in terms of accuracy is insufficient—performance must be examined not only as a function of signal-to-noise ratio (SNR), but through consideration of what the DNN is learning, however elusive a prospect. A variety of techniques have been developed to try to visualize what DNNs are learning. When the activation maps of the neuron with largest activation is examined at various layers through the networks, it may be observed that different networks deal with spurious signals differently. In one study[‡], the spurious signal was seen throughout the layers of the CNN but never appeared in the activation maps of the convolutional autoencoder and was much fainter when transfer learning from VGGnet or GoogLeNet was applied. GoogLeNet also appears to have more faithfully represented the class signature, while the spurious signal was slightly stronger in the activation maps of VGGnet. Again, it may be observed that different network architectures behave differently in terms of the ability to model a given class and reject unwanted artifacts in the data—why?
- **Case 3:** Generative adversarial networks (GANs) have become quite popular in computer vision for generating synthetic imagery or improving the realism of synthesized images. Yet, when applied to radar micro-Doppler data, a significant percentage of GAN-generated synthetic signatures are in fact *misleading* because they correspond to kinematic events that are either impossible or deviate from the motion class being synthesized[§]. Thus, for example, while a "walking" signature is desired, the GAN may instead generate a "walk-stop-walk" signature. Or where a "falling" signature is desired, the GAN may instead generate a "progressive fall"—motion corresponding to first falling on the knees, then to the ground, not an impulsive, immediate fall to the ground. While GANs have issues in generating synthetic RF data with desired kinematic fidelity, they are effective in capturing other artifacts of the data due to sensor imperfections or clutter. This ties in to the broader challenge of how best to integrate knowledge of radar phenomenology into the DNN architecture and training so that the network weights are optimized with respect to concepts and metrics that reflect qualities meaningful to radar signal understanding.
- **Case 4:** Another related issue is better understanding of how DNNs deal with ground clutter. In traditional ML, a moving target indication filter or some other form of clutter suppression is used to improve the SNR of the data and the accuracy of extracted features. Interestingly, in DNNs, such preprocessing actually can degrade the resulting classification accuracy. This is both a demonstration of the power of DNNs to extract signal information and an indication that radar engineers must also reconsider the conventional wisdom in radar signal processing

[‡]M. S. Seyfioğlu, A. M. Ozbayoglu and S. Z. Gurbuz, "Deep Convolutional Autoencoder for Radar-Based Classification of Similar Aided and Unaided Human Activities," in IEEE Transactions on Aerospace and Electronic Systems, vol. 54, no. 4, pp. 1709–1723, 2018.

[§]B. Erol, S. Z. Gürbüz and M. G. Amin, "Motion Classification Using Kinematically Sifted ACGAN-Synthesized Radar Micro-Doppler Signatures," in IEEE Transactions on Aerospace and Electronic Systems, vol. 56, no. 4, pp. 3197–3213, Aug. 2020.

when classifying data with DNNs. In some cases, simply letting the DNN deal with a less adulterated form of the data may prove to be beneficial‖.

As the previous examples illustrate, we must avoid direct "copy–paste" implementation of designs from other applications but instead use advancements in DNNs to inspire RF-specific solutions to RF applications. Taking a deeper look into the differences between radar data and natural images in the context of the previous examples, we conjecture on some possible causes of these observations and suggest possible research directions.

From the example in Case 1, we see that deeper more complicated networks are not necessarily better than a shallower simpler network when it comes to radar applications. Radar data is very different than that of natural images; physically, they occupy significantly lower frequencies in the electromagnetic spectrum, reflects (scatters) off materials in very different ways, and of course is not perceivable by humans. One of the key notions of DNNs is to extract hierarchical information, which exists at different resolutions. While this same notion applies to RF data, it manifests in different ways and clearly necessitates alternative approaches. Consider the fact that radar measurements are a superposition of time-delayed and Doppler-scaled (or shifted) replicas of different waveforms spread over time and frequency, and clutter is the addition of other signals coexisting in time and frequency. When observing the mathematical formulation, the obvious differences between RF and electro-optical images become clear.

Furthermore, architectural design formulation based on exploiting symmetries of data using algebraic group¶ convolutions** has been developed recently. This approach to architecture has achieved state-of-the-art performance in challenging data domains, such as classifying 3D imagery. Observing that certain symmetries and properties present in radar data can be modeled using groups provides an exciting direction of research. For example, the delay and Doppler effects seen in received radar echoes can be modeled as convolutions over the Affine and Heisenberg group, a property that could be useful in designing networks invariant to transmitter motion, useful in waveform design and classification problems.

In summary, we need alternative network architectures designed to incorporate inductive bias relevant to radar signal structure. This is a very fascinating, relevant, and important research direction needed to fully realize the power of DNNs for RF data.

P.2 Physics – aware ML that exploits RF data richness

Radar data is acquired as a time stream of complex I/Q data and is thus not normally in the form of an image. It is only after a number of radar signal processing steps

‖X. Yang, W. Yin and L. Zhang, "People Counting Based on CNN Using IR-UWB Radar," in 2017 IEEE/CIC International Conference on Communications in China (ICCC), Qingdao, 2017, pp. 1–5.

¶Foregoing a rigorous mathematical explanation, in this case a group can be viewed as filter endowed with an operation that preserves certain geometric symmetries.

**https://www.quantamagazine.org/an-idea-from-physics-helps-ai-see-in-higher-dimensions-20200109

in that radar data can be converted into a form appropriate for input to DNNs. For example, SAR reconstruction algorithms, together with algorithms to filter speckle noise and correct for platform motion, are required to generate airborne SAR imagery. Time–frequency transforms are typically applied to generate an image of a moving target's micro-Doppler signature. But there are many more possibilities for presenting radar data to DNNs in the form of 2D and 3D representations, based on decomposition into multiple time/frequency-scales. In recent years, researchers have become more creative in their explorations of data representations, computing point clouds, applying saliency analysis to generate heatmaps, or generating range-Doppler videos of moving target profiles, to name just a few examples. If, additionally, we consider that radar hardware may be constructed to form interferometric, dual-receive channel radars, multichannel systems, or multi-static radar networks, radar signal processing algorithms may be used to extract not just target range, radial and cross-track velocity, but also direction-of-arrival, and power spectral density. Observations from multiple channels or RF sensors at different perspectives can be fused, used to estimate joint domain representations, or processed and assessed separately.

Not only is the optimal way to represent and preprocess RF data prior to input to DNNs an open question, but so is the question of the optimal input dimensionality. In image processing, this question could be framed as how large an image should be supplied to a DNN? In the context of radar, we may instead consider the FFT length and window size used when computing spectrograms of the RF data. There is a trade-off between dimensionality and resolution, but no fixed approach for determining the optimum is known at this time. Methods for dimension reduction that will not skew or corrupt the underlying relationship of the data samples are critically important when applying DNNs, which are computationally intense to train. This is especially a problem since radar signals are not necessarily stationary and posses correlations within and between different subsequences.

These challenges are especially emphasized by the varying timescales present in radar data. For example, in SAR we have two timescales: slow and fast, which are determined by the platform motion and sampling rate. If, additionally, the antenna has a scanning capability, either physically or electronically, the rate of scan represents a third timescale. As with time–frequency analysis, there is a trade-off between reducing sampling while still capturing critical events and features. How to represent the varying timescales for a single-data sample and deal with the varying sampling rates is an important and open problem in applying DNNs to many radar problems of interest.

Another rich aspect of radar data is that data is typically represented as complex signal following I/Q demodulation. Complex representations of real-valued data have garnered some attention in the DL community, as they capture more information via representation of signal phase as well as amplitude. Several approaches can be taken for processing complex data with neural networks. One approach is to separate the real and imaginary parts of the signal into separate channels, which are either treated independently the entire way through the network or mixed to emulate complex multiplication. This approach exploits existing tools to feed the network both components of the complex data but does not capture the totality of the relationship between the real and imaginary components of a measurement. A more sophisticated approach[††] that was recently developed uses weighted Fréchet means to define a modified convolutional operator. This has been shown to preserve equivariance with respect to shifts over the complex plane.

Despite these current advances, a unified approach based on Wirtinger derivatives has yet to be developed using automatic differentiation, as well as development and study of non-holomorphic activation functions. While there is no formal automatic differentiation software for Wirtinger derivatives, they were used in recent work on DL for inverse problems and SAR imaging[‡‡].

P.3 RF sensing problems that can benefit from DL

The last two sections have focused on areas of research that pertained to the development of DNNs specifically designed for RF applications. In the last couple years there has been a surge of publications on using DL for almost every radar problem under the sun: many of which do not make sense. There have been some attempts at replacing fundamental signal processing blocks, e.g. time–frequency transforms, with DNNs. Here, we would recommend caution, as DNNs are a data greedy method, which may not be as robust as conventional methods when challenges from real-world environmental factors, such as clutter and interference, are also considered.

Of course there are the fundamental radar problems that seem well suited for DL, for example, automatic target recognition (ATR) with SAR data. SAR ATR is fundamentally a classification problem, where features can be very hard to extract due to the complex physical scattering behavior. There are challenges, specifically, generalization issues pertaining to limited training data and the habit of DNN to learn background statistics, such as clutter.

Beyond classification problems, DNN can serve as a good surrogate for optimization. In the passive radar problem, many components are unknown, such as the transmitted waveform, the location of transmitter, and a myriad of other uncertainties. The conventional approach is to use optimization to solve for these unknowns and gracefully handle the uncertainties. In the case of unknown transmitted waveform and transmitter location we are left with a bilinear inverse problem to recover the reflectivity and waveform. When the position of the transmitter is unknown, part of Green's function is missing and the problem becomes even more difficult for conventional optimization.

This book provides an in-depth examination of applications of DL to radar that have been pursued to-date. Surely, as research advances, many more innovative ways of applying DL to radar sensing challenges will arise.

[††] R. Chakraborty, J. Wang and S. Yu, "SurReal: Fréchet Mean and Distance Transform for Complex-Valued Deep Learning," in Proc. IEEE Computer Vision and Pattern Recognition Workshop on Perception, June 16, 2019.

[‡‡] B. Yonel, E. Mason and B. Yazıcı, "Deep Learning for Passive Synthetic Aperture Radar," in IEEE Journal of Selected Topics in Signal Processing, vol. 12, no. 1, pp. 90–103, 2018.

P.4 Overview of this book

It is our hope that this book will serve as a first introduction to the opportunities and challenges of designing DNNs for the analysis of radar datasets. RF applications draw researchers and engineers from several different areas of expertise, including electrical engineering, computer science, mathematics, and physics.

Part I—Fundamentals is intended to serve as a brief introduction to RF sensing, signal processing (Chapter 1), ML (Chapter 2), and DL (Chapter 3), presented at an introductory level assuming just a general scientific and mathematical background.

Part II—Special topics includes chapters targeted toward the solution of fundamental challenges in applying DL to radar datasets: how we present radar data to DNNs and subsequently train them, especially when little training data is available. Chapter 4 discusses different forms of 2D and 3D data representations in the context of indoor radar for the classification of activities of daily living. Chapters 5 and 6 both consider the challenges of DNN training, especially the problem of synthetic training data generation. Chapter 5 focuses on training for micro-Doppler classification, while Chapter 6 is focused on SAR image synthesis and efficacy.

Part III—Applications comprises chapters highlighting five different applications of DL to radar datasets, including micro-Doppler signature classification (Chapter 7), SAR and inverse SAR (Chapter 8), passive SAR imaging (Chapter 9), and fusion in radar networks (Chapter 10). Chapter 11 is a broad survey of the role of DL in a wide range of radar remote sensing applications for earth observation, environmental science, land use classification, agriculture, geophysical parameter estimation, meteorology, ice sounding, and radar aeroecology, among others.

DL is a rapidly advancing field, and by each day, new innovative ideas are brought forward. This book will provide a foundation on the current state-of-the-art, but it is up to you, the reader, to continue to empower next-generation radar systems with breakthroughs in AI.

Part I

Fundamentals

Chapter 1
Radar systems, signals, and phenomenology

Sevgi Zubeyde Gurbuz[1], Shunqiao Sun[1], and David Tahmoush[2]

Radar, short for "radio detection and ranging," was first conceived in the 1930s as means to provide early warning of approaching aircraft. It operates by transmitting an electromagnetic (EM) wave and processing the received echoes to measure the distance, velocity and scattering properties of objects and their surroundings. In the early years of its development (1930s and 1940s), radar was used primarily for target detection; however, since then, its uses have grown to include classification, tracking and imaging. While birthed from military applications, radar has become central to many commercial technologies that we as a society have come to rely on. Some of the more common uses include police radar to enforce traffic regulations, meteorological radars to monitor weather conditions and identify the type of precipitation expected, and air traffic control radars to guide commercial aircraft in the vicinity of airports. Airborne and spaceborne radar systems have been used for a wide range of environmental remote sensing applications, including mapping, land use classification, forest monitoring, earthquake damage assessment, soil moisture estimation, ice sounding, and much more. As radar transceivers themselves have become smaller, lighter in weight, and lower in cost due to advances in solid-state microelectronics, so has radar emerged as a key enabler in newer fields, such as automotive sensing and self-driving vehicle technology, gesture recognition for human–computer interfaces, and biomedical applications of remote health monitoring and vital sign detection. The proliferation of radar is indeed remarkable and presents a vast arena where deep learning, cybernetics, and artificial intelligence can enable disruptive technological advancements.

1.1 Physics of electromagnetic scattering

The radio frequency (RF) EM waves transmitted by a radar system propagate through a medium, interact with the environment, as a result of which a certain portion of

[1]Department of Electrical and Computer Engineering, University of Alabama, Tuscaloosa, AL, USA
[2]Information and Telecommunication Technology Center, University of Kansas, Lawrence, KS

the transmission will propagate back to the radar receiver. EM waves are formed by time-varying, oscillating electric, and magnetic fields, which are perpendicular to each other, as depicted in Figure 1.1. Mathematically, the oscillation of the E and M fields are sinusoidal, and the *period* of the wave is defined as the time that must elapse for the wave to go through one cycle. When the period T_0 is expressed in seconds, the *frequency* is the inverse of the period and represents the number of cycles incurred during 1 s:

$$f = \frac{1}{T_0} \tag{1.1}$$

In a vacuum, EM waves move at the speed of light (c), or 300,000 km/s. The wavelength (λ) and frequency of an EM wave depend upon its speed of propagation, which for free space can be expressed as

$$\lambda f = c. \tag{1.2}$$

The *polarization* of an EM wave describes the motion and orientation of the electric field. An EM wave is *linearly polarized* if the wave propagates along a straight line, e.g., such as one of the axis. If x represents the horizontal axis, a wave propagating along the x-axis would be *horizontally polarized*, while a wave would be *vertically linearly polarized* if it propagated along the y-axis. More general polarizations can be defined by representing the electric wave in general by its x and y components: $\vec{E} = E_x\hat{\imath} + E_y\hat{\jmath}$. If the x and y components of the electric field are equal, and the phases differ by an odd multiple of $\pi/2$, the electric field vector will move along a circle as the wave propagates. In this case, the wave is said to be *circularly polarized*.

Many properties of the way an EM wave interacts with the environment are affected by the frequency of a radar transmission, as a result, the transmission frequencies used strongly depend on the application. It is typical to refer to different frequency intervals, or bands, according to standard letter-based names, as summarized in Table 1.1. Here, it is important to note that all RF transmissions are regulated

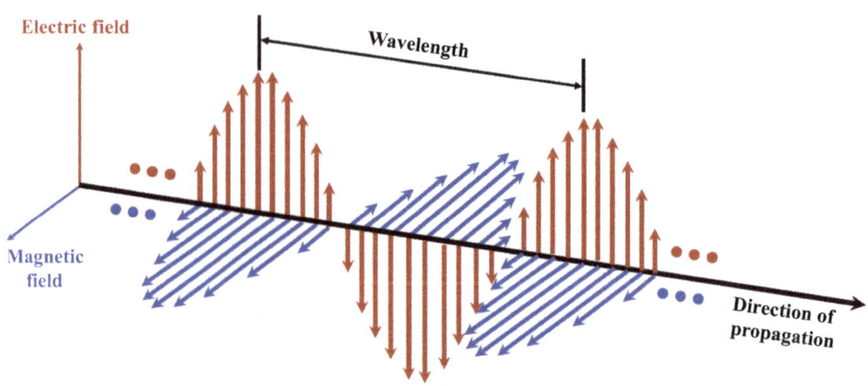

Figure 1.1 Electromagnetic wave

Table 1.1 RF/radar bands and common applications

Band	Frequency	Wavelength	ITU frequency	Example applications
HF	3–30 MHz	100–10 m		OTH[a] radar
VHF	30–300 MHz	10–1 m	138–144 MHz 216–225 MHz	Ground penetrating radar
UHF	0.300–1 GHz	1–30 cm	420–450 MHz 890–942 MHz	Ionospheric radar
L	1–2 GHz	30–15 cm	1.215–1.4 GHz	En-route air traffic control
S	2–4 GHz	15–7.5 cm	2.3–2.4 GHz 2.7–3.7 GHz	Terminal air traffic control and Long-range weather radar
C	4–8 GHz	7.5–3.75 cm	5.250–5.925 GHz	Long-range tracking
X	8–12 GHz	3.75–2.5 cm	8.50–10.68 GHz	Missile guidance
Ku	12–18 GHz	2.5–1.67 cm	13.4–14.0 GHz 15.7–17.7 GHz	High-resolution mapping
K	18–27 GHz	1.67–1.11 cm	24.05–24.25 GHz 24.65–24.75 GHz	Handheld police radar and Long-range automotive radar
Ka	27–40 GHz	1.11–0.75 cm	33.4–36.0 GHz	High-resolution EO[b]
V	40–75 GHz	7.5–4 mm	59.0–64.0 GHz	Hand gesture recognition
W	75–110 GHz	4–2.7 mm	76.0–81.0 GHz 92.0–100.0 GHz	Short-range automotive radar Cloud profiling radar
mm	100–300 GHz	3–1 mm	126.0–142.0 GHz 144.0–149.0 GHz 231.0–235.0 GHz 238.0–248.0 GHz	Experimental

[a]OTH, over-the-horizon.
[b]EO, Earth Observation.

by the International Telecommunications Union (ITU) and, in the United States the Federal Communications Commission, which works closely with the ITU. So, for example, while "C-band" refers to frequencies between 4 and 8 GHz, over that band, only transmissions between 5.250 and 5.925 GHz are authorized by the ITU.

Military radar applications have traditionally used the Ku-bands or lower, while civilian radar applications can be found across virtually all radar bands, even the K-band (18–27 GHz), where water vapor tends to preclude airborne or spaceborne sensing applications. K-band has been effective for ground-based radar applications, such as automotive or police radars. Earth observation (EO) missions have also exploited the higher as well as lower frequencies of the RF spectrum, with cloud profiling radars operating as high as 95 GHz. Short-range automotive radars operate in the low W-band, while biologists have used this band for studying insects and measuring insect wing flapping frequency.

1.1.1 Effects of propagation medium

The speed at which an EM wave propagates depends on the medium. When propagating through the atmosphere, part of the energy being transported by the EM wave gets absorbed in the air. This results in *attenuation* of the transmitted signal. Atmospheric attenuation is very little below 1 GHz under clean-air conditions. Above 1 GHz, the absorption of the atmosphere steadily increases, with significant peaks occurring at 22 GHz, due to water vapor absorption, and 60 GHz, due to oxygen absorption. Above 10 GHz, there are windows of lower atmospheric absorption in the Ka-bands and W-bands, making these bands preferable for airborne and spaceborne radars operating in the very high frequencies. The amount of atmospheric attenuation incurred depends not only on frequency but also on the weather conditions and type of precipitation. Thus, more attenuation is incurred during heavy rain than during light rain. Although such losses weaken the transmitted signal, radar systems are still operational in cloudy or harsh weather. This represents a distinct advantage over electro-optical imagers, which are completely blocked by such conditions.

When the EM wave passes through other mediums, such as through materials of different dielectric constants, the waves will bend due to the change in propagation speed at the interface. This is known as *refraction* and is encountered in remote sensing applications, such as ground penetrating radar. Over-the-horizon radar systems exploit refraction in the Earth's ionosphere to detect targets several thousand miles away.

1.1.2 Radar cross section

When the propagated wave comes into contact with any object or surface, part of the energy of the EM wave is absorbed, while the remaining energy is scattered in varying directions. If the angle of incidence of the transmitted wave is equal to the angle of reflection of the backscatter, as would be a mirror reflects light, then the scattering is said to be *specular*. Very few objects are purely specular, although many man-made objects cause strong specular returns. In contrast, if the transmitted wave is backscattered so that there is some energy at every angle of reflection, the scattering is said to be *diffuse*. This is the type of scattering that would be expected, for instance, if the EM wave interacted with a rough surface, such as perhaps the Earth's surface.

The backscatter of an object can be quantified through the concept of *radar cross section (RCS)*. Assuming the backscatter from the object is isotropic, the backscattered power density may be expressed as

$$Q_b = \frac{P_b}{4\pi R^2}, \tag{1.3}$$

where P_b is the total backscattered signal power and R is the distance between the object and the radar transceiver. The RCS (σ) is formulated as the fictional area over which the transmitted power density Q_t must be summed to account for the total backscattered power, P_b. Thus, $P_b = \sigma Q_t$, and

$$\sigma = 4\pi R^2 \frac{Q_b}{Q_t}. \tag{1.4}$$

Formally, RCS is defined not in terms of power density, but in terms of the magnitude of the electric field as follows:

$$\sigma = 4\pi \lim_{R\to\infty} \left[R^2 \frac{|E^b|^2}{|E^t|^2}\right] \tag{1.5}$$

where $|E^b|$ and $|E^t|$ are the amplitudes of the backscattered and transmitted electric fields, respectively. Thus, RCS does not simply quantify the total backscatter but, in fact, characterizes how much of the transmitted signal was intercepted by the object, and how much of that energy is directed back towards the radar's receiver. The RCS of an object depends not only on material properties, but also on the frequency, polarization, orientation, geometry, and angle of incidence of the transmitted wave. The RCS of the same object will therefore differ for *monostatic* and *bistatic* radar configurations. In monostatic radar configurations, the transmitter and receiver are colocated, whereas in bistatic radar, the transmitter is at a different location than the receiver. The backscatter power in different directions can vary; therefore, for example, it may be easier to detect a stealth aircraft with bistatic radar than with a monostatic radar.

To give another example, consider the monostatic RF reflectivity of a flat metal plate. Based on our normal intuition derived from daily experience with optical reflections, we may think a flat plate to be a great reflector. Indeed, when the surface of the flat plate directly faces radar transceiver, a strong specular return is recorded. But, should the plate be even slightly tilted, the returns will be directed away from the transceiver, and very little power will be received, as shown in Figure 1.2. In contrast, geometrical shapes having corners or edges formed from the intersection of two planes tend to have a larger RCS over a wider range of angles. Thus, most radar

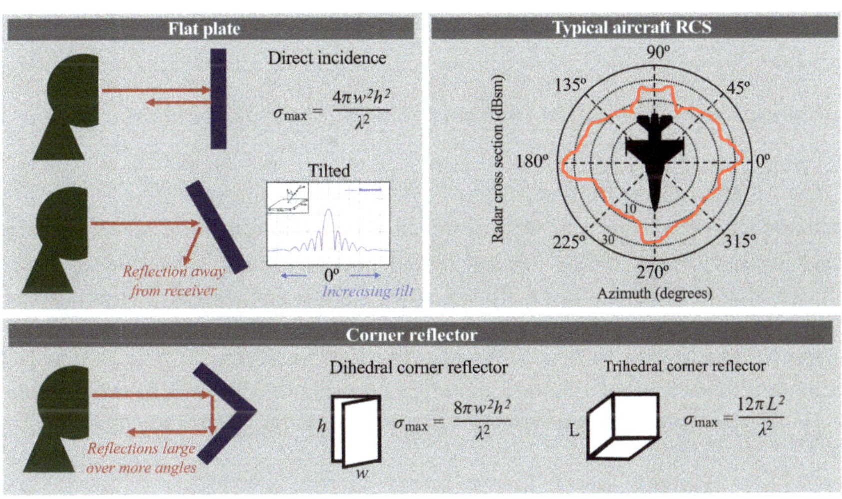

Figure 1.2 Radar cross section of a flat plate, corner reflector, and typical aircraft as a function of angle

calibration targets are chosen to be some type of corner reflector. For the same reason, the corners and edges formed by an aircraft's wings and tail are stronger reflectors than the surface of the aircraft itself. Large physical size does not necessarily result in a large RCS.

1.1.3 Radar range equation

A basic, deterministic model of the power received by a radar system is given by the *radar range equation (RRE)*. The RRE is very useful when doing basic radar system design, link budget analysis, or feasibility studies. There are several forms of the equation, dependent upon whether the target is a single discrete scatterer, also known as a *point target*, an area scatterer, or a volumetric scatterer.

Let us first assume that the scatterer is a point target, the signals received are narrowband and that the transmitting antenna is directional with a maximum gain of G. Then, the *peak-transmitted power density (W/m²)* can be found by multiplying the isotropic power density with the peak gain:

$$Q_t = \frac{P_t G}{4\pi R^2} \tag{1.6}$$

given that the transmitter output power is P_t and the distance between the radar and the target is R.

If the RCS of the point target is σ and we assume that the backscatter is also isotropic, then the re-radiated power density can be expressed as

$$Q_b = \frac{P_t G \sigma}{(4\pi)^2 R^4} \tag{1.7}$$

The power actually received by the radar depends on the *effective aperture* of the receiving antenna. It can be shown that the effective aperture (A_e) of an antenna is related to its gain and the wavelength of the transmitted signal (λ) as

$$A_e = \frac{\lambda^2 G}{4\pi} \tag{1.8}$$

Substituting into (1.7), the total received power may be found to be

$$P_r = \frac{P_t G \sigma A_e}{(4\pi)^2 R^4} = \frac{P_t G^2 \lambda^2 \sigma}{(4\pi)^3 R^4} \tag{1.9}$$

Equation (1.10) describes the ideal received power, if there were no losses. However, there are system losses due to antenna inefficiency and propagation through the radome and waveguides, and electronic noise, which are often a function of radar transmit frequency, as well as atmospheric attenuation, which depends on range and frequency. There may also be additional sources of interference, such as caused by jamming, or unwanted reflections from other objects or surfaces in the environment, also known as *clutter*. When these factors are added, the more typical form of the RRE for a single-point target is obtained:

$$P_r = \frac{P_t G^2 \lambda^2 \sigma}{(4\pi)^3 R^4 L_{sys} L_a} \tag{1.10}$$

where L_{sys} represents system losses and L_a represents atmospheric losses.

Oftentimes, we are more interested in the *signal-to-noise ratio (SNR)*, as the performance of radar detection algorithms directly relates to SNR, as does the performance of radar target trackers, classifiers, and imaging systems. If considering only the electronic noise floor, the noise power (P_n) may be computed as $P_n = kT_0\beta F$, where k is the Boltzmann's constant, T_0 is the standard reference temperature in Kelvin (290K), β is the transmitted signals bandwidth, and F is the noise figure. Then, the SNR can be computed as

$$SNR = \frac{P_r}{P_n} = \frac{P_t G^2 \lambda^2 \sigma}{(4\pi)^3 R^4 L_{sys} L_a k T_0 \beta F} \tag{1.11}$$

If the clutter power P_c is also taken into account, then the signal-to-interference-plus-noise ratio (SINR) may computed by first adding to the noise power and then taking the radio with the received power: $SINR = P_r/(P_c + P_n)$.

This analysis may be generalized to distributed targets by representing the entire target as the integral over a differential volume defined in polar coordinates as $d\phi d\theta dr$, and considering not just the maximum antenna gain, but the entire antenna beam pattern $\mathscr{P}(\theta, \phi)$. Here, θ and ϕ are the elevation and azimuth angles, respectively. Then, the total received power can be expressed as

$$P_r = \frac{P_t \lambda^2}{(4\pi)^3 L_{sys}} \int_V \frac{\mathscr{P}(\theta, \phi)^2}{r^4 L_a(r)} d\sigma(r, \theta, \phi) \tag{1.12}$$

The atmospheric losses are kept inside the integral due to their dependence on range, one of the integration variables.

An important special case of (1.12) is *area scattering*, best exemplified by scattering from the Earth's surface. In remote sensing and Earth observation applications, the goal is to characterize the scattering properties of the surface. In contrast, many surveillance and recognition application focus on the sensing of a discrete target, in which case the scattering from the ground is a source of clutter. Both the extraction of environmental parameters from radar remote sensing data and the characterization of ground clutter are subjects that have been extensively studied.

One practical approach is to calculate the radar backscatter in terms of the *area reflectivity*, σ^o, defined using the differential RCS as $d\sigma = \delta(r - r_0)\sigma^o$, where r_0 is the distance to a particular differential area. Computing the double integral over the area over the surface illuminated by the radar, the received power from the surface can be calculated for two different cases: (1) the beam-limited case, where the area illuminated is determined by the beamwidth, and (2) the pulse-limited case, where the area illuminated is determined by the pulse-width. These two cases are illustrated in Figure 1.3.

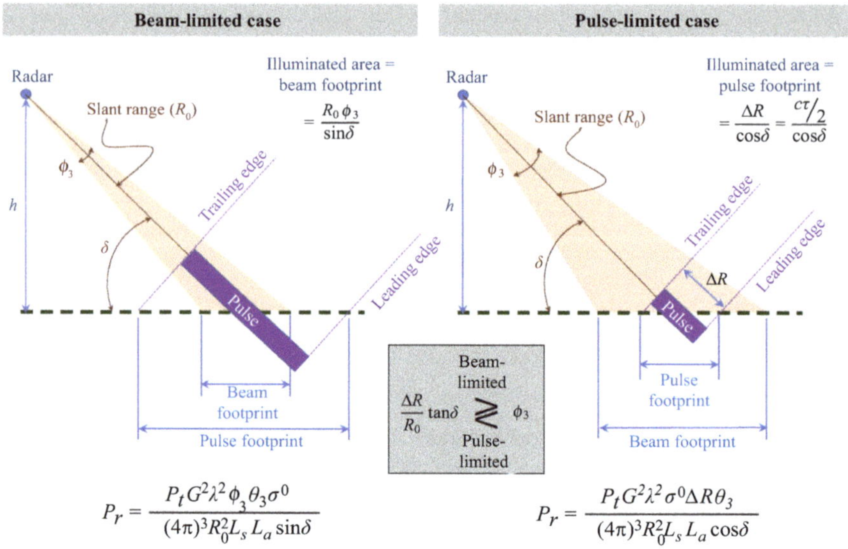

Figure 1.3 Beam-limited versus pulse-limited area scattering

1.2 Basic radar measurements and waveforms

1.2.1 Continuous wave radar

The most basic waveform that can be transmitted by a radar system is a continuous wave (CW), defined as a purely sinusoidal signal. Then, the received signal will be of similar form, but with amplitude (A), frequency (f) in Hertz, and phase (θ) in radians:

$$x_r(t) = A(t)\sin(2\pi ft + \theta(t)) \tag{1.13}$$

However, one of the key issues the radar receiver must resolve is what exactly the phase of the received signal is. Knowledge of the sine (or cosine) alone is not sufficient to uniquely ascertain the angle. However, by mixing the received signal with the sine and cosine of the transmit frequency, the components necessary to uniquely find the phase can be formed. Consider the products shown next:

$$A\sin(2\pi ft + \theta) \times 2\sin(2\pi ft) = A\cos[\theta] - A\cos[2(2\pi f)t + \theta] \tag{1.14}$$

$$A\sin(2\pi ft + \theta) \times 2\cos(2\pi ft) = A\sin[\theta] + A\sin[2(2\pi f)t + \theta] \tag{1.15}$$

After low-pass filtering, each product yields either the sine or the cosine of the phase, which together unambiguously specifies the phase. As a result, one channel is typically labeled the *in-phase* or "I" channel, while the other is 90° phase shifted and hence termed the *quadrature* or "Q" channel.

Once both channels are digitized, a complex representation of the data is typically formed: the I-channel is assigned to be the real part, while the Q-channel is assigned as the imaginary part. Thus, the data acquired by the radar receiver can be represented as a single complex signal

$$x(t) = I(t) + Q(t) = A(t)e^{j(2\pi ft + \theta(t))} \tag{1.16}$$

The overall structure of the I/Q or quadrature receiver is shown in Figure 1.4.

The amplitude of the received signal, $A(t)$, is directly related to the received power, $P_r(t)$, as modeled by the RRE. Although A_t will depend on the range, measurement of the amplitude will not give an accurate estimate of the range. Range is typically determined by measuring the round-trip time-of-flight, t_r, yielding $R = ct_r/2$. However, in CW radar systems, the receiver has no means for marking time-of-flight, so range measurements cannot be made.

CW radars can measure a target's radial velocity. If the target is moving, the frequency of the received signal (f_r) will slightly shifted in accordance to the *Doppler effect*:

$$f_r = \left(\frac{1 + v/c}{1 - v/c}\right) f_t \tag{1.17}$$

where v is the radial velocity of the target, c is the speed of light, and f_t is the frequency of the transmitted signal. Assuming that the speed of the target is much smaller than the speed of light, which for all man-made targets will be the case, this expression is well approximated by

$$f_D = \frac{2v}{c} f_t = \frac{2v}{\lambda_t} \tag{1.18}$$

where f_D is the Doppler shift, defined as the difference between the frequencies of the transmitted and received waveforms. It is important to underscore that the velocity determining the Doppler shift is not the absolute speed of the target, but the component of the velocity falling in the direction of the radar. This represents a fundamental, physics-based constraint on the capabilities of radar systems, which all RF sensing systems must take into account.

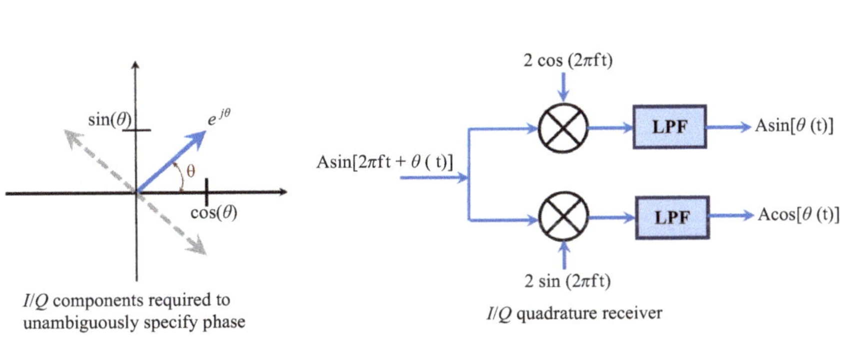

Figure 1.4 I/Q signal decomposition and quadrature receiver

1.2.2 Pulsed CW radar

The ability to measure range as well as velocity can be gained through the transmission of a pulsed CW waveform, illustrated in Figure 1.5. The *pulse width* is defined as the duration over which the CW signal is transmitted, while the *pulse repetition interval (PRI)* indicates the time between pulses. Typically, a monostatic radar transceiver is used to transmit pulsed waveforms, so that while the radar is transmitting, short-range backscatter cannot be received. The minimum range at which targets can be perceived is

$$R_{\min} = \frac{c\tau}{2} \tag{1.19}$$

In general, the range to a target is computed relative to the start of each PRI, so that for a target located at a range of R_0, the round-trip flight time can be found as

$$t_0 = \frac{2R_0}{c} - \tau \approx \frac{2R_0}{c} \tag{1.20}$$

If, however, the range corresponds to a flight time longer than a PRI, the radar system will mistakenly measure the distance of the target as $R_{true} - (2/c)(m \times PRI)$, where m is the number of pulses that have been transmitted before reception of the target return. This is known as *range aliasing* and results in pulsed radar systems having a *maximum unambiguous range* of

$$R_{\max,ua} = \frac{2PRI}{c} \tag{1.21}$$

In pulsed radar systems, the round-trip flight time t_0 is usually easily found by applying a *matched filter* on the received signal. The matched filter is a filter that has the special property of maximizing the output SNR. It may be shown that, in general, the matched filter of a signal $x(t)$ is simply

$$h_M(t) = \alpha x^*(- t) \tag{1.22}$$

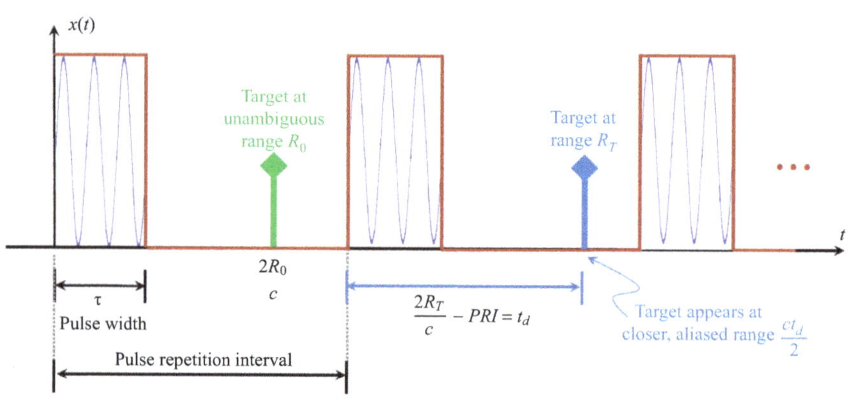

Figure 1.5 Pulsed CW waveform illustrated with unambiguous and aliased targets

where α is a constant scaling factor. This particular filter is known as a matched filter because its response is "matched" to its input: the filter itself is just a time-reversed copy of the input signal. The matched filter output for the pulsed CW waveform has a peak that corresponds to the flight time corresponding to the target range. Thus, by determining the time of the peak, the range of the target can be ascertained.

1.2.3 Range and Doppler resolution

An important factor in radar system design is *resolution*—the ability to differentiate two targets of equal strength in either range or velocity (i.e., Doppler). A useful tool for signal analysis is the Fourier transform that allows interpretation of the radar signals in the frequency domain. Consider, for example, the Fourier transform of matched filter output, shown in Figure 1.6 for two CW pulses from returns slightly separated in range. To differentiate the two pulses, we require that the peak of one response aligns with the null of the other. As the peak-to-null bandwidth of the Fourier response is $1/\tau$ (Hz), it corresponds to a difference in time of τ. Converting to range, the range resolution for the pulsed CW waveform can be found as

$$\Delta R = \frac{c\tau}{2} \tag{1.23}$$

The Doppler resolution can similarly be found through Fourier analysis of the transmission of N CW pulses. In this case, the peak-to-null bandwidth of the Fourier transform determines the Doppler resolution, which can be found to be the inverse of the total observation duration (also known as the *dwell time*):

$$\Delta f_D = \frac{1}{N \times PRI} \tag{1.24}$$

The velocity resolution can be found from this result using the Doppler shift relation as follows:

$$\Delta v = \frac{c}{2f_t}\Delta f_D. \tag{1.25}$$

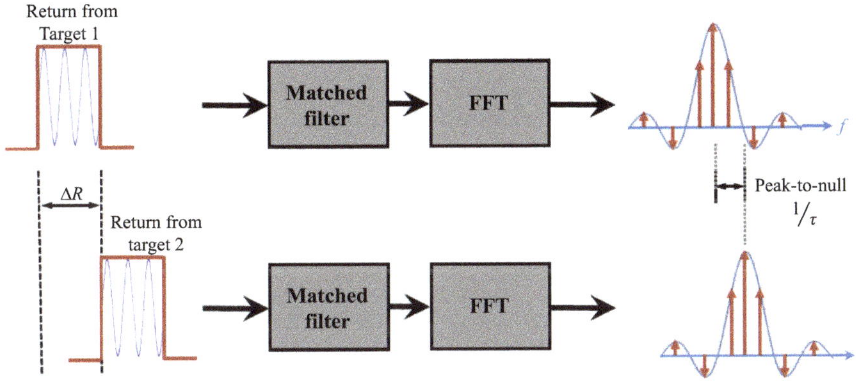

Figure 1.6 Frequency domain analysis of range resolution of pulsed CW waveform

The pulsed CW waveform enables both range and velocity measurements; however, there is a trade-off between average signal power and resolution. While range resolution improves as the pulse width gets smaller, the average transmit power also is reduced. The probability of detecting a target increases with SNR, and, hence, transmit power. Thus, reducing the pulse width improves range resolution to the detriment of the probability of detection. One remedy for this situation is to transmit a signal with linear frequency modulation (LFM), discussed next.

1.2.4 Pulsed-Doppler radar

An LFM, or chirp, waveform is defined by a sinusoid whose frequency linearly increases as

$$x(t) = Re\{e^{j\pi\beta t^2} \tau\} = \cos\left(\pi\frac{\beta}{\tau}t^2\right) \tag{1.26}$$

where β is the bandwidth of the signal and τ is the pulse width. The instantaneous frequency of a chirp waveform is the derivative of the phase and equal to $\beta t/\tau$. The slope of the linear increase in frequency is also known as the *chirp rate*, $\gamma = \beta/\tau$.

A typical chirp waveform is shown in Figure 1.7. The range resolution is determined by the minimum distance between targets that allows their differentiation. In the case of the chirp waveform, the matched filter output is a sinc function in the time-domain. Thus, the minimum distance between two targets corresponds to the situation when the peak of one target aligns with the null of the other. Since the peak-to-null duration is $1/\beta$, the range resolution may be found as

$$\Delta R = \frac{c}{2\beta} \tag{1.27}$$

Notice that now the range resolution is decoupled from the pulse width, allowing for independent design of transmit power requirements and waveform duty cycle.

1.2.5 Frequency-modulated continuous wave (FMCW) radar

Pulsed-Doppler radars are most often used for long-range applications, where the minimum detectable range imposed by the pulse width is not an issue. For short-range applications, frequency-modulated CW (FMCW) can be used in conjunction with *stretch processing* to reap the advantages of LFM without having any minimum range restrictions. A typical FMCW waveform is shown in Figure 1.7. FMCW radar

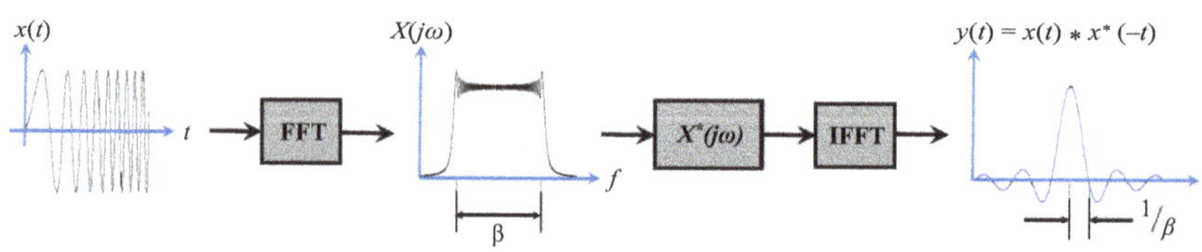

Figure 1.7 Chirp signal and its matched filter output

typically has very high bandwidth, resulting in fine range resolution. The maximum bandwidth of an FMCW radar is constrained by the pulse duration τ and the chirp rate γ. FMCW radars typically have separate antennas for transmit and receive, and hence no minimum detection range. The maximum range is limited by the PRI, which in this case is equal to the pulse duration, τ. Hence, FMCW is best suited for applications that require high resolution at short ranges.

To determine target range using stretch processing, consider the target return in relation to the transmitted signal, as illustrated in Figure 1.8. The round-trip flight time t_d can be found if the instantaneous difference in frequency, f_b, also known as the beat frequency, could be determined. This can be accomplished by splitting the transmit signal and using it as a reference signal that is mixed with the received signal. The mixing operation will generate a low frequency component at the beat frequency, as well as a high frequency component that is easily suppressed through application of a low-pass filter (LPF). The architecture of an FMCW radar is depicted in Figure 1.8. On the transmit chain, an exciter is required to generate a sinusoid of the desired center frequency. Voltage-controlled oscillators (VCOs), whose instantaneous frequency is controllable via an input voltage, are often used for this purpose. When the VCO is driven with a function generator, a linearly increasing input signal will result in an LFM waveform at the output. Typically, the output power of the VCO is boosted with a low-noise amplifier before the signal is transmitted. The received waveform is mixed with the transmitted reference signal as well as with a 90° phase-shifted copy of the reference to generate a complex I/Q output. Once the I/Q channels are digitized, a fast Fourier transform (FFT) can be applied in the digital signal processor (DSP),

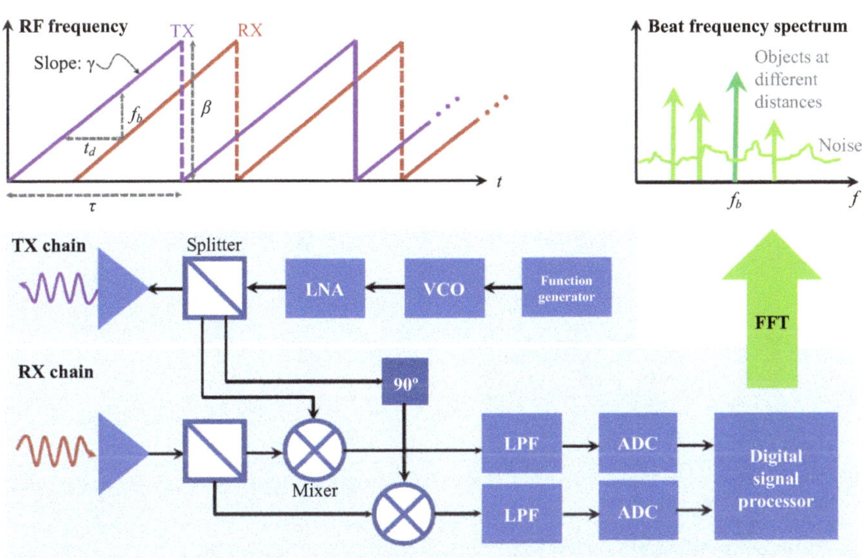

Figure 1.8 Typical FMCW radar and illustration of stretch processing

e.g., a computer or embedded processing platform, to measure the beat frequency. Hence, the target range found as

$$\frac{\beta}{\tau} = \frac{f_b}{t_d} = \frac{f_b}{2R/c} \rightarrow R = \frac{\tau c}{2\beta} f_b \tag{1.28}$$

Note that the analog-to-digital converter (ADC) bandwidth required can be reduced by adjusting the duration of the reference chirp used in the mixing process and dechirping about a nominal reference range.

1.2.6 Ambiguity functions and range-Doppler coupling

The matched filter yields the maximum output SNR when the amplitude and phase of the filter matches that of the received waveform. If we knew in advance the velocity of the target, the maximum output SNR would be attained. However, without prior knowledge, the velocity, and, hence, Doppler shift of targets result in unintentional mismatch between the received waveform and matched filter. *Ambiguity functions* provide a way to characterize the effect of target Doppler shift on the matched filter output. The ambiguity function is defined as

$$A(t, f_d) = \left| \int x(\alpha) \exp(j2\pi f_d \alpha) x^*(\alpha - t) d\alpha \right| \tag{1.29}$$

where $x(\alpha)\exp(j2\pi f_d\alpha)$ is the Doppler-shifted receive signal, and α is the dummy integration variable. The ideal ambiguity function, illustrated in Figure 1.9, is shaped like a thumbtack pointing upward: the matched filter output is maximum for the target range and Doppler, while everywhere else the ambiguity function is zero, so that there is no uncertainty in the radar measurement. In contrast, the ambiguity function for an FMCW waveform is a sheared ridge. Thus, any Doppler mismatch between the target and matched filter results in a shift in the range measurement. This is referred to as *range-Doppler coupling* and is a consequence of the range measurement being accomplished through the measurement of difference in received frequency in FMCW radars. Because of range-Doppler coupling, the range measurement for a stationary target versus a moving target located the same distance from the radar slightly differs.

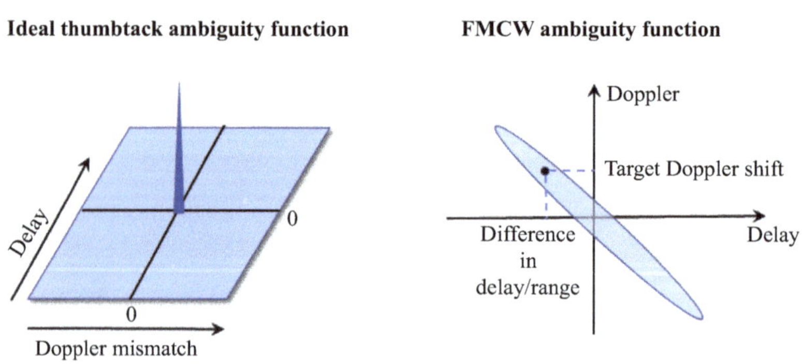

Figure 1.9 Ideal and FMCW ambiguity functions

1.3 Real and synthetic aperture radar processing

The raw I/Q data acquired by a radar system is typically segmented and stored as a 3D stack of matrices in fast-time and slow-time for each antenna element (also known as receive channel). *Fast-time* refers to the samples of the received signal as digitized by an ADC; thus, the time interval between each fast-time sample, Δt, is the inverse of the sampling frequency, f_s, in Hz: $\Delta t = 1/f_s$.

The interval between pulses, i.e., the PRI, also known as *slow-time*, is so termed as typically the PRI is much longer a duration than the ADC sampling interval. Figure 1.10 illustrates the radar data cube that results from this form of data stacking.

A variety of signal processing algorithms can be applied on 1D and 2D cross sections of this data cube to accomplish principle radar functions, such as detection, tracking, classification, and imaging. For example, *pulse compression* [1] involves application of a 1D matched-filter across fast-time and results in a peak at ranges corresponding to target locations. Hypothesis testing [2] is typically used to detect target returns of sufficiently large SNR above the noise floor. Similarly, *Doppler processing*, or the application of a discrete-time Fourier transform across slow-time will reveal peaks at the frequencies corresponding to the Doppler shift of the target motion. From the Doppler shift, the target's radial velocity can be extracted.

In subsequent sections, we describe in more detail some of the radar signal processing algorithms that can be applied on single transceivers, antenna arrays,

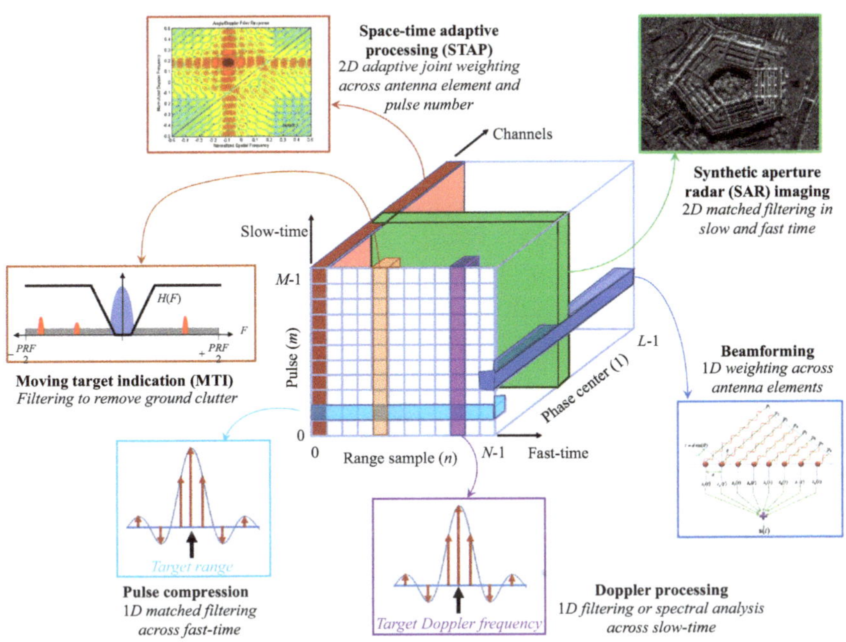

Figure 1.10 Radar data cube and common radar signal processing algorithms

and synthetic apertures to mitigate clutter returns, boost target SNR, and generate synthetic aperture radar (SAR) images.

1.3.1 Moving target indication

Often times we are interested in detecting and identifying moving targets in radar data, such as vehicles or people. In this context, all stationary radar returns comprise clutter. Moving target indication (MTI) [3] is a process for filtering out stationary ground clutter to improve the SNR, and, hence, detection or classification performance. One common way for designing MTI filters is based on the observation that the radar return from stationary targets does not change significantly from pulse to pulse. In contrast, the phase from moving targets does change. So, by subtracting successive pulses, the clutter component can be removed while moving target returns are preserved. The frequency responses for a two-pulse and three-pulse canceller are shown in Figure 1.11. Notice that the two-pulse canceller has a sharp null at 0 Hz. Having a broader null, such as obtained using a three-pulse canceller, is advantages for removing clutter. But, moving targets may be attenuated or amplified depending on the speed relative to the PRF.

Higher order N-pulse cancellers may be obtained by cascading two or three pulse cancellers and have a transfer function of

$$H_N(z) = (1 - z^{-1})^{N-1} \tag{1.30}$$

Other digital high-pass filters could also be designed for MTI; however, filters with infinite impulse response, such as Butterworth filters, incur errors due to truncation, so finite impulse response filters are more commonly used.

1.3.2 Array processing techniques

In a radar with multiple phase centers, such as in an antenna array, each additional array element enables spatial sampling along the azimuth that can be used for direction of arrival (DOA) estimation, boosting of the received target SNR and clutter suppression. Let us consider the linear antenna depicted in Figure 1.12 as an elementary example. The distance between each array element is a fixed value, d; thus, the difference in

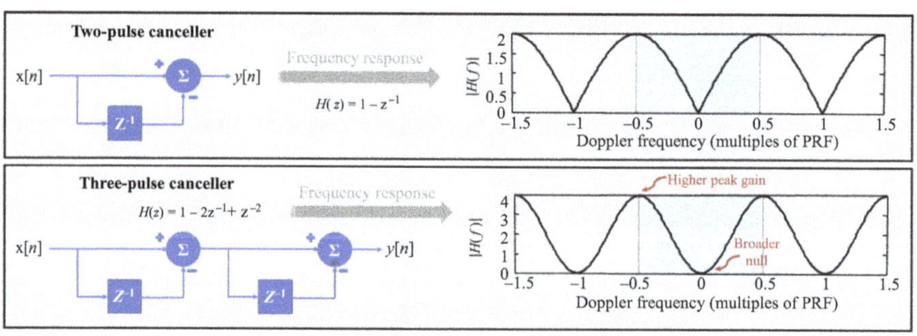

Figure 1.11 Pulse-cancellers for MTI

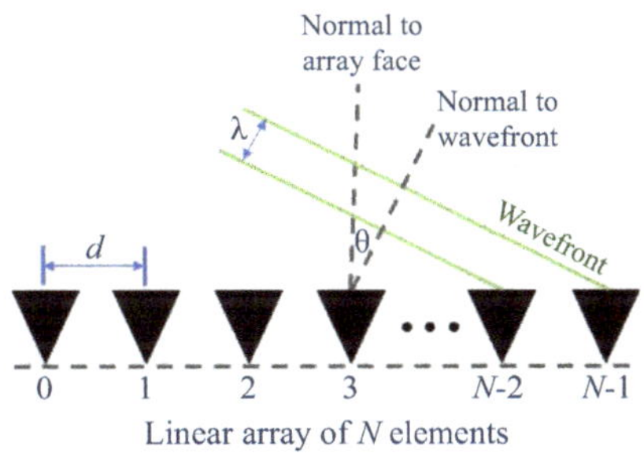

Figure 1.12 Illustration showing array-wavefront geometry for a linear array

time from the wavefront reaching adjacent elements is $d \sin(\theta)/c$. The signal observed at the nth array element may be written as

$$y_n(t) = Ae^{j[\Omega(t - nd \sin(\theta)/c) + \phi_o]} \tag{1.31}$$

Beamforming [4] is a technique that exploits constructive interference to increase the target SNR. Because the phase shift between each element is known given the DOA, θ, appropriate weighting may be applied to each array element so as to maximize the coherent integration of each element's received signal. Mathematically, a snapshot of the signals received in the array may be written as

$$\vec{y} = [y[0] \; y[1] \; \cdots \; y[N-1] \;]^T = A\vec{a}_s \tag{1.32}$$

where $\vec{a}_s$ is the *steering vector*

$$\vec{a}_s = [1 \;\; e^{-jK_\theta} \;\; e^{-j(n-1)K_\theta}]^T \tag{1.33}$$

and K_θ is the spatial frequency $2\pi d \sin(\theta)/\lambda$.

For a beamformer steered to θ_o, the filter output $z(\theta)$ is

$$z(\theta) = \vec{(h)}^T y = A \sum_{n=0}^{N-1} \alpha_n e^{-j(K_\theta - K_{\theta_o})n} \tag{1.34}$$

This is just the discrete Fourier transform (DFT) of the weights h_n. So, by taking the DFT, beamforming converts the data cube from a representation in terms of range-Doppler-channel to one of range-Doppler-*angle*, as shown in Figure 1.13.

In this space, the different components of the received signal, namely, the target, clutter, and interference, can be more distinctly visualized and often exhibit separation, which can be exploited for boosting SNR and mitigating ground clutter. This is illustrated in Figure 1.14, which shows how discrete targets, clutter, and jamming would appear in this *space-time* signal environment. *Space-time adaptive processing* [5–7] is an adaptive filtering method that can selectively design nulls to suppress unwanted signal components, such as clutter and jammers, thereby revealing target returns that might otherwise fall under the detection threshold. Figure 1.14 illustrates

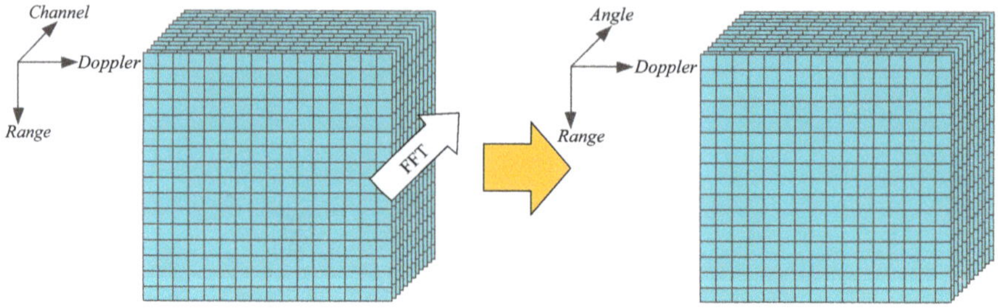

Figure 1.13 Digital beamforming (DBF)

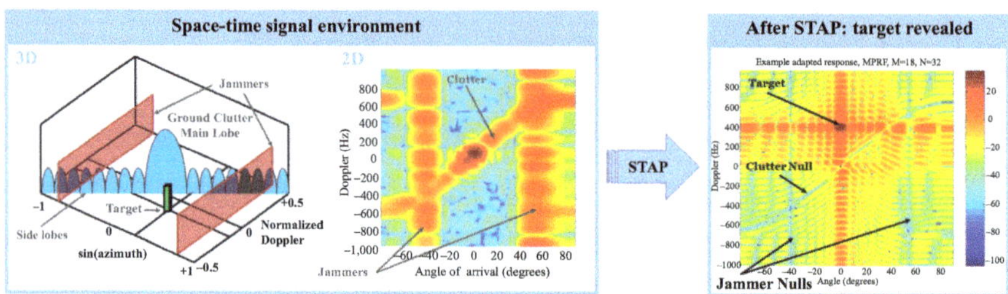

Figure 1.14 Space-time adaptive processing in angle-Doppler representation

the efficacy of STAP to suppress clutter and interference by showing a sample space-time signal environment before and after STAP. Notice that a null is directly positioned over unwanted signal components and that the target is now distinctly visible.

1.3.3 The micro-Doppler effect and time–frequency analysis

The micro-Doppler effect [8] is caused by the vibration or rotation of any part of a target resulting in frequency modulation in addition to the Doppler shift caused by the target's translational motion. For example, the rotation of the wheels of a vehicle, the treads of a tank, and the blades of a helicopter all result in radar micro-Doppler. In humans, the time-varying motion of the torso, arms, legs, hands, and feet results in a unique *micro-Doppler signature* that can be visually separated from that of other targets, including animals. Mathematically, the micro-Doppler signature is a time–frequency (TF) transform of the complex I/Q radar data, which reveals how the received Doppler frequencies vary with time.

TF analysis methods can be classified into linear transforms and quadratic time–frequency distributions (QTFDs) [9,10]. The former includes the short-time Fourier transform (STFT) [9], Gabor transform [11], fractional Fourier transform [12], and wavelet transform [13]. The most commonly applied TF transform is the spectrogram, denoted by $S(t, \omega)$, which is the square modulus of the STFT, and a special case of (1.37). It can be expressed in terms of the employed window function, $w(t)$, as

$$S(t, \omega) = \left| \int_{-\infty}^{\infty} w(t-u)x(u)du \right|^2 \tag{1.35}$$

While reaping the benefits of linearity, STFTs must trade off TF resolution due to fixed length windows. In contrast, alternative transforms can capture local behavior of the signal in differing ways.

The instantaneous autocorrelation function (IAF) of a continuous-time signal $x(t)$ is [9]

$$C(t, \tau) = x(t+\tau)x^*(t-\tau) \tag{1.36}$$

where τ is the time lag, and $(:)^*$ denotes the complex conjugate. The Wigner–Ville distribution (WVD) is the Fourier transform of the IAF with respect to τ, whereas the Fourier transform over the time delay t yields the ambiguity function, $A(\theta, \tau)$. The WVD has better TF resolution in comparison to the STFT; however, the presence of cross-term interference limits its practical application to micro-Doppler analysis. As a result, the Gabor–Wigner transform, defined as the product of the STFT with the Wigner distribution, has been proposed as a means for removing cross-term interference while maintaining the clarity and resolution of the result [14].

QTFDs, on the other hand, aim at concentrating the signal power along the instantaneous frequency of each signal component. In the case of radar backscattering from a moving target, such concentration accurately reveals the target velocity, acceleration, and higher order terms. The Cohen's class of QTFDs of signal $x(t)$ is defined as [15,16]

$$D(t, \omega) = FT_2[\Phi(\theta, \tau)A(\theta, \tau)] \int_{-\infty}^{\infty} \int_{-\infty}^{\infty} \Phi(\theta, \tau)A(\theta, \tau)e^{-j(\omega\tau + \theta t)}d\tau d\theta \tag{1.37}$$

where $FT_2[:]$ denotes 2D Fourier transform. The kernel $\Phi(\theta, \tau)$ has typically LPF characteristics to suppress signal cross-terms and preserve auto-terms, leading to reduced interference distributions [15]. The radar signal backscatterings from humans in motion are viewed as a deterministic signal in additive noise rather than random processes [17].

1.3.4 Synthetic aperture radar

The cross-range resolution (ΔCR) of a real aperture radar system is determined by the azimuth beamwidth (θ_{az}) of the antenna:

$$\Delta CR = 2R_o \sin\left(\frac{\theta_{az}}{2}\right) \approx R_o\theta_{az} \tag{1.38}$$

where R_o is the nominal range. The cross-range dimension is orthogonal to the range. Thus, the range resolution, ΔR and the cross-range resolution, ΔCR, together determine the minimum resolution cell in real distance on the ground. The azimuth beamwidth, in turn, is related to the size of the antenna as $\theta_{az} = k(\lambda/D_{az})$, where λ is the wavelength of the transmission and D_{az} is the physical length of the antenna in the azimuth direction. The scalar k is a factor that can vary according to antenna type.

Thus, to construct an antenna array with a narrow beam in azimuth (and hence finer cross-range resolution) a physically large antenna array is required. This would prove impractical for many remote sensing applications, such as satellite-based imagers, for which the nominal range R_o is large and the sensor platform itself cannot carry a large physical array. The idea behind a synthetic aperture is the generation of a large array by instead moving the platform carrying a single antenna across a large distance in space. At the transmission of each pulse, the position of the antenna will be different by (ideally) a fixed distance. By coherently combining the signals received at each position, the operation of a large physical array of elements located at each transmit position can be emulated. This concept is illustrated in Figure 1.15.

The Rayleigh beamwidth is defined as $\theta = \lambda/d$. When applied to the synthetic aperture, this yields a synthetic beamwidth of

$$\theta_{SAR} = \frac{\lambda}{2D_{SAR}} = \frac{\lambda}{2 \times v \times N \times PRI} \tag{1.39}$$

where v is the platform velocity, and N is the number of pulses transmitted (i.e., number of synthetic array elements). So, the SAR cross-range resolution is

$$\Delta CR = \frac{\lambda R_o}{2 \times v \times N \times PRI} \tag{1.40}$$

and the round-trip fast-time delay for a particular synthetic aperture location u is

$$t = \frac{2\sqrt{x^2 + (y-u)^2}}{c}. \tag{1.41}$$

Mathematically, the previous equation is that of a half-hyperbola. Consequently, prior to SAR focusing, the target will leave a hyperbolic signature at its location. There

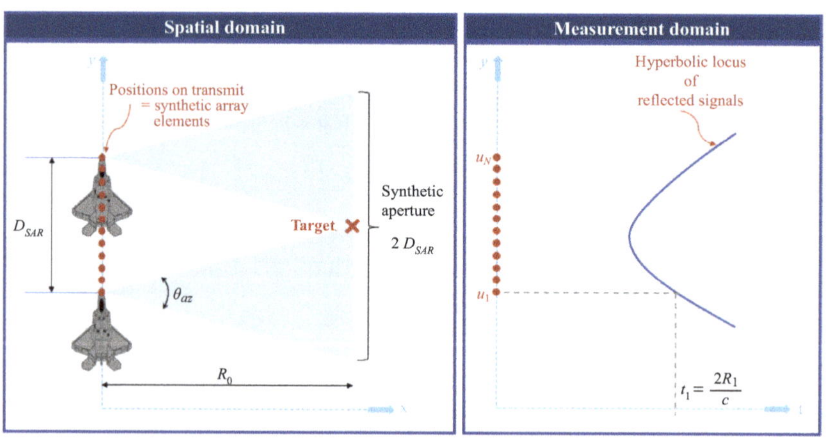

Figure 1.15 Synthetic aperture radar concept

are various techniques for coherently combining pulses, such as backprojection and the Omega–K algorithm, among others [18,19], which trade off computational complexity with image reconstruction accuracy. In practice, effective focusing of SAR images will additionally require compensation for platform motion. Coarse compensation can be done using GPS data, but more accurate methods for eliminating phase errors can be done using a variety of autofocus techniques. SAR imaging is a powerful processing technique with a wide range of applications in remote sensing, including space-based and airborne systems and ground penetrating radar.

1.4 Radar data denoising for machine learning

One of the nice attributes of deep convolutional nets is that the approach finds the features that help one to improve the classification. This means that all of the information in the radar imagery is taken into account. However, there are particular parts of the radar image that are known to be unrelated to the target, which can be a distraction. Sensitivity to distracting data unrelated to the target will produce a classifier that fails to generalize well. The main parts of a radar image that are not related to the target are the clutter and the noise. The diversity of the data collection in areas such as speed, direction, and orientation should also be considered. By carefully avoiding the issues in radar data and targets, machine learning can be highly effective on radar data.

It is recommended that radar imagery use a consistent clutter canceller before training and during deployment. This will ensure that the trained imagery is similar to the test imagery. Clutter cancellation is a large field in radar analysis, but for stationary radars with mild clutter conditions, the simple blanking of the clutter line or a ridge around the clutter line can be sufficient. For complex clutter applications, it is recommended that the potential clutter variations be injected into the imagery. This assumes that the training data is not available across all variations in clutter, which is almost always the case. This clutter injection significantly increases the number of training images that have to be analyzed, but since the clutter varies across imagery with the same target, the classifier will focus on the target characteristics instead of the clutter characteristics. This increases the intra-class variability in the clutter signals, which minimizes the importance of clutter signals to the classifier. By injecting clutter variations into multiple classes, the interclass variations are minimized, which also reduces the importance of clutter signals to the classifier. The analog for clutter in an optical image when you are looking for cats would be everything except the cats in the image, so the technique for including many clutter variations for the radar imagery is the same idea as using imagery of cats in many different backgrounds to help the classifier focus on the cat features between images.

Just as clutter varies with the environment, the noise varies with the radar and with the range to the target. To help the classifier avoid focusing on the noise, the imagery can be denoised by raising the minimum in the dB scale of the imagery. This technique has been used throughout this book both to make the imagery appear cleaner and also to prevent the classifier from focusing on the texture of the noise pattern for classification. By raising the noise floor slightly, most of the variations of the noise floor will not be visible in the imagery, while all of the signals will still be visible. This prevents the classifier from focusing on the signals from the noise and instead to focus on the signals from the target.

The target signature does not vary with the distance to the sensor in radar imagery the way that it does in optical images. In optical images, the target signature grows or shrinks with the distance to the sensor, which means that multiple resolutions have to be included in the dataset. This can often be done in software that minimizes the amount of data that needs to be collected. The radar signature in micro-Doppler imagery does not vary at all with range, though this is not true in SAR imagery. In micro-Doppler imagery, a change in range means a change in the SNR for the imagery. Using the range equation, the signal is degraded by the distance to the sensor. However, the noise is typically a product of the sensor itself and does not change with range. In order to make a classifier for radar data that is useful at multiple ranges, injecting noise into the collected imagery will give a more general classifier. Because multiple images with different noise levels are presented to the classifier with the same label, the classifier will tend to avoid using the noise signals in the classification and instead focus on the targets, which is our goal in setting up our dataset.

To demonstrate how important noise injection can be for the performance of a classifier, an example comparison can be done with one classifier trained without noise injection versus one with noise injection. The first network is trained with only unaltered data, while the second network was trained with samples from the unaltered data and the data with multiple levels of artificially induced noise. The classes can include different target classes with similar characteristics, confuser classes, and one empty class. The empty class is included to prevent what are called "false alarms" or classifications when there is nothing in the data. This is a fundamental issue for radars, so including an empty class is highly recommended. For the network trained only with the noise-free data, the performance typically falls significantly on imagery with lower SNR. As the SNR falls further, the performances of the classifiers typically vary significantly. By including the empty data imagery, it is often found that deep learning classifiers are exceptional at detecting targets regardless of the class.

Overall, the most important treatments to give data before using a classifier include clutter rejection or augmentation, diversity of the target motions or diversification in software, and most importantly the denoising or noise injection of the imagery. All of these techniques are designed to enhance the diversity of the dataset and to enable the classifier to generalize well to new data in new environments. Overall, the robustness of the classifier can be significantly increased by employing these techniques.

1.5 Radar data representations for machine learning

The previous section shows that there are a wide range of radar signal processing algorithms that may be employed to better reveal target characteristics from the overall radar return signal. Thus, one of the first decisions that must be made prior to applying any machine learning algorithm on the data is *how to best represent the radar data*. One approach could be to directly classify the 1D complex time stream acquired by the radar. This would require application of a complex neural network, or representing each I/Q channel as a separate channel of real data. Moreover, this relies on the deep neural network (DNN) to automatically learn relevant features while forcing the DNN itself to sort out the target return from clutter and interference. Not only are there losses involved with processing inherently complex data as multichannel real data, but the DNN is being forced to learn an unnecessarily high dimensional mapping of data to target classes. This drives up the computation cost and complexity of the resulting network while also precluding utilization of radar signal processing approaches that can mitigate clutter and simplify the DNN architecture—these are techniques that typically also result in networks with more robust performance, simpler training, and optimization.

In short, the optimal representation of radar data is still a subject of research and is application dependent, as well as a key factor in subsequent design of the architecture (complex versus real, 1D/2D/3D, or recurrent) of the DNN utilized. Sometimes, multiple representations can be exploited for different purposes in a given sensing problem. One application that involves not only the full range of radar signal processing algorithms but also exploits different representations is the challenge of remote sensing with automotive radar. As a case study, we discuss data representations in automotive radar in more detail, next.

1.5.1 State-of-the-art automotive radar

Since the late of 1990s, radar sensors have found widespread applications in advanced driver assistance systems (ADAS), such as adaptive cruise control (ACC) and automatic emergency braking (AEB). More recently, radar has emerged as one of the key technologies in autonomous driving systems, providing environmental perception under all weather conditions. Some of today's self-driving cars, such as Zoox, have over ten radars, providing a 360° surround sensing. Different from ground-based or airborne surveillance radars, automotive radars have small size (multi-inch by multi-inch), short range (within multi-hundred meters), low power (multi-watt), and low cost, integrated behind the vehicle bumper or windshield, operating in a highly dynamic environment with multipath [20].

Figure 1.16 shows a typical radar perception configuration for autonomous driving. Short-range radar (SRR) sensors with an azimuth field of view (FOV) $[-75°, 75°]$ and a detection range of 45 m are deployed at the four corners of the vehicle for blind spot detection, cross traffic alert, etc. Midrange radar (MRR) sensors with an azimuth FOV of $[-40°, 40°]$ and a detection range of 100 m are deployed at both front and rear sides for lane change assist and AEB. A long-range radar (LRR) sensor with an azimuth FOV of $[-15°, 15°]$ and a detection range of 250m is deployed at the front for ACC. All radar sensors in a vehicle are connected to an electronic control unit for further processing, such as radar tracking and sensor fusion.

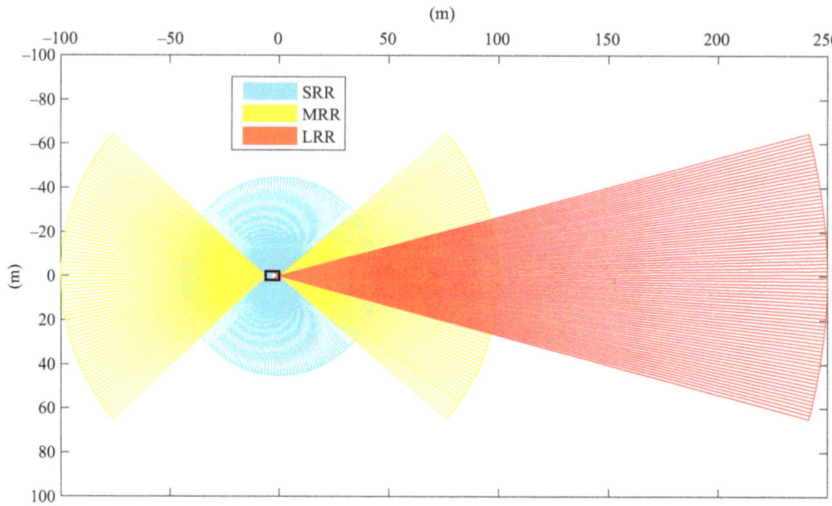

Figure 1.16 A typical radar-perception configuration for autonomous driving contains four short-range radars, deployed at the four corners of the car and providing 360° coverage; two midrange radars for both forward and rearward looking; and one long-range radar for forward looking

1.5.2 Automotive radar with MIMO radar technology

For control critical functions, such as AEB, distant obstacles need to be detectable with high angular resolution. Therefore, automotive radar for ADAS and autonomous driving needs to have high angle discrimination capability. In state-of-the-art automotive FMCW radar, the range and Doppler parameters of targets can be estimated by using single receive antenna. However, to estimate the angle parameter of targets, a receive antenna array is needed. Multi-input–multi-output (MIMO) radar [21–23] can synthesize virtual arrays with large aperture using only a small number of transmit and receive antennas. This advantage has been exploited by almost all major automotive suppliers, in their different types of radar products, such as SRR, MRR, and LRR.

In MIMO radar, the transmit antennas transmit FMCW sequences in a way that guarantees their orthogonality. At each receive antenna, the contribution of each transmit antenna is extracted by exploiting waveform orthogonality. For M_t transmit antennas and M_r receive antennas, a virtual array with M_tM_r elements can be synthesized. The array response of the *synthesized* array, i.e., the array corresponding to a MIMO radar can be expressed as $\mathbf{a}_t(\theta_t) \otimes \mathbf{a}_r(\theta_r)$, where $\mathbf{a}_t(\theta_t)$ and $\mathbf{a}_r(\theta_r)$ are the transmit and receive array steering vectors, corresponding to direction of departure θ_t and DOA θ_r, respectively. Here, $\otimes$ denotes the Kronecker product.

Virtual array synthesis in automotive FMCW radar using MIMO radar technology relies on the separability of the transmit signals of the different antennas.

The separation is easier when the transmit signals of different antennas are orthogonal. In the following, we review techniques to achieve waveform orthogonality while transmitting FMCW, such as time division multiplexing (TDM), Doppler division multiplexing (DDM), and frequency division multiplexing [20].

Figure 1.17 shows one MIMO radar configuration with $M_t = 2$ transmit and M_r receive antennas with $d_t = M_r d_r$ and $d_r = \lambda/2$. Here, $\lambda = c/f_c$ is the wavelength. It can be verified that the synthetic virtual arrays are equivalent to uniform linear array (ULA) with $M_t M_r$ elements and spacing of d_r when $\theta_t = \theta_r$.

In automotive MIMO radar with M_t transmit and M_r receive antennas, a virtual ULA of $M_t M_r$ elements can be synthesized with inter-element spacing d. The array response can be written as

$$\mathbf{y} = \mathbf{A}(\theta)\,\mathbf{s} + \mathbf{n}, \tag{1.42}$$

where $\mathbf{A}(\theta) = [\mathbf{a}(\theta_1), \ldots, \mathbf{a}(\theta_K)]$ is the virtual array steering matrix with

$$\mathbf{a}(\theta_k) = \left[1, e^{j(2\pi/\lambda)d\sin(\theta_k)}, \ldots, e^{j(2\pi/\lambda)(M_t M_r - 1)d\sin(\theta_k)}\right]^T. \tag{1.43}$$

Here, $\mathbf{n}$ is a noise term and $\mathbf{s} = [\beta_1, \ldots, \beta_K]^T$, where β_k denotes the target reflection coefficient for the kth target. The array response at a particular time instance consisting of data obtained at all the virtual receivers and corresponding to the same range-Doppler bin is the *array snapshot*. In highly dynamic automotive scenarios, usually, only a small number of array snapshots, or even a single snapshot in the worst case is available.

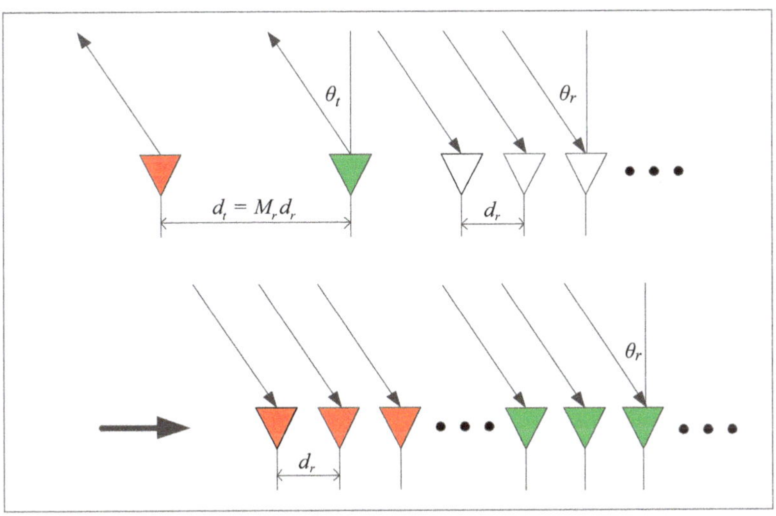

Figure 1.17 Illustration of one MIMO radar virtual array configuration using TDM or DDM scheme with $M_t = 2$ transmit antennas and M_r receive antennas with $d_t = M_r d_r$ and $d_r = \lambda/2$. Different colors indicate the transmit antenna either transmits different time slots or codes.

In automotive MIMO radar with virtual ULA, angle finding can be done with digital beamforming (DBF) by performing FFT on snapshots taken across the array elements, i.e., $\mathbf{y}$ in (1.42) (see Figure 1.13). DBF can be implemented efficiently in an embedded DSP with a single snapshot. However, DBF is not a high-resolution angle finding method. Higher resolution angle finding can be achieved with subspace-based methods, such as MUSIC [24] and ESPRIT [25], sparse sensing-based methods [26–30], or the iterative adaptive approach [31,32].

1.5.3 High-resolution automotive radar and point clouds

For Levels 4 and 5 (L4/L5) autonomous driving, both azimuth and elevation angular resolution of automotive radar need to be less than 1°. Current generation of automotive radar for ADAS has rather limited ability to resolve closely spaced targets. High-resolution imaging radar with hundreds of virtual array elements are currently being developed to produce the so-called *point clouds* that are groups of points detected by radar representing the object's shape [33].

LiDAR systems have better angular resolution (less than 1°) and have been introduced in L4/L5 autonomous driving systems. LiDAR can provide point clouds. Via the use of DNNs, such as PointNet [34] and PointNet++ [35], the point clouds can lead to target identification. However, due to its use of light spectrum wavelength, LiDAR is susceptible to bad weather conditions, such as fog, rain, snow, and dust in the air. In addition, the cost of LiDAR is high.

On the other hand, automotive radar with millimeter waveform technology has the potential to provide point clouds at much lower cost than LiDAR, and with more robustness to weather conditions. Such radar is referred to as high-end radar, or *imaging radar* [38]. Computer vision techniques that were previously reserved for high-resolution camera sensors and LiDAR systems can be applied to imaging radar data to identify targets. For example, a car can be identified based on 2D radar points of an imaging radar using PointNet. Figure 1.18 shows an example of a high-resolution automotive radar 2D detection points in highway scenario in one frame. The radar used in this experiment has 8 transmit, 16 receive antennas, and a bandwidth of 3 GHz, shown in Figure 1.19 [36,37]. With MIMO radar technology, a virtual linear array of 128 elements can be synthesized for azimuth angle finding. As a further example, Figure 1.20 shows a comparison of imaging radar detection points, LiDAR, and camera with 3D annotation.

General Motors and Texas Instruments successfully demonstrated that up to 4 Texas Instruments AWR1243 radar chips can be cascaded together to provide 12 transmit and 16 receive antennas, enabling a synthesis of 192 virtual array elements [40]. Uhnder has developed a radar system-on-chip that has 12 transmit and 16 receive antennas, enabling synthesis of 192 virtual array elements [41].

1.5.3.1 Micro-Doppler representation of vulnerable road users

In autonomous driving, the vulnerable road users (VRUs), including pedestrians and cyclists, need to be detected and classified. With automotive radar data, a variety of signal processing tools, e.g., FFT, STFT, filtering, and beamforming have been

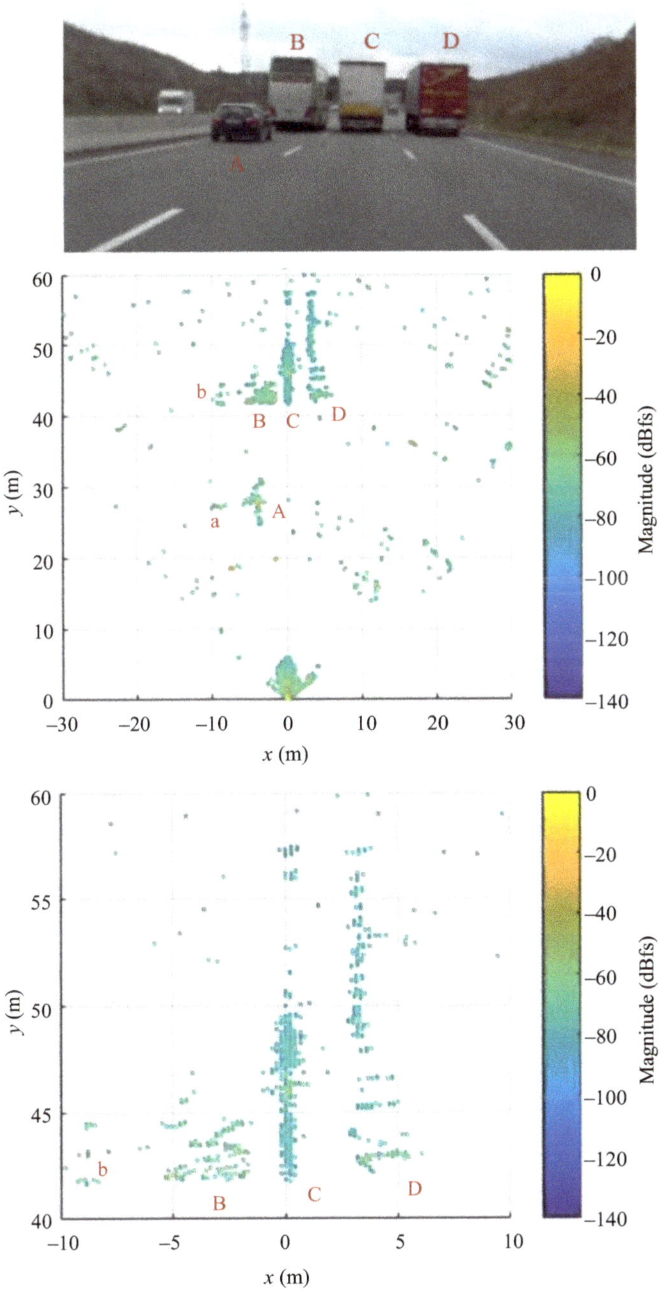

Figure 1.18 A frame of high-resolution automotive radar performance in the highway scenario (top) and corresponding radar detection points (middle) with zoom on the relevant objects (bottom). Here, due to multipath propagation introduced by the barrier, a and b are ghost targets, which are mirrors of real objects A and B (Source: Robert Bosch GmbH [36]).

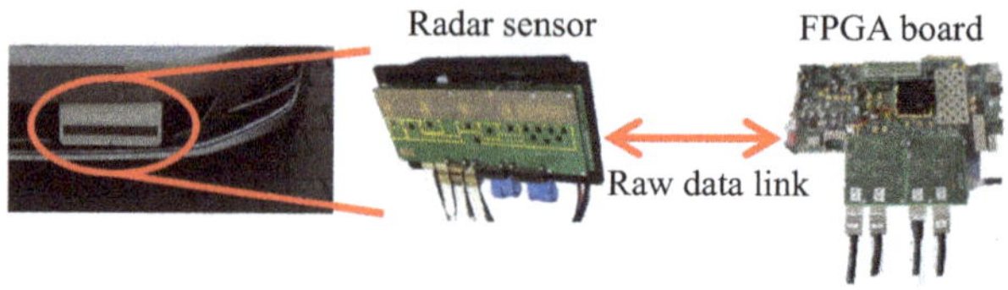

Figure 1.19 A high-resolution automotive radar system of 8 transmit, 16 receive antennas, and a bandwidth of 3 GHz by Robert Bosch GmbH [37]

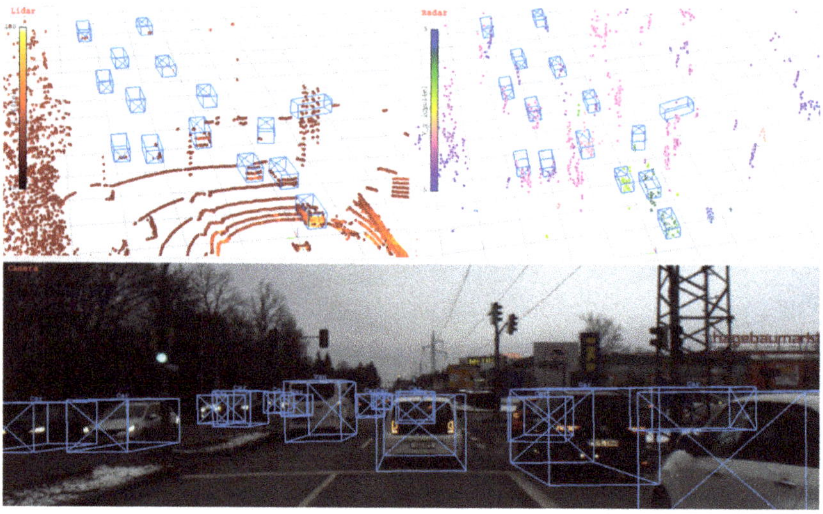

Figure 1.20 A frame of high-resolution imaging radar, LiDAR, and camera with 3D annotation (Source: Astyx [39])

adopted to obtain target features, such as micro-Doppler spectrum of pedestrians. Machine learning algorithms and DNNs have also been applied in automotive radar for VRU recognition and classification.

It is well-known that the relative movement between target and radar generates Doppler effect. There are two types of relative movements. The first type is between the main body such as pedestrian torso and radar. The second type is between parts of VRUs, such as legs or arms of pedestrian, wheels, or pedals of a bike, which would generate the so-called micro-Doppler signature. Many parts of the VRUs have periodic actions and as a result, the micro-Doppler signature has pattern of cycles. Different VRUs are characterized by distinguishable micro-Doppler features that can be exploited for target classification. Figure 1.21 shows micro-Doppler pattern of a pedestrian walking to the Texas Instruments automotive radar AWR1642 evaluation board running at 77 GHz [42].

One of the challenges of using micro-Doppler signatures for VRU detection and classification is that the RCSs of VRUs are relatively low. Design of automotive radars with high dynamic range is of great interest. Further, since radar measures the radial

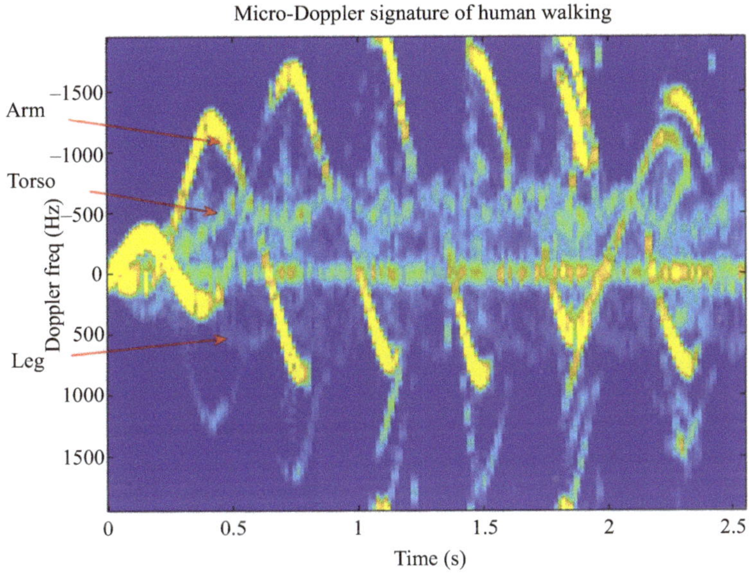

Figure 1.21 Micro-Doppler pattern of a walking pedestrian using real measurement of a Texas Instruments AWR1642 automotive radar evaluation board in 77 GHz [42]

Doppler effect, micro-Doppler signature changes dramatically as the DOA of VRUs to automotive radar varies. In the extreme case of 90° DOA, there is no Doppler effect. In general, multiple automotive radars incorporated in a single vehicle, such as two SRRs at front left and right corners of a vehicle, can be connected and collaboratively carry out VRUs detection and classification.

1.6 Additional reading

There are a number of excellent text books regarding the fundamentals of radar systems, signals, and signal processing. A 3-volume book series that covers all aspects of radar systems at an introductory level is *Principles of Modern Radar: Basic Principles*, edited by Richards *et al.* [43]. The second and third volumes of this series, *Advanced Techniques* [44] and *Radar Applications* [45], cover advanced topics such as radar waveforms and waveform design, SAR, array processing and interference mitigation, passive radar, and post-processing techniques, such as target tracking. Applications detailed include CW radar, weather radar and air traffic control, pulse-Doppler radar, fire control radar, ground MTI and ground penetrating radar. A fundamental textbook that is the basis for many graduate radar signal processing classes is *Fundamentals of Radar Signal Processing* authored by Richards [46]. A textbook that utilizes MATLAB® in a more hands-on approach to introducing radar is *Radar Systems Analysis and Design Using MATLAB* by Mahafza [47]. The text *Microwave Radar and*

Radiometric Remote Sensing by Ulaby *et al.* [48] provides an in-depth description of EM scattering from a variety of natural surfaces for application to radar remote sensing of the environment. Solid introductory texts, especially for undergraduate students, include *Introduction to Airborne Radar* by Stimson *et al.* [49] and *Small and Short-Range Radar Systems* by Charvat [50], which is also a companion text for MIT's Coffee Can Radar Project [51]. While there are many more outstanding texts on radar, it is our hope that this short list will serve as an effective starting point for engineers interested in learning more about radar.

References

[1] Barton DK. *Pulse Compression, Radars*, Volume 3. Artech Radar Library, Artech Publishing, Boston; 1975.

[2] Kay S. *Fundamentals of Statistical Signal Processing, Volume II: Detection Theory*. Prentice Hall, Upper Saddle River, NJ, USA; 1998.

[3] Schleher DC. *MTI and Pulsed Doppler Radar With MATLAB*. Artech House Remote Sensing Library, Artech Publishing, Boston; 2009.

[4] Johnson DH and Dudgeon DE. *Array Signal Processing: Concepts and Techniques*. Prentice Hall, Upper Saddle River, NJ, USA; 1993.

[5] Klemm R. *Principles of Space-Time Adaptive Processing*. Radar Sonar and Navigation Series. London, The Institution of Engineering and Technology; 2006.

[6] Ward J. *Space-Time Adaptive Processing for Airborne Radar*. Lincoln Laboratory Technical Report No. 1015 / ESC-TR-94-109; 1994.

[7] Guerci J. *Space-Time Adaptive Processing for Radar, 2nd Edition*. Artech House, Boston; 2014.

[8] Chen VC. *The Micro-Doppler Effect in Radar, 2nd Edition*. Artech House, Boston; 2019.

[9] Cohen L. Time-frequency distributions-a review. *Proceedings of the IEEE*. 1989;77(7):941–981.

[10] Chen V and Ling H. *Time-Frequency Transforms for Radar Imaging and Signal Analysis*. Artech House, Boston; 2002.

[11] Tivive FHC, Phung SL, and Bouzerdoum A. Classification of micro-Doppler signatures of human motions using log-Gabor filters. *IET Radar, Sonar Navigation*. 2015;9(9):1188–1195.

[12] Almeida LB. The fractional Fourier transform and time-frequency representations. *IEEE Transactions on Signal Processing*. 1994;42(11):3084–3091.

[13] Daubechies I. The wavelet transform, time-frequency localization and signal analysis. *IEEE Transactions on Information Theory*. 1990;36(5):961–1005.

[14] Sridharan K, Thayaparan T, SivaSankaraSai S, *et al.* Gabor-Wigner transform for micro-Doppler analysis. In: *9th Int. Radar Symp*. Bangalore, India; 2013.

[15] Choi H and Williams WJ. Improved time-frequency representation of multicomponent signals using exponential kernels. *IEEE Transactions on Acoustics, Speech, and Signal Processing*. 1989;37(6):862–871.

[16] Amin MG. Spectral decomposition of time-frequency distribution kernels. *IEEE Transactions on Signal Processing*. 1994;42(5):1156–1165.

[17] Boashash B. *Time-Frequency Signal Analysis and Processing: A Comprehensive Reference (Eurasip and Academic Press Series in Signal and Image Processing) 2nd Edition*. Academic Press, London, UK; 2016

[18] Soumekh M. *Synthetic Aperture Radar Signal Processing: With MATLAB Algorithms*. John Wiley & Sons, New York, NY; 1999.

[19] Cumming IG and Wong FH. *Digital Processing of Synthetic Aperture Radar Data*. Artech House, Boston, MA; 2005.

[20] Sun S, Petropulu AP, and Poor HV. MIMO radar for advanced driver-assistance systems and autonomous driving: advantages and challenges. *IEEE Signal Processing Magazine*. 2020;37(4):98–117. doi: 10.1109/MSP.2020.2978507.

[21] Li J and Stoica P. MIMO radar with colocated antennas. *IEEE Signal Processing Magazine*. 2007;24(5):106–114.

[22] Li J and Stoica P. *MIMO Radar Signal Processing*. Hoboken, NJ, Wiley; 2009.

[23] Bergin J and Guerci JR. *MIMO Radar: Theory and Application*. Boston, MA, Artech House; 2018.

[24] Schmidt R. Multiple emitter location and signal parameter estimation. *IEEE Transactions on Antennas and Propagation*. 1986;34(3):276–280.

[25] Roy R and Kailath T. ESPRIT – Estimation of signal parameters via rotation invariance techniques. *IEEE Transactions on Acoustics, Speech, and Signal Processing*. 1989;17(7):984–995.

[26] Yu Y, Petropulu AP, and Poor HV. MIMO radar using compressive sampling. *IEEE Journal of Selected Topics in Signal Processing*. 2010;4(1):146–163.

[27] Yu Y, Sun S, Madan RN, *et al.* Power allocation and waveform design for the compressive sensing based MIMO radar. *IEEE Transactions on Aerospace and Electronic Systems*. 2014;50(2):898–909.

[28] Strohmer T and Friedlander B. Analysis of sparse MIMO radar. *Applied and Computational Harmonic Analysis*. 2014;37:361–388.

[29] Sun S, Bajwa WU, and Petropulu AP. MIMO-MC radar: a MIMO radar approach based on matrix completion. *IEEE Transactions on Aerospace and Electronic Systems*. 2015;51(3):1839–1852.

[30] Sun S and Petropulu AP. Waveform design for MIMO radars with matrix completion. *IEEE Journal of Selected Topics in Signal Processing*. 2015;9(8):1400–1411.

[31] Yardibi T, Li J, Stoica P, *et al.* Source localization and sensing: a nonparametric iterative adaptive approach based on weighted least squares. *IEEE Transactions on Aerospace and Electronic Systems*. 2010;46(1):425–443.

[32] Roberts W, Stoica P, Li J, *et al.* Iterative adaptive approaches to MIMO radar imaging. *IEEE Journal of Selected Topics in Signal Processing*. 2010;4(1):5–20.

[33] Bilik I, Bialer O, Villeval S, *et al.* Automotive MIMO radar for urban environments. In: *Proc. IEEE Radar Conference*. Philadelphia, PA; 2016.

[34] Qi CR, Su H, Mo K, *et al.* PointNet: deep learning on point sets for 3D classification and segmentation. In: *Proc. the IEEE Conference on Computer Vision and Pattern Recognition (CVPR)*. Honolulu, HI; 2017.

[35] Qi CR, Yi L, Su H, *et al.* PointNet++: deep hierarchical feature learning on point sets in a metric space. In: *Proc. 31st Conference on Neural Information Processing Systems (NIPS)*. Long Beach, CA; 2017.

[36] Meinl F, Stolz M, Kunert M, *et al.* An experimental high performance radar system for highly automated driving. In: *Proc. Intl. Conference on Microwaves for Intelligent Mobility (ICMIM)*. Nagoya, Japan; 2017.

[37] Li M, Stolz M, Feng Z, *et al.* An adaptive 3D grid-based clustering algorithm for automotive high resolution radar sensor. In: *Proc. IEEE Intl. Conference on Vehicular Electronics and Safety (ICVES)*. Madrid, Spain; 2018.

[38] Brisken S, Ruf F, and Hohne F. Recent evolution of automotive imaging radar and its information content. *IET Radar, Sonar & Navigation*. 2018;12(10):1078–1081.

[39] Meyer M and Kuschk G. Automotive radar dataset for deep learning based 3D object detection. In: *Proc. European Radar Conference (EuRAD)*. Paris, France; 2019.

[40] Bilik I, Villeval S, Brodeski D, *et al.* Automotive multi-mode cascaded radar data processing embedded system. In: *Proc. IEEE Radar Conference*. Oklahoma City, OK; 2018.

[41] Giannini V, Goldenberg M, Eshraghi A, *et al.* A 192-virtual-receiver 77/79GHz GMSK code-domain MIMO radar system-on-chip. In: *Proc. IEEE Intl. Solid-State Circuits Conference (ISSCC)*. San Francisco, CA; 2019.

[42] Zhang R and Cao S. Real-time human motion behavior detection via CNN using mmWave radar. *IEEE Sensors Letters*. 2019;3(2):1–4. doi: 10.1109/LSENS.2018.2889060.

[43] Richards M, Sheer J, and Holm W. *Principles of Modern Radar: Basic Principles*. SciTech Publishing; 2010.

[44] Melvin WL and Sheer J. *Principles of Modern Radar: Advanced Techniques*. SciTech Publishing, Raleigh, NC; 2012.

[45] Melvin LW and Scheer JA. *Principles of Modern Radar: Radar Applications*. SciTech Publishing, Raleigh, NC; 2013.

[46] Richards M. *Fundamentals of Radar Signal Processing*. McGraw Hill, New York, NY; 2014.

[47] Mahafza B. *Radar Systems Analysis and Design Using MATLAB*. Chapman and Hall/CRC, Boca Raton, FL; 2016.

[48] Ulaby F and Long D. *Microwave Radar and Radiometric Remote Sensing*. University of Michigan Press, Ann Arbor, MI; 2013.

[49] Stimson GW, Griffiths HD, Baker CJ, *et al. Introduction to Airborne Radar*. SciTech Publishing, Raleigh, NC; 2014.

[50] Charvat G. *Small and Short-Range Radar Systems*. CRC Press, Boca Raton, FL; 2014.

[51] MIT. Build a Small Radar System Capable of Sensing Range, Doppler, and Synthetic Aperture Radar Imaging; 2020. https://ocw.mit.edu/resources/res-ll-003-build-a-small-radar-system-capable-of-sensing-range-doppler-and-synthetic-aperture-radar-imaging-january-iap-2011/.

Chapter 2
Basic principles of machine learning

Ali Cafer Gurbuz[1] *and Fauzia Ahmad*[2]

A fundamental feature of advanced cognitive processing is the ability to seek and distinguish patterns. In fact, pattern recognition is key to various unique aspects of human intelligence, such as creativity, invention, language, communication, decision making, and reasoning. As such, pattern recognition and machine learning (ML) play a central role in artificial intelligence. ML can be broadly defined as computational methods that exploit experience and/or available data in order to improve a defined performance metric to accomplish a specific task. In traditional computing, a predefined fixed set of rules or computations is applied to the data to generate an output. In essence, the mapping from the data domain to the output domain is predefined. Unlike traditional computing, ML uses training data and the desired outputs to learn the rules for mapping. Thereafter, when presented with new data, the ML algorithm exploits the learned mapping to generate predictions that are able to scale well with the size of the data.

With the ever increasing amounts and types of data, ML has emerged over the past decades as a major approach in data science and signal processing with applications in nearly all aspects of life. This chapter is specifically focused on applications of ML in radar. Although radar technology has traditionally been used in military and security, it has recently found many civilian applications, including automotive and autonomous systems, remote sensing, smart cities, and healthcare. The increased accessibility to radar data and the wide range of radar-related tasks together with the inherent challenges associated with processing and interpreting radar measurements render ML, smart data analysis, and signal processing necessary ingredients for many radar problems.

This chapter provides an overview of the basic principles of ML, outlining the fundamental concepts that need to be applied correctly for a broad range of radar applications. We expect the reader to have background knowledge of basic linear algebra and probability theory, which form the foundations of ML. In Section 2.1, we describe the concept of learning from data and introduce the main categories of ML, namely, supervised and unsupervised learning. We also present different tasks that ML can tackle under each category and provide relevant radar-based examples.

[1]Department of Electrical and Computer Engineering, Mississippi State University, Starkville, MS, USA
[2]Department of Electrical and Computer Engineering, Temple University, Philadelphia, PA, USA

In Section 2.2, we briefly describe the various components of an ML algorithm. We present several fundamental techniques of supervised and unsupervised learning in Section 2.3. In Section 2.4, we define various performance assessment metrics and describe the design and evaluation of a learning algorithm. More recent learning approaches, such as variants of deep neural networks (DNNs), and more specific ML tools related to the various radar applications will follow in subsequent chapters of this book.

2.1 Learning from data

The concept of learning from data is the core principle of an ML algorithm. What we imply by learning is succinctly defined in [1] as follows:

> A computer program is said to learn from experience E with respect to some class of tasks T and performance measure P, if its performance at tasks in T, as measured by P, improves with experience E.

The task T may include detection, estimation, or classification, all of which are highly relevant to the radar modality. For example, an automotive radar, in conjunction with other sensors, can be used to classify the scene being observed into drivable road or targets (automobiles, pedestrians, etc.) [2,3]. Synthetic aperture radar (SAR) images can be used to classify a target of interest, say a tank, into several intended role-based or weight-based tank classes [4]. Radar micro-Doppler signatures can be used to distinguish between different human activities [5–7]. The presence or absence of a target in a given range-Doppler bin or a crossrange–downrange image pixel can be determined in the presence of clutter using ML algorithms [8].

The experience E can consist of a dataset by itself, a dataset associated with labels, or an online interaction with the environment. These respective experiences lead to the categorization of ML algorithms as *unsupervised, supervised*, or *reinforcement learning (RL)*. More specifically, unsupervised learning refers to extracting structure or useful information from an unlabeled dataset. On the other hand, if the transformation between input and output spaces is learned using a dataset where each data sample is associated with a labeled output, the resultant algorithm is supervised. In RL, the agents take actions in an environment they interact with and learn to take optimal actions in order to maximize a cumulative reward function. Next, we describe supervised and unsupervised learning in more detail. RL is omitted for brevity. For more details on RL, the reader is referred to [9].

2.1.1 Supervised learning

In supervised learning, we have a labeled dataset $\mathbb{D}$, commonly known as the training dataset, which is used to learn a mapping from each input data sample $\mathbf{x}_i$ to the corresponding output $\mathbf{y}_i$. In this case, the dataset $\mathbb{D}$ can be represented as

$$\mathbb{D} = \{(\mathbf{x}_i, \mathbf{y}_i),\ i = 1, \ldots, N\} \tag{2.1}$$

where N is the number of samples in the dataset. Input and output data sample pairs $(\mathbf{x}_i, \mathbf{y}_i)$ can be discrete or continuous depending on the task. They can also be in the form of a scalar, vector, matrix, or a tensor of arbitrary size. In order to set up the learning process, we require two functions. First, we need a decision function (model), parameterized by unknown parameters $\boldsymbol{\theta}$, which maps the input $\mathbf{x}$ to an estimated output $\widehat{\mathbf{y}}$ as

$$\widehat{\mathbf{y}} = f_{\boldsymbol{\theta}}(\mathbf{x}). \tag{2.2}$$

We want to learn the mapping function $f_{\boldsymbol{\theta}}(\mathbf{x})$, thereby learning the parameters $\boldsymbol{\theta}$, using the training dataset $\mathbb{D}$. To this end, the second required function is a loss function that informs the ML algorithm as to how good or bad it is performing in its objective of learning the mapping. The loss function $\ell(\widehat{\mathbf{y}}, \mathbf{y}_i) \in \mathbb{R}$ measures some kind of distance between the estimated output $\widehat{\mathbf{y}}$ and the labeled expected output $\mathbf{y}_i$ for the corresponding input $\mathbf{x}_i$. The choice of the decision and loss functions is dictated by the specific technique adopted for supervised learning. With the two functions in hand, the goal of the supervised learning is to determine the parameter vector $\boldsymbol{\theta}$ that minimizes the total loss over the training dataset. That is,

$$\boldsymbol{\theta}^* = \underset{\boldsymbol{\theta}}{\text{argmin}} \sum_{i=1}^{N} \ell(f_{\boldsymbol{\theta}}(\mathbf{x}_i), \mathbf{y}_i). \tag{2.3}$$

A solution to (2.3) can be obtained using various techniques, such as *stochastic gradient descent* (SGD) [10]. The main idea behind the training process with SGD is to update the parameters iteratively in such a manner that the total loss function is reduced, that is, by taking small steps opposite to the gradient.

$$\boldsymbol{\theta}^{(t+1)} = \boldsymbol{\theta}^{(t)} - \eta_t \nabla \ell(\mathbf{f}_{\boldsymbol{\theta}}(\mathbf{x}_i), \mathbf{y}_i) \tag{2.4}$$

The learning rate η_t in (2.4) is a hyperparameter that controls the step size in the update iterations of $\boldsymbol{\theta}$. The hyperparameters in a learning algorithm should either be selected prior to the learning process or can be designed using a validation dataset. We will revisit this topic in Section 2.4. In order to ensure faster and robust learning of the parameters, several variations of the SGD have been utilized, such as using averaged gradients over mini-batches to update the parameters making the learning process faster, or momentum that uses a linear combination of the gradient and previous updates [11].

We note that although the general structure of the supervised learning techniques is as described earlier, the specific choices of the loss and decision functions and the discrete or continuous nature of the output can vary by task or from one technique to another. Next, we discuss some of the common tasks that supervised learning can handle and provide specific radar examples.

2.1.1.1 Classification

Classification is a common task in ML where the ML algorithm should decompose the input dataset into K distinct predefined classes. To this end, a mapping between the input data domain and numeric code of each class is learned. That is, $f_{\boldsymbol{\theta}} : \mathbb{R}^M \to 1, \ldots, K$ where M is the dimension of input data samples, i.e., $\mathbf{x}_i \in \mathbb{R}^M$.

Detection and classification are highly desirable tasks in a majority of radar applications [12]. Deciding whether a measurement corresponding to a specific radar range-Doppler bin arises due to a target or noise plus clutter is a fundamental detection question for radar systems. Detection task can be considered as a binary classification problem between two classes of target being present and absent. In [13], target detection in the presence of non-Gaussian clutter is considered for cognitive radar using ML approaches. ML techniques are also employed for land cover identification in polarimetric SAR images [14–16], target classification in SAR images [17,18], and detecting landslides in spaceborne SAR images [19]. Different human activities and hand gestures are recognized using ML approaches applied to time-frequency domain radar data [20–23]. ML approaches are also used to classify targets embedded in stratified dielectric media using ground penetrating and through-the-wall radar systems [8,24,25]. ML-based classification is also considered in electronic countermeaure systems for recognition of intra-pulse modulations and low probability of intercept waveforms [26,27].

2.1.1.2 Estimation/regression

In the task of regression, ML predicts an estimate of a parameter, thereby creating a numerical value as output rather than the class indices. Doppler weather radar measurements are used within an ML framework to estimate weather parameters, such as precipitation [28,29]. Soil moisture can be estimated using reflected microwave signals utilizing an artificial neural network (ANN) [30]. Radar arrays can be used to estimate the direction-of-arrival or the number of targets using ML techniques [31,32]. Heart rate estimation is achieved with ultrawideband radars and convolutional neural networks for healthcare applications [33].

2.1.1.3 Denoising

In general, raw radar returns are corrupted by noise, clutter, or in some cases by radio-frequency (RF) interference. Recovering desired signals from the corrupted measurements is an important denoising task for radar. Sea clutter has been suppressed using ML-based approaches in [34]. In [35], interference in through-the-wall radar systems is mitigated with autoencoders. Speckle removal approaches for SAR images have been developed using convolutional neural networks [36]. A dictionary learning approach is utilized for the removal of RF interference from SAR signals [37].

2.1.2 Unsupervised learning

Unsupervised learning departs from its supervised counterpart by the lack of data label information. Rather, the main goal in unsupervised learning is to determine previously unknown patterns in the input data. Two of the typical unsupervised learning tasks are clustering and finding reduced dimensionality representations of data.

2.1.2.1 Dimensionality reduction

Large datasets are increasingly common in radar, which not only makes their processing very demanding but also leads to difficulty in interpreting the data. The fundamental idea of the dimensionality reduction task is to preserve as much information as possible about the data while keeping the representation as simple or sparse as possible. Such dimensionality reduction of large datasets, while minimizing the loss of information, eases the processing demands and enhances data interpretability [38]. In [39], dimensionality reduction is performed on high-range resolution profiles to significantly improve the speed of subsequent processing for target recognition. A 2-D subspace is identified to represent Doppler velocity data, corresponding to atmospheric vortices, measured from a single plan position indicator scan in [40]. Manifold learning methods are used in [41] to deal with the curse-of-dimensionality problem, thereby enabling automatic target recognition in wide-angle SAR.

2.1.2.2 Clustering

When class labeling of the training dataset is not available, the task assumes the form of revealing the organization of input samples into clusters that permit discovery of similar and different attributes among the training dataset. Such clustering can lead to useful conclusions about the input data samples. Identification of aircraft from high-resolution radar range profiles is performed via unsupervised clustering in [42]. In [43], clustering is used to assist with 3-D reconstruction of buildings using tomographic SAR point clouds of urban areas, while natural media classification in remote sensing is performed via clustering methods applied to polarimetric SAR imagery in [44]. Clustering techniques are also considered in [45] to aid in clutter reduction, resulting in enhanced radar images that, in turn, permit improved detection of stationary targets inside buildings.

2.2 Ingredients of an ML algorithm

While this chapter focuses on applications of ML techniques in different radar problems, in this section, we present the essential elements for formulation of an ML approach from a more general perspective.

The recipe for supervised learning from data requires the following ingredients:

- ***Input data samples:*** The inputs comprise the raw data or its transform-domain representations. For example, in the classification task of tagging an image as "cat" or "dog," we can use the raw images as input data samples. On the other hand, to classify a human activity as "walking" or "running" using radar, the inputs are typically the time-frequency representations of the raw radar data.
- ***Expected outputs:*** The outputs of an ML algorithm depend on the specific task. For example, the output could be a class label, such as "walking" or "running" for the radar-based human activity classification, or an estimated parameter value for a regression problem.
- ***Decision function or model:*** This ingredient represents the general structure of the parametric function that maps the input samples to the output. The specific model of this mapping, whether it is a decision tree or a neural network, is a

design choice. The ML algorithm learns the unknown parameters of the selected model to optimize a loss function.

- ***Loss function:*** The loss function is used to evaluate how well an ML algorithm is learning the model parameters. It could assume the form of a distance between the ML algorithm's current output and the expected output or an award related to the action of the ML system. The loss function provides feedback to the ML algorithm, thereby enabling it to adjust the parameters of its model. It is this adjustment step that enables what is often viewed as *learning*.

Input data samples are actually a collection of information measured from some process. In classical supervised learning approaches, the input data is defined as a feature vector where a domain expert transforms the raw sensor data into a lower dimensional vector of features. Each feature carries a unit of information that is assumed to be representative of the underlying process. Such a transformation can be regarded as *feature engineering*, and the job of the supervised learning algorithm is then to learn the mapping from the engineered features to the expected outputs. However, the giant leap forward of ML, specifically in the last decade, is primarily due to the models that have the ability to learn optimal features in the sense of defined loss functions from the raw data. This *feature learning* capability of (deep) learning models permits raw sensor data, such as images, sound, and raw radar data, or at least much simpler representations of raw data to be used directly within an ML framework. The decrease in need for domain expertise to design features also allows ML to be applicable to a much broader range of applications.

The provided recipe for a supervised ML algorithm indicates that we learn the parameters of an ML model to optimize the loss function. However, the main challenge in ML is not only to fit the given input data but also to perform similarly well on the previously unseen, new data samples. This is known as the *generalization* of the ML approach. In this regard, it is very important to properly evaluate the designed ML algorithms. We will discuss the general ML workflow and evaluation approaches in Section 2.4 to establish a standard evaluation procedure.

Compared to supervised learning, the ingredients for unsupervised ML neither include the model nor the expected outputs. Instead, the recipe entails discovering information and patterns in the input data through optimization of a loss or objective function.

2.3 Basic techniques of supervised and unsupervised learning

We briefly describe several fundamental supervised and unsupervised learning approaches. More specifically, we first introduce ANNs and support vector machines (SVMs), both being supervised learning approaches. Then, we present principal component analysis (PCA) and K-means clustering, both of which are unsupervised learning techniques. More recent deep learning techniques are described in later chapters of this book.

2.3.1 Supervised learning approaches

2.3.1.1 Artificial neural networks

The concept and model of an ANN is inspired from the structure of human neural system, including the brain where electrical events produced by neurons are used for information computation, processing, and communications with the human sensory organs [46].

An ANN consists of hierarchically connected artificial neurons called nodes. These artificial neurons are the basic building blocks of a neural networks, each receiving information from its incoming connections, applying a simple processing to the information, and generating an output information that can be transferred to another artificial neuron. Figure 2.1 depicts a schematic representation of an artificial neuron, indicating its following main components:

- The incoming information $\mathbf{x} = [x_1, x_2, \ldots, x_M]^\top$.
- The weights $\mathbf{w} = [w_1, w_2, \ldots, w_M]^\top$ at the incoming connections and the bias b for the neuron, both of which constitute the parameters of the neuron.
- An intermediate processing step within the body of the neuron implemented simply as a linear combination of the incoming data $\mathbf{x}$ using the weights $\mathbf{w}$ and bias b as $z = \sum_{i=1}^{M} w_i x_i + b = \mathbf{w}^\top \mathbf{x} + b$.
- The activation function $\sigma(\cdot)$ of the neuron which applies a nonlinear transformation to the intermediate result z, thus creating the output information, y, of the neuron.

More compactly, the output of the artificial neuron model can be represented as

$$y = \sigma(\mathbf{w}^\top \mathbf{x} + b). \tag{2.5}$$

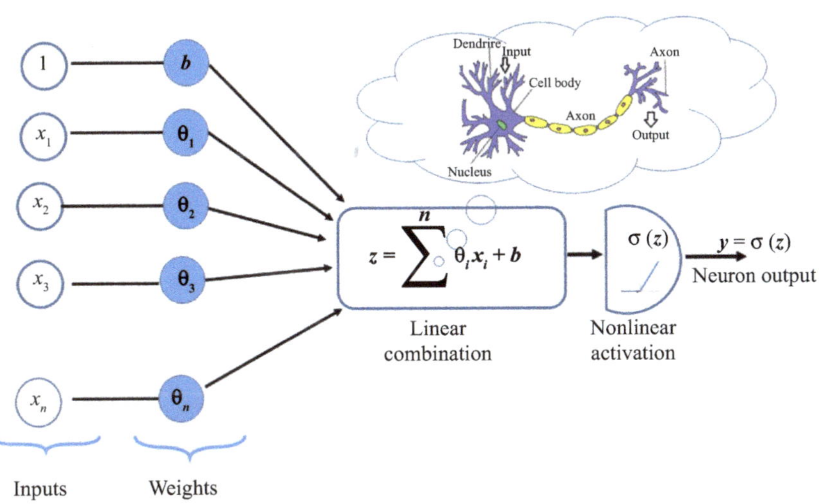

Figure 2.1 A single artificial neuron model

Some highly utilized activation functions, namely, the sigmoid, the tangent hyperbolic, and the rectified linear unit [46] are plotted in Figure 2.2, which also provides their definitions. The weights and bias of each neuron are the parameters to be learned to approximate the optimal mapping between the input and output data samples. The nonlinear activation functions, in particular, are important for permitting more complex nonlinear mappings to be represented by a network of simple neurons.

An ANN consists of an interconnection of neurons in a layered structure. While the first layer of neurons takes data samples as their inputs, the final layer generates the output samples whether they be classification or regression results. There are one or more hidden layers that progressively transform the first-layer neuron activations into the final output samples. Figure 2.3 shows an example of ANN with two hidden layers.

Using the input–output relation of a single neuron in (2.5), we can define the output samples $\mathbf{x}^{\ell} \in \mathbb{R}^m$ of all m neurons grouped in layer ℓ as

$$\mathbf{x}^{\ell} = \sigma(\mathbf{W}^{\ell}\mathbf{x}^{\ell-1} + \mathbf{b}^{\ell}), \tag{2.6}$$

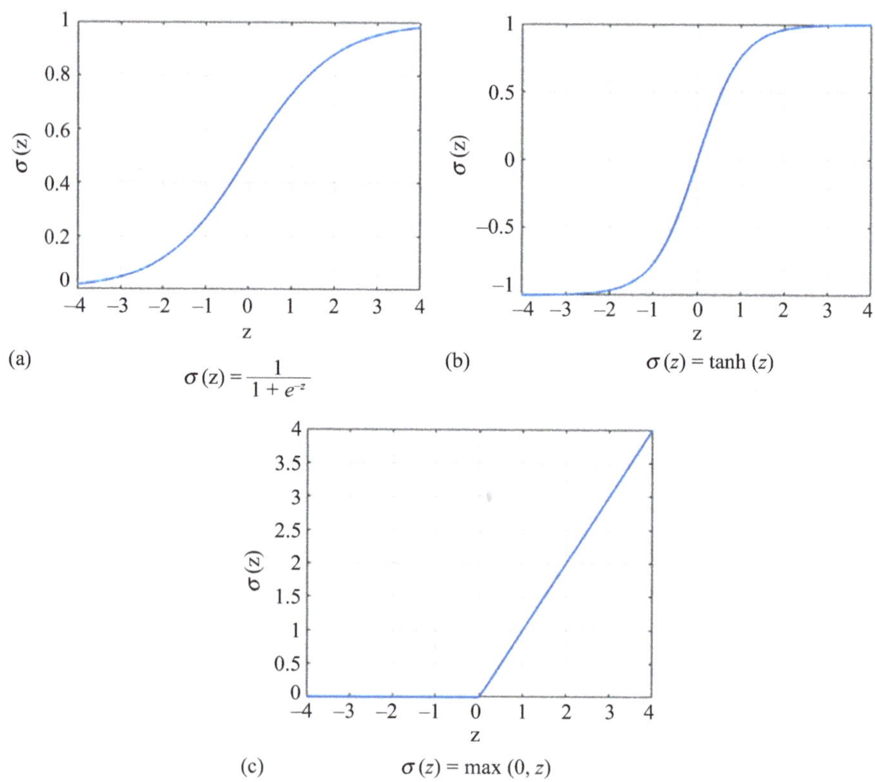

Figure 2.2 Three of the most commonly used activation functions: (a) sigmoid $\sigma(z) = \frac{1}{1+e^{-z}}$, (b) tangent hyperbolic $\sigma(z) = \tanh(z)$, and (c) rectified linear unit (ReLU) $\sigma(z) = \max(0, z)$

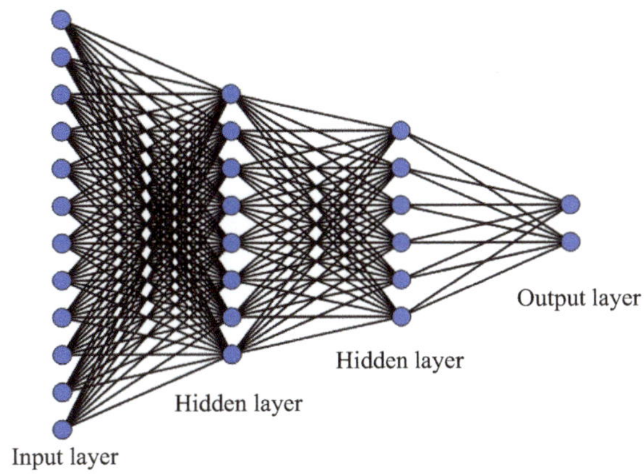

Figure 2.3 An example ANN structure with two hidden layers

with $x^{\ell-1} \in \mathbb{R}^k$ denoting the output from the previous $(\ell - 1)$th layer, $\mathbf{b}^\ell \in \mathbb{R}^m$ representing the bias vector, and $\mathbf{W}^\ell \in \mathbb{R}^{k \times m}$ denoting the weight matrix for all links between the two layers. Considering all L layers, the final output of an ANN will be a highly complex nonlinear function of the input data samples and $\{(\mathbf{W}^\ell, \mathbf{b}^\ell), \ell = 1, \ldots, L\}$. Given the supervised setting and the training data labels, learning for ANNs entails determining the parameters (weights and biases for each layer) such that the preselected loss function is minimized over the training dataset. Both SGD and its variants are common approaches for learning the parameters of ANNs. When an ANN is trained properly, it should be able to map any new input sample to a new output sample. As such, the generalization capability of a learned model should be assessed correctly as discussed in Section 2.4.

As the number of hidden layers increases, the ANN structure becomes increasingly deep, thereby turning into a DNN. Although ANNs with a large number of hidden layers have been named DNNs in the literature [47], deep learning and DNN structures are more related to feature learning capability from raw sensor data and models and structures that facilitate such capability.

2.3.1.2 Support vector machines

SVM is highly practical and relatively easy-to-use supervised learning approach [48, 49]. SVMs can construct complex models, such as subclasses of neural nets, radial basis functions (RBFs), and polynomial classifiers that can handle practical learning challenges. The mathematical formulation of SVM is straightforward as it is a linear technique in a higher dimensional feature space.

SVM classifiers are based on hyperplanes. As such, the decision boundary between two classes will take the form

$$\mathbf{w}^\top \mathbf{x} + b = 0 \quad \mathbf{w}, \mathbf{x} \in \mathbb{R}^M,\ b \in \mathbb{R}. \tag{2.7}$$

For the simplest case of binary classification, SVM aims to find the hyperplane (i.e., decision boundary) that linearly separates two groups of points in the training dataset by the largest margin, as illustrated in Figure 2.4. We consider the decision function related to the boundary in (2.7) to be $f(\mathbf{x}) = \text{sign}(\mathbf{w}^\top \mathbf{x} + b)$. Here, $f(\mathbf{x}) = +1$ represents the class of points above the boundary, while $f(\mathbf{x}) = -1$ corresponds to the points below the boundary. However, there could be several different hyperplanes between two classes of linearly separable data points, and we would like to find the one that yields the largest margin. Mathematically, this objective can be expressed as a simple quadratic programming problem [48],

$$\min_{\mathbf{w},b} \mathbf{w}^\top \mathbf{w} \quad \text{subject to} \quad \mathbf{y}_i(\mathbf{w}^\top \mathbf{x}_i + b) \geq 1 \tag{2.8}$$

where $\mathbf{y}_i$ denotes the label of the input data sample $\mathbf{x}_i$. The constraint in (2.8) indicates correct classification for input sample $\mathbf{x}_i$ if both the label $\mathbf{y}_i$ and the decision function $(\mathbf{w}^\top \mathbf{x}_i + b)$ are $+1$ or -1 at the same time. The solution of (2.8), $(\mathbf{w}, b)$, defines the decision hyperplane, and the weights $\mathbf{w}$ can be linearly expressed in terms of a subset of the input samples $\mathbf{x}_i$ that lies on the margin, as shown in Figure 2.4. The distance between these points is, in fact, $2/\sqrt{\mathbf{w}^\top \mathbf{w}}$ and, hence, minimization of the quadratic term $\mathbf{w}^\top \mathbf{w}$ in (2.8) corresponds to maximizing the distance between the margins. The input samples on the margin, called the support vectors, carry the relevant information about the classification problem. For the case where the data is not linearly separable, the quadratic programming in (2.8) can be relaxed by introducing slack variables to allow some data points of one class to appear on the other side of the boundary [49].

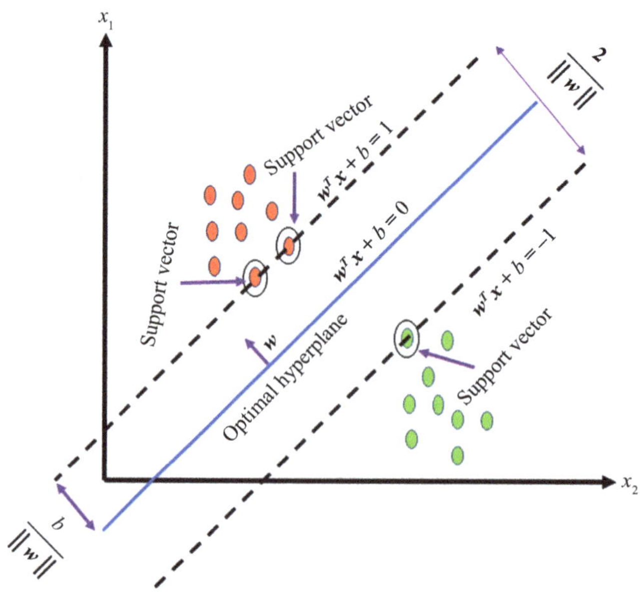

Figure 2.4 Linear decision boundary in the feature space for the support vector machine (SVM) algorithm

SVMs can also be used to generate nonlinear decision boundaries [50]. One important property of SVMs is that both the quadratic optimization problem in (2.8) and the decision function $f(\mathbf{x})$ depend on dot products between data samples. As a result, mapping the data samples to a higher dimensional space only requires the evaluation of dot products in that space. This simple kernel trick may generate linearly separable classes in a higher dimensional space, while the data samples may not be linearly separable in the original space. Different choices of kernel functions, such as polynomial, RBF, or sigmoidal kernels, correspond to different architectures for nonlinear classifiers.

2.3.2 Unsupervised learning approaches

2.3.2.1 Principal component analysis

Assume N input samples are available with $\mathbf{x}_i \in \mathbb{R}^M$ denoting the ith sample. Note that in case of two-dimensional data points, such as images, each sample $\mathbf{x}_i$ is obtained by column-wise vectorization of the corresponding matrix. Alternatively, row-wise vectorization of each matrix could be considered. The N input samples are organized as a zero-centered data matrix

$$\mathbf{X} = [\mathbf{x}_1 - \mathbf{m}, \mathbf{x}_2 - \mathbf{m}, \ldots, \mathbf{x}_N - \mathbf{m}] \in \mathbb{R}^{M \times N}, \tag{2.9}$$

where the mean vector $\mathbf{m}$ is defined as

$$\mathbf{m} = \frac{1}{N} \sum_{i=1}^{N} \mathbf{x}_i. \tag{2.10}$$

Given $\mathbf{X}$, PCA seeks a small number of orthogonal directions that define a subspace wherein the presence of the input samples is maximized. Mathematically, PCA approximates the data matrix $\mathbf{X}$ by the low-rank matrix $\mathbf{Q}\mathbf{Y}^\top \in \mathbb{R}^{M \times N}$, where $\mathbf{Q}^\top\mathbf{Q} = \mathbf{I}_d$ and $d \le \text{rank}(\mathbf{X}) \le \min(N, M)$ so that the ℓ_2-norm of the approximation error is minimized. That is, PCA is formulated as

$$(\mathbf{Q}_{L2}, \mathbf{Y}) = \underset{\mathbf{Q}\in\mathbb{R}^{M\times d};\ \mathbf{Q}^\top\mathbf{Q}=\mathbf{I}_d;\ \mathbf{Y}\in\mathbb{R}^{N\times d}}{\text{argmin}} \left\| \mathbf{X} - \mathbf{Q}\mathbf{Y}^\top \right\|_2^2, \tag{2.11}$$

where the squared ℓ_2-norm ($\| \cdot \|_2^2$) returns the sum of the squared entries of its matrix argument [51]. Observing that, for any given $\mathbf{Q}$, $\mathbf{Y} = \mathbf{X}^\top\mathbf{Q}$ minimizes the error in (2.11), $\mathbf{Q}_{L2}$ can be determined as the solution to the equivalent projection maximization problem

$$\mathbf{Q}_{L2} = \underset{\mathbf{Q}\in\mathbb{R}^{M\times d};\ \mathbf{Q}^\top\mathbf{Q}=\mathbf{I}_d}{\text{argmax}} \left\| \mathbf{Q}^\top\mathbf{X} \right\|_2^2. \tag{2.12}$$

Accordingly, the desired data with the reduced dimension is given as $\mathbf{Y} = \mathbf{X}^\top\mathbf{Q}_{L2}$. Principal component matrix $\mathbf{Q}_{L2}$ consists of the d-dominant left singular vectors of $\mathbf{X}$, obtained with standard singular value decomposition [52]. Note that several methods are available for estimating the value of the hyperparameter d [53].

2.3.2.2 *K*-Means clustering

Clustering methods partition the input samples into classes. The classes are generally unknown and are explored based on the data by using a similarity measure. Given N input samples $\{\mathbf{x}_i, i = 1, 2, \ldots, N\}$, the K-means clustering method partitions the samples into K clusters or classes by minimizing the sum of the within-cluster variances (WCSS) [54],

$$\text{WCSS} = \sum_{k=1}^{K} \sum_{\mathbf{x}_i \in C_k} \|\mathbf{x}_i - \boldsymbol{\mu}_k\|_2^2, \tag{2.13}$$

where C_k is the kth cluster with centroid $\boldsymbol{\mu}_k$. The pseudocode for the K-means clustering is given as follows:

1. Initialize the number of classes K and centroids $\boldsymbol{\mu}_k, k = 1, 2, \ldots, K$.
2. Assign each sample $\mathbf{x}_i, i = 1, 2, \ldots, N$ to the cluster whose centroid is the closest.
3. After all N input samples have been assigned, recalculate the K centroids.
4. Repeat steps 2 and 3 until the centroids no longer change.

Note that the number of classes needs to be specified *a priori*. In the absence of this knowledge, one may resort to measures that could estimate the hyperparameter K automatically [55].

2.4 Evaluation of a machine learning algorithm

In supervised ML, the objective is to learn models that can generalize that is, perform similarly well as training data on never-before-seen data. Hence, it is critical to reliably measure the generalization capability of the learned model. Unlike supervised learning, there are no expected outputs to guide the unsupervised process of learning underlying relations among the input data samples. However, similar to supervised learning, unsupervised learning algorithms generate estimates that are functions of the input samples. In this case, it is the generalization capability of these estimates that is of interest. Generalization in supervised learning has been extensively examined, while only a few works have addressed this aspect in the context of unsupervised learning. For PCA and K-means clustering, in particular, good generalization has been shown to be linked to the choice of the number of principal components and the number of clusters, respectively [56]. Several methods have been proposed to estimate these numbers, which were mentioned in Section 2.3. As such, the primary focus of this section is on providing a blueprint for developing and evaluating a supervised ML algorithm. However, we briefly remark on items that are also relevant to unsupervised methods, when appropriate.

2.4.1 General workflow of an ML algorithm

Constructing an ML algorithm consists of several important steps. A flowchart of these steps is shown in Figure 2.5. The first step in developing an ML algorithm is establishing a problem definition. One needs to define the task in hand to start

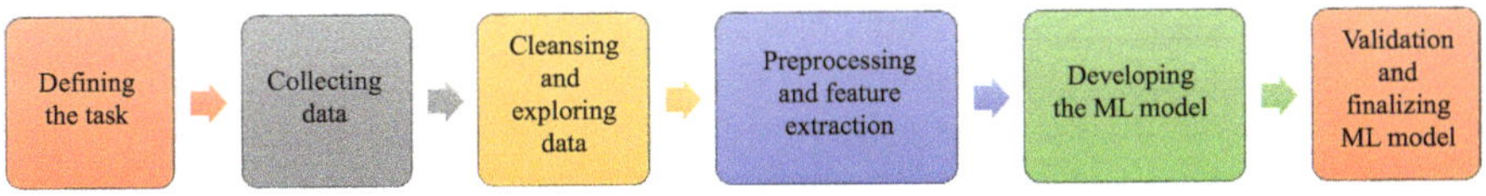

Figure 2.5 Flowchart of data-driven ML model development

building an ML approach. Next, data should be collected depending on the defined task. This includes determining the input samples and expected outputs. Availability of labeled vs unlabeled data will determine whether to opt for supervised or unsupervised learning techniques. On the other hand, the output depends on the type of problem being considered. For example, classification and regression problems have different outputs. In addition to the quantity of data samples, quality of the data in terms of being diverse enough to be a good representative of various conditions in real world is very important. Generalization of the ML approach depends on the representation power of the collected data for training and simply adding more data samples does not always aid an ML algorithm in achieving better performance.

Before direct development of the ML model can commence, one needs to first cleanse and explore the data. In most cases, the collected data is in a form that cannot be directly used by an ML algorithm. Data cleansing is essentially the task of detecting, removing, or correcting any corrupted, incorrect, incomplete, duplicated, or improperly formatted data. Radar data is also not an exception in this sense. The next step after cleansing and exploration is preprocessing and feature extraction. Data preprocessing is generally required to make the data suitable for use by the ML algorithm. Preprocessing in case of radar data may include operations, such as filtering, denoising, clutter suppression, and normalization. Later, features that will be used by the ML algorithm should be generated. Despite the fact that more recent techniques of deep learning permit feature learning from raw sensor measurements, preprocessing and suitable data representation are nonetheless an important part of such techniques. Dedicating efforts on data cleansing and preprocessing is definitely a good investment for efficient ML model development.

Identifying the problem type and calculation of input features will provide guidance in terms of selection of both the proper model and the loss function to be optimized in case of supervised learning and the loss function only in case of unsupervised learning. In the process of learning from the input data, the goal is to learn a model or obtain estimates that can generalize. In other words, selection of the model and its hyperparameters should not provide good results only when applied to the training data. In the following sections, we discuss the performance metrics and the validation approaches for measuring generalization performance of an ML algorithm.

2.4.2 Performance metrics

The performance of a learned model can be assessed using different metrics. The choice of metrics that can be used depends on the specific task the ML algorithm is designed to accomplish. The metrics for a classification task differ from those for a regression task. These metrics provide feedback so that improvements to the learned model can be made by tuning its hyperparameters. Another important aspect of performance metrics is their ability to discriminate among results obtained with different models. The best model can be selected as the one which achieves the highest performance in a set of metrics. In general, the use of a single metric does not always provide a good indication of model performance. Next, we briefly review the most common performance metrics used in supervised ML algorithms, along with a discussion of when these are most effective.

2.4.2.1 Classification metrics

The confusion matrix is one of the most intuitive and useful tools to calculate many classification metrics. A confusion matrix is a $K \times K$ matrix, where K is the number of classes being predicted. Each row of this matrix represents the instances in a predicted class, while each column represents the instances in an actual class. An example confusion matrix for binary classification, $K = 2$, is illustrated in Figure 2.6.

We can define following metrics using the confusion matrix:

1. *Accuracy*: Accuracy of the model usually denotes classification accuracy. It is the ratio of the number of correct predictions to the total number of input samples. For a given confusion matrix $\mathbf{C}$ where predicted classes constitute the rows and true classes constitute the columns of $\mathbf{C}$ as in Figure 2.6, the accuracy is given as

$$Accuracy = \frac{\sum_k \mathbf{C}(k,k)}{\sum_{k,l} \mathbf{C}(k,l)} \tag{2.14}$$

   where $\mathbf{C}(k,l)$ represents the (k,l) index of the matrix $\mathbf{C}$. Accuracy is a good measure only when the target classes in the data are nearly balanced. Consider the case where class A comprises 98% of the input samples, while only 2% of the samples correspond to class B. If we consider accuracy to be the determining metric for this case, then even a classifier that always declares a class A prediction will have 98% accuracy.

		True classes		
		Class 1 (positive)	**Class 2 (negative)**	
Predicted classes	**Class 1 (positive)**	a (true positive)	b (false positive)	Precision $\frac{a}{a+b}$
	Class 2 (negative)	c (false negative)	d (true negative)	Negative predictive value $\frac{d}{d+c}$
		Recall, P_D $\frac{a}{a+c}$	P_{FA} $\frac{b}{b+d}$	Accuracy $\frac{a+d}{a+b+c+d}$

Figure 2.6 Confusion matrix

2. *Precision*: The proportion of correct predictions over all predictions for a given class is defined as the precision metric for that class. Precision by itself is not sufficient to evaluate performance. If there is only one positive instance and the algorithm correctly predicts that one positive instance, it will have 100% precision. The precision for class k in a multi-class classification problem is given by

$$Precision_k = \frac{\mathbf{C}(k,k)}{\sum_l \mathbf{C}(k,l)}. \tag{2.15}$$

3. *Recall*: The proportion of all class k-labeled cases that were correctly predicted as class k is the recall or sensitivity metric for that class. For a binary classification or a detection problem, recall is the same as the probability of detection or power of the detection test. The recall for class k in a multi-class classification problem is expressed as

$$Recall_k = \frac{\mathbf{C}(k,k)}{\sum_l \mathbf{C}(l,k)} \tag{2.16}$$

4. F_1*-score*: It is clear that precision and recall together capture the performance of a classifier better than each one individually. F_1-score is defined as the harmonic mean of precision and recall and is given by

$$F_{1,k} = 2 \cdot \frac{Precision_k \times Recall_k}{Precision_k + Recall_k} \tag{2.17}$$

5. *Specificity*: Specificity is defined as the ratio of true negatives to the total number of true-negative and false-positive samples. For a detection problem in radar, a metric of interest is the probability of false alarm (P_{FA}). Specificity is the complement of P_{FA}, i.e., $1 - P_{FA}$.
6. *Area under the curves—receiver operating characteristics (ROC) or precision–recall*: ROC curve is the main statistical tool used in radar to analyze the performance of a detection algorithm describing how probability of detection P_D varies with changes in P_{FA}. Precision–recall curves describe similar operating characteristics of an ML algorithm. When an ML algorithm directly predicts the class values, we are in essence operating at a single (P_D, P_{FA}) point on an ROC curve or at a single precision–recall point. Essentially, an ML algorithm or a statistical detection test has thresholds on the class probabilities or test statistics and changing the decision thresholds moves the operating point on ROC or precision–recall curves. As the threshold is moved through its minimum to maximum values, ROC and precision–recall curves are constructed. Figure 2.7 shows example ROC and precision–recall curves corresponding to logistic regression and no skill random selection for the same binary classification problem. The area under these curves is an important metric indicating average performance of an algorithm. In addition, using these curves, an algorithm and its operating point can be chosen depending on the requirements.

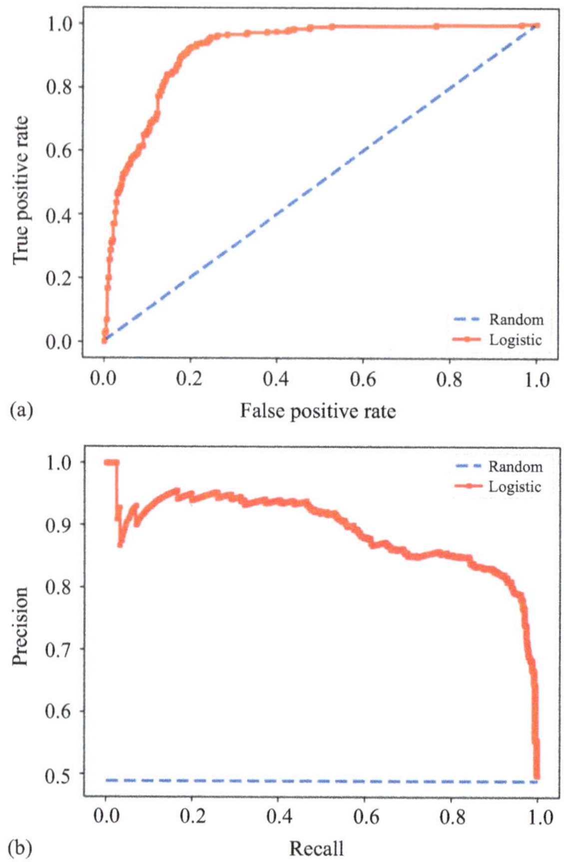

Figure 2.7 (a) Receiver operating characteristics (ROC) and (b) precision–recall curves for performances of different models

2.4.2.2 Regression metrics

Regression algorithms predict a continuous output value rather than a class label or probability. The most popular evaluation metrics for regression task are the ones that measure distances in terms of different norms.

1. *Mean squared error (MSE)*: MSE is defined as the average of the square of the difference between the actual and the predicted values. It provides a measure of how far the predictions are from the actual output in an ℓ_2-norm sense.

$$MSE = \frac{1}{N}\sum_{i=1}^{N}(y_i - \widehat{y_i})^2 \tag{2.18}$$

2. *Mean absolute error (MAE)*: MAE is defined as the average of the difference between the actual values and the predicted values. It is the distance measured in an ℓ_1-norm sense.

$$MAE = \frac{1}{N}\sum_{i=1}^{N} |y_i - \widehat{y_i}| \tag{2.19}$$

2.4.3 Measuring generalization: training, validation, and test sets

Evaluating an ML model and measuring its generalization capability essentially reduce to splitting the available data in such a manner that permits training of the model, validation and tuning of the hyperparameters, and testing the model's ability to make accurate predictions on "new" data. The proper approach is to divide the dataset into three parts: training, validation, and test sets. One could ask why not have only the training and test sets, which is often done in the literature. The reason is that, in addition to the model parameters or the model weights, every model has hyperparameters. For example, in ANNs, the number and sizes of layers or learning rates constitute the hyperparameters. The model weights are learned over a training dataset for a given set of hyperparameters, and the learned model is evaluated, using a different test dataset, to determine the performance metric. Generally, the model development cycle also involves tuning of the hyperparameters to obtain the highest performance in this second dataset. In fact, the search for the optimal hyperparameters is another type of learning. As a result, tuning the hyperparameters of the model only to obtain the highest performance metric value in the test dataset could lead to overfitting of this data. Although the model weights are not trained on the test dataset, the hyperparameters and the structure of the model are tuned to obtain the best performance from this set. More precisely, what we imply by generalization is the performance of the learned and tuned model on data that is completely new to the model. To this end, the available dataset should be divided into three parts: training, validation, and test datasets. The model weights are learned using the training dataset, while evaluation and tuning of the hyperparameters are performed on the validation dataset. After learning and tuning the model using training and validation sets, we should end up with a model that is performing well in validation dataset. The actual generalization performance is calculated one final time on the test dataset, which is never before seen or never used in the learning process of the model.

With information on how to split the dataset in hand, the next question that arises is in regard to determining the optimal split ratio for training, validation, and test sets. This is not straightforward to accomplish, and the split is specific to the use case. However, the split ratio is, in general, dependent on the total number of samples in the dataset and the model complexity. Models with very few hyperparameters typically require much less validation data for tuning, while complex models with a large number of weights need larger training sets. A specific split may introduce bias, and in the case of limited dataset, a three-way split may also hamper effective learning of the model.

A computational procedure commonly used for splitting the data and evaluating the model is K-fold cross validation. Model evaluation via cross validation or data split into training, validation, and test sets is illustrated in Figure 2.8. The central idea of cross validation procedure is as follows. If we have less data, we should keep aside a certain percentage of the dataset for testing and split the remaining data into K different subsets. We use $K-1$ subsets, also called folds, for training and validating the model with the remaining fold. We average the model against each of the folds and then finalize the model. After the model is finalized, it can be tested on the test set. If data is sufficiently large, it is possible to learn the model and hyperparameters without a K-fold validation scheme. Although varying on a case-by-case basis, a three-way split of the whole dataset into 60%, 20%, and 20% for training, validation, and testing, respectively, is a possible way of generating these sets. Cross validation is also generally applied with K=5–10 folds, although many other variants are also available.

2.4.4 Overfitting and underfitting

The main goals of cross validation approach and splitting the data into three sets were to evaluate the generalization of an ML approach and to select the hyperparameters. We now discuss in more detail the selection of the model and its hyperparameters. Which dataset is more important for the evaluated metric? If the model is underperforming, how should we move forward? Do we need to increase or decrease the

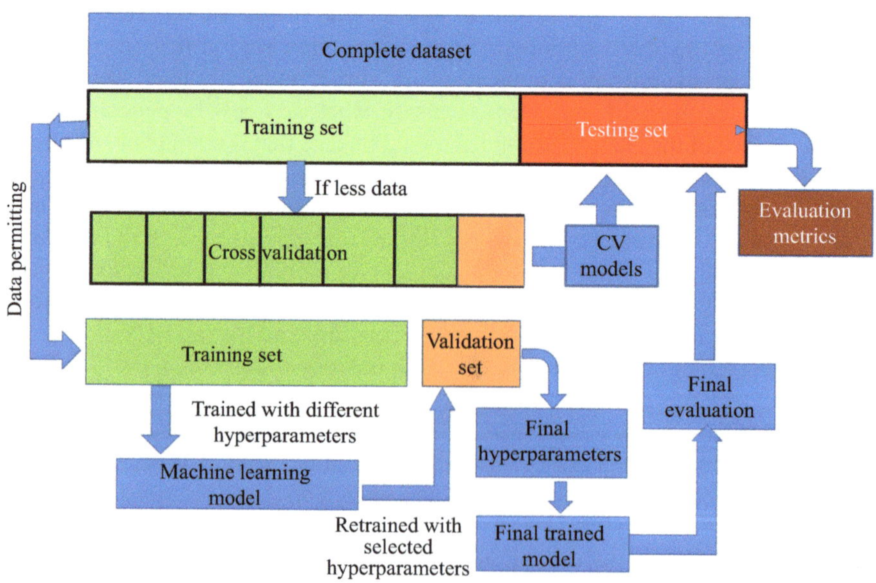

Figure 2.8 Model evaluation and testing via cross validation or using training, validation, and testing datasets

model complexity? Do we need more training data? All of these are important questions for practical applications of ML. The ability to determine the steps required to find answers to these questions will not only improve the generalization capability of the model but also enable successful practice of ML in varying applications.

Fundamentally, answers to such questions in order to find the "best model" and its hyperparameters lie in bias and variance analyses of the models. Consider Figure 2.9 where the objective entails learning a polynomial fit to a set of training samples. Three different models with varying model complexities fit the same training dataset. Here, the number of parameters in the polynomial fit model is the hyperparameter of the best model search. It can be observed that neither the simple line model in Figure 2.9(a) nor the high-degree polynomial fit in Figure 2.9(c) is a good fit for the underlying function. However, they fail in different ways. The model in Figure 2.9(a) is a straight line whereas the true function generating the training data is more complicated. This scenario is called *underfitting* or *high bias*, since the proposed model lacks sufficient complexity to represent the data.

On the other hand, the model in Figure 2.9(c) provides a near perfect fit to the training data. However, we observe that the model is different from the true function. This case fits more with noise and outliers in the training dataset than the intrinsic true function generating the data. This scenario is called *overfitting* or *high variance*.

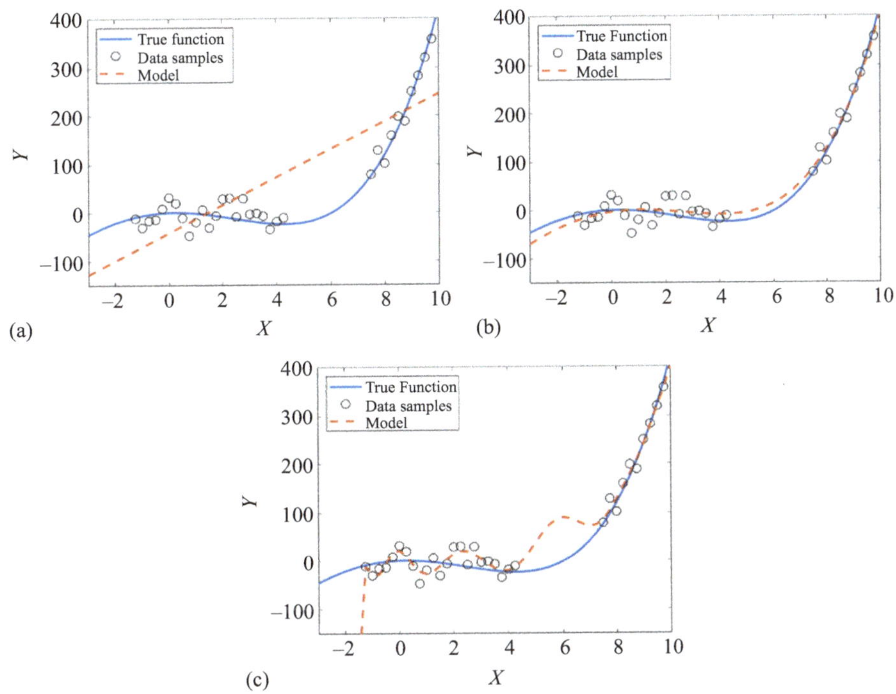

Figure 2.9 Model fits to the underlying data samples under three cases: (a) underfitting, (b) fitting with correct model order, and (c) overfitting

In this case, the model has a higher complexity than the actual function, which causes it to deviate from the true function in order to account for the noise or outliers in the data. The model in Figure 2.9(b) is the sweet spot for the model complexity between the underfit and overfit scenarios.

With the knowledge of underfitting and overfitting in hand, we want to understand the scenario the model corresponds to so that appropriate actions can be taken to improve the model. One such approach is to plot validation and learning curves. Validation curve plots the associated cost of the model in both training and validation datasets over a varying hyperparameter. For the example of polynomial fit, the hyperparameter is the polynomial degree, which equals the model complexity. The obtained validation curve for this example is shown in Figure 2.10. Clearly, while the loss function for the training set decreases as the model complexity increases, the loss in the validation dataset spans out a convex curve indicating an optimal model complexity choice.

The other tools for model selection are the learning curves that are illustrated in Figure 2.11. In learning curves, the costs of both training and validation datasets are shown as a function of training set size at a fixed hyperparameter value. For the case of underfitting, while the loss functions for training and validation datasets become similar as training set size increases, they both have high errors. On the other hand, for the overfitting case, the training loss will remain well below the validation set loss. One important feature of a learning curve is the convergence of training and validation losses as the training set size increases. If these loss functions converge at a particular training set size, adding more training data will not improve the model performance. If this convergence has occurred at a high loss value, the model is underfit and we

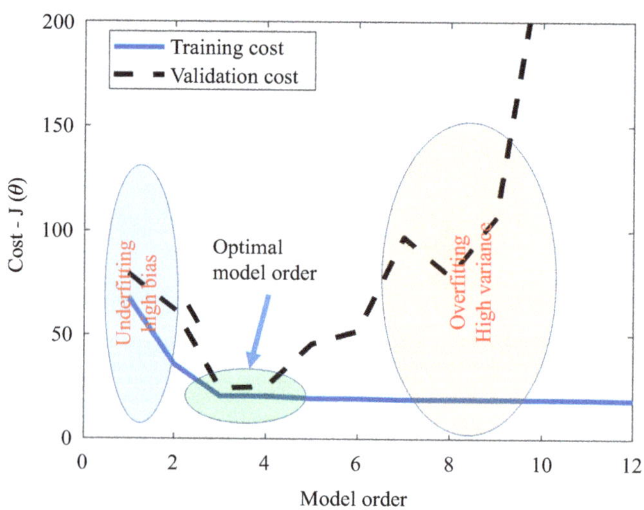

Figure 2.10 Validation curve: cost as a function of hyperparameter both for training and validation sets

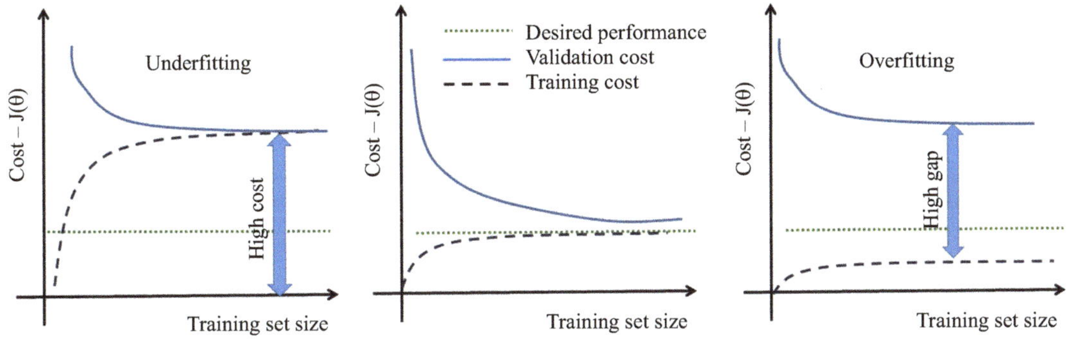

Figure 2.11 Learning curves showing loss (cost) function as a function of training set size for underfitting, optimal scenario, and overfitting cases

need to increase the model complexity to deal with it. If the two curves have not converged, as in the case of overfitting, there are several possible solutions, such as collecting more data if possible, applying regularization, or using a simpler model.

2.5 Conclusion

This chapter provided an overview of the basic principles of ML. Both supervised and unsupervised ML categories were discussed, and commonly used methods under each category were presented. Also, different tasks that ML can tackle under each category were described. Representative examples from radar applications were provided, when appropriate. Moreover, a detailed blueprint for designing and evaluating an ML algorithm with a good generalization capability was outlined, which can be readily adapted for a broad range of radar applications.

References

[1] Mitchell TM. *Machine Learning*. New York, NY: McGraw-Hill; 1997.

[2] Cai X and Sarabandi K. A machine learning based 77 GHz radar target classification for autonomous vehicles. In: *2019 IEEE International Symposium on Antennas and Propagation and USNC-URSI Radio Science Meeting*, IEEE, Atlanta, GA, USA; 2019. p. 371–372. doi: 10.1109/APUSNCURSINRSM.2019.8888647.

[3] Patel K, Rambach K, Visentin T, Rusev D, Pfeiffer M and Yang B. Deep learning-based object classification on automotive radar spectra. In: *2019 IEEE Radar Conference (RadarConf)*, Boston, MA, USA; 2019. p. 1–6. doi: 10.1109/RADAR.2019.8835775.

[4] Pei J, Huang Y, Huo W, *et al.* SAR Automatic Target Recognition Based on Multiview Deep Learning Framework. *IEEE Transactions on Geoscience and Remote Sensing*. 2018;56(4):2196–2210.

[5] Seyfioglu MS, Erol B, Gurbuz SZ, *et al.* DNN Transfer Learning From Diversified Micro-Doppler for Motion Classification. *IEEE Transactions on Aerospace and Electronic Systems*. 2019;55(5):2164–2180.

[6] Markopoulos PP, Zlotnikov S, and Ahmad F. Adaptive Radar-Based Human Activity Recognition With L1-Norm Linear Discriminant Analysis. *IEEE Journal of Electromagnetics, RF and Microwaves in Medicine and Biology*. 2019;3(2):120–126.

[7] M G Amin (Ed). *Radar for Indoor Monitoring: Detection, Classification, and Assessment.* Boca Raton, FL: CRC Press; 2017.

[8] Kilaru V, Amin MG, Ahmad F, *et al.* Gaussian Mixture Modeling Approach for Stationary Human Identification in Through-the-Wall Radar Imagery. *Journal of Electronic Imaging.* 2015;24(1):1–11. Available from: https://doi.org/10.1117/1.JEI.24.1.013028.

[9] Sutton RS and Barto AG. *Reinforcement Learning: An Introduction.* 2nd ed. The MIT Press; 2018. Available from: http://incompleteideas.net/book/RLbook2018.pdf.

[10] Robbins H and Monro S. A Stochastic Approximation Method. *Annals of Mathematical Statistics*. 1951;22(3):400–407. Available from: https://doi.org/10.1214/aoms/1177729586.

[11] Rumelhart DE, Hinton GE, and Williams RJ. In: *Learning Internal Representations by Error Propagation.* Cambridge, MA: MIT Press; 1986. p. 318–362.

[12] Gini F and Rangaswamy M. *Knowledge Based Radar Detection, Tracking and Classification.* Adaptive and Learning Systems for Signal Processing, Communications and Control Series. Wiley; 2008. Available from: https://books.google.com/books?id=bkYfAQAAIAAJ.

[13] Metcalf J, Blunt SD, and Himed B. A machine learning approach to cognitive radar detection. In: *2015 IEEE Radar Conference (RadarConf)*, Arlington, VA; 2015. p. 1405–1411. doi: 10.1109/RADAR.2015.7131215

[14] Chen W, Gou S, Wang X, *et al.* Complex Scene Classification of PoLSAR Imagery Based on a Self-Paced Learning Approach. *IEEE Journal of Selected Topics in Applied Earth Observations and Remote Sensing.* 2018;11(12):4818–4825.

[15] Guo J, Wang L, Zhu D, *et al.* Subspace Learning Network: An Efficient ConvNet for PolSAR Image Classification. *IEEE Geoscience and Remote Sensing Letters.* 2019;16(12):1849–1853.

[16] Chen S and Tao C. PolSAR Image Classification Using Polarimetric-Feature-Driven Deep Convolutional Neural Network. *IEEE Geoscience and Remote Sensing Letters.* 2018;15(4):627–631.

[17] Geng J, Wang H, Fan J, *et al.* SAR Image Classification via Deep Recurrent Encoding Neural Networks. *IEEE Transactions on Geoscience and Remote Sensing.* 2018;56(4):2255–2269.

[18] Lang H and Wu S. Ship Classification in Moderate-Resolution SAR Image by Naive Geometric Features-Combined Multiple Kernel Learning. *IEEE Geoscience and Remote Sensing Letters.* 2017;14(10):1765–1769.

[19] Mahrooghy M, Aanstoos JV, Nobrega RAA, *et al.* A Machine Learning Framework for Detecting Landslides on Earthen Levees Using Spaceborne SAR Imagery. *IEEE Journal of Selected Topics in Applied Earth Observations and Remote Sensing.* 2015;8(8):3791–3801.

[20] Lin Y, Le Kernec J, Yang S, *et al.* Human Activity Classification With Radar: Optimization and Noise Robustness With Iterative Convolutional Neural Networks Followed With Random Forests. *IEEE Sensors Journal.* 2018;18(23):9669–9681.

[21] Seyfioğlu MS, Özbayoğlu AM, and Gürbüz SZ. Deep Convolutional Autoencoder for Radar-Based Classification of Similar Aided and Unaided Human Activities. *IEEE Transactions on Aerospace and Electronic Systems.* 2018;54(4):1709–1723.

[22] Amin MG, Zhang YD, Ahmad F, *et al.* Radar Signal Processing for Elderly Fall Detection: The Future for In-Home Monitoring. *IEEE Signal Processing Magazine.* 2016;33(2):71–80.

[23] Kim Y and Toomajian B. Hand Gesture Recognition Using Micro-Doppler Signatures With Convolutional Neural Network. *IEEE Access.* 2016;4:7125–7130.

[24] Sun Y and Li J. Time-Frequency Analysis for Plastic Landmine Detection via Forward-Looking Ground Penetrating Radar. *IET Proceedings – Radar, Sonar and Navigation.* 2003;150(4):253.

[25] Giovanneschi F, Mishra KV, Gonzalez-Huici MA, *et al.* Dictionary Learning for Adaptive GPR Landmine Classification. *IEEE Transactions on Geoscience and Remote Sensing.* 2019;57(12):10036–10055.

[26] Qu Z, Wang W, Hou C, *et al.* Radar Signal Intra-Pulse Modulation Recognition Based on Convolutional Denoising Autoencoder and Deep Convolutional Neural Network. *IEEE Access.* 2019;7:112339–112347.

[27] Kong S, Kim M, Hoang LM, *et al.* Automatic LPI Radar Waveform Recognition Using CNN. *IEEE Access.* 2018;6:4207–4219.

[28] Kon S, Tanaka T, Mizutani H, and Wada M. A machine learning based approach to weather parameter estimation in Doppler weather radar. In: *2011 IEEE International Conference on Acoustics, Speech and Signal Processing (ICASSP)*, Prague; 2011. p. 2152–2155. doi: 10.1109/ICASSP.2011.5946753.

[29] Chen H, Chandrasekar V, Cifelli R, *et al.* A Machine Learning System for Precipitation Estimation Using Satellite and Ground Radar Network Observations. *IEEE Transactions on Geoscience and Remote Sensing.* 2019:1–13.

[30] Eroglu O, Kurum M, Boyd D, *et al.* High Spatio-Temporal Resolution CYGNSS Soil Moisture Estimates Using Artificial Neural Networks. Remote Sensing. 2019;11(19). Available from: https://www.mdpi.com/2072-4292/11/19/2272.

[31] Terabayashi K, Natsuaki R, and Hirose A. Ultrawideband Direction-of-Arrival Estimation Using Complex-Valued Spatiotemporal Neural Networks.

IEEE Transactions on Neural Networks and Learning Systems. 2014;25(9): 1727–1732. Available from: http://ieeexplore.ieee.org/document/6797874/.

[32] Gardill M, Fuchs J, Frank C, and Weigel R. A multi-layer perceptron applied to number of target indication for direction-of-arrival estimation in automotive radar sensors. In: *2018 IEEE 28th International Workshop on Machine Learning for Signal Processing (MLSP)*, Aalborg; 2018. p. 1–6. doi: 10.1109/MLSP.2018.8516952.

[33] Wu S, Sakamoto T, Oishi K, *et al.* Person-Specific Heart Rate Estimation With Ultra-Wideband Radar Using Convolutional Neural Networks. *IEEE Access.* 2019;7:168484–168494.

[34] Callaghan D, Burger J, and Mishra AK. A machine learning approach to radar sea clutter suppression. In: *2017 IEEE Radar Conference (RadarConf)*, Seattle, WA; 2017. p. 1222–1227. doi: 10.1109/RADAR.2017.7944391.

[35] Vishwakarma S, Ummalaneni V, Iqbal MS, Majumdar A, and Ram SS. Mitigation of through-wall interference in radar images using denoising autoencoders. In: *2018 IEEE Radar Conference (RadarConf18)*, Oklahoma City, OK; 2018. p. 1543–1548. doi: 10.1109/RADAR.2018.8378796.

[36] Ma D, Zhang X, Tang X, Ming J, and Shi J. A CNN-based method for SAR image despeckling. In: *IGARSS 2019 – 2019 IEEE International Geoscience and Remote Sensing Symposium*, Yokohama, Japan; 2019. p. 4272–4275. doi: 10.1109/IGARSS.2019.8899122.

[37] Nguyen LH and Tran TD. Separation of radio-frequency interference from SAR signals via dictionary learning. In: *2018 IEEE Radar Conference (RadarConf18)*, Oklahoma City, OK; 2018. p. 0908–0913. doi: 10.1109/RADAR.2018.8378681.

[38] Duda RO, Hart PE, and Stork DG. *Pattern Classification.* 2nd ed. New York, NY: Wiley; 2001.

[39] Wu H, Dai D, and Wang X. A Novel Radar HRRP Recognition Method with Accelerated T-Distributed Stochastic Neighbor Embedding and Density-Based Clustering. *Sensors.* 2019;19(23):5112.

[40] Harasti PR and List R. Principal Component Analysis of Doppler Radar Data. Part I: Geometric Connections Between Eigenvectors and the Core Region of Atmospheric Vortices. *Journal of the Atmospheric Sciences.* 2005;62: 4027–4042.

[41] Ertin E. Manifold learning methods for wide-angle SAR ATR. In: *2013 International Conference on Radar*, Adelaide, SA; 2013. p. 500–504. doi: 10.1109/RADAR.2013.6652039.

[42] Pham DT. Applications of unsupervised clustering algorithms to aircraft identification using high range resolution radar. In: *Proceedings of the IEEE 1998 National Aerospace and Electronics Conference.* NAECON 1998. Celebrating 50 Years (Cat. No.98CH36185), Dayton, OH, USA; 1998. p. 228–235. doi: 10.1109/NAECON.1998.710121.

[43] Liu H, Pang L, Li F, *et al.* Hough Transform and Clustering for a 3-D Building Reconstruction With Tomographic SAR Point Clouds. *Sensors.* 2019;19(24):5378.

[44] Kersten PR, Lee J-S, and Ainsworth TL. Unsupervised Classification of Polarimetric Synthetic Aperture Radar Images Using Fuzzy Clustering and EM Clustering. *IEEE Transactions on Geoscience and Remote Sensing.* 2005;43(3):519–527.

[45] Seng CH, Amin MG, Ahmad F, *et al.* Image Segmentations for Through-the-Wall Radar Target Detection. *IEEE Transactions on Aerospace and Electronic Systems.* 2013;49(3):1869–1896.

[46] Aggarwal CC. *Neural Networks and Deep Learning.* Cham: Springer; 2018.

[47] Schmidhuber J. Deep Learning in Neural Networks: An Overview. Neural Networks. 2015;61:85–117. Available from: http://www.sciencedirect.com/science/article/pii/S0893608014002135.

[48] Boser BE, Guyon IM, and Vapnik VN. A training algorithm for optimal margin classifiers. In: *Proceedings of the Fifth Annual Workshop on Computational Learning Theory.* COLT '92. New York, NY: ACM; 1992. p. 144–152. Available from: http://doi.acm.org/10.1145/130385.130401.

[49] Cortes C and Vapnik V. Support-Vector Networks. Machine Learning. 1995;20(3):273–297. Available from: https://doi.org/10.1023/A:1022627411411.

[50] Cristianini N and Shawe-Taylor J. *An Introduction to Support Vector Machines: And Other Kernel-based Learning Methods.* New York, NY: Cambridge University Press; 2000.

[51] Hotelling H. Analysis of a Complex of Statistical Variables Into Principal Components. *Journal of Educational Psychology.* 1933;24:417–441.

[52] Eckart C and Young G. The Approximation of One Matrix by Another of Lower Rank. *Psychometrika.* 1936;1:211–218.

[53] Jolliffe IT. *Principal Component Analysis.* 2nd ed. Springer Series in Statistics. New York, NY: Springer-Verlag; 2002.

[54] Seber G. *Multivariate Observations.* Hoboken, NJ: Wiley; 1984.

[55] Pham D, Dimov S, and Nguyen C. Selection of k in k-Means Clustering. Proceedings of the Institution of Mechanical Engineers, Part C: *Journal of Mechanical Engineering Science.* 2005;219(1):103–119.

[56] Hansen LK and Larsen J. Unsupervised learning and generalization. In: *Proceedings of International Conference on Neural Networks (ICNN'96)*, Washington, DC, USA. vol. 1; 1996. p. 25–30. doi: 10.1109/ICNN.1996.548861.

Chapter 3

Theoretical foundations of deep learning

Stefan Brüggenwirth[1] and Simon Wagner[1]

In this chapter, we will derive the theoretical foundations of deep neural network architectures. In contrast to shallow neural topologies, deep neural networks comprise more than one hidden layer of neurons. Even though the concept and theory has been around since many decades, efficient deep learning methods were developed in the last years and made the approach computationally tractable. This chapter will hence begin with a short review of historical and biological introduction to the topic. Then, we will address the mathematical model of the perceptron that still forms the basis of multilayer architectures. We will introduce the backpropagation algorithm as a state-of-the-art method for training of neural networks. We will briefly cover some popular optimization strategies that have been successfully applied and are quite relevant for radar applications that are sometimes quite different from the optical domain (e.g., scarce training data sets). These include stochastic gradient decent, cross-entropy (CE), regularization and the optimization of other hyperparameters. We will then cover convolutional neural networks (CNNs), some specific processing elements and learning methods for them. We will also look at some well-known and successfully applied architectures. The focus in this chapter lies on supervised learning for classification problems. To round up this chapter, an unsupervised method for representation learning called autoencoder (AE) is illustrated. The structure and theoretical derivations in this chapter follow standard-textbooks [1–3] and online sources [4,5].

3.1 Introduction

Biological investigations revealed quite early that information processing in the human brain is established as an emergent effect of the interplay of a multitude of simple neurons, an approach that cognitive science terms as connectionism. The first network of artificial neurons has been described in the 1940s by [6]. Opposite to symbol-based approaches that make working memory content and reasoning explicit,

[1]Fraunhofer FHR, Wachtberg, Germany

and have been popular both in cognitive psychology and artificial intelligence, neural networks operate with a subsymbolic representation. It often criticized that this makes the systematic design, understanding of the output and certification of neural networks (e.g., for safety critical applications) hard. However, it is without doubt that subsymbolic approaches have the great strength of generalizing concepts from training data and have proven successful in various domains. An interesting approach that tries to overcome this weakness is explainable-Artificial Intelligence (AI) [7].

A neuron is a special type of cell that can be electrically excited and hence receive and process signals in the human body. It consists of the cell body, dendrites that receive and an axon that transmits signals to other neurons. It is estimated that the human brain consists of 90 billion neurons, each of them connected on average to 10,000 other nerve cells as shown in Figure 3.1. The information is encoded in the transmission of an action potential, a discrete spike (known as depolarization) in the electric charge of the cell membrane that propagates along the axon toward the synaptic gap. Mostly, chemical synapses translate the electric impulse to a chemical transmitter that is released and received by neighboring cells. Synapses can increase

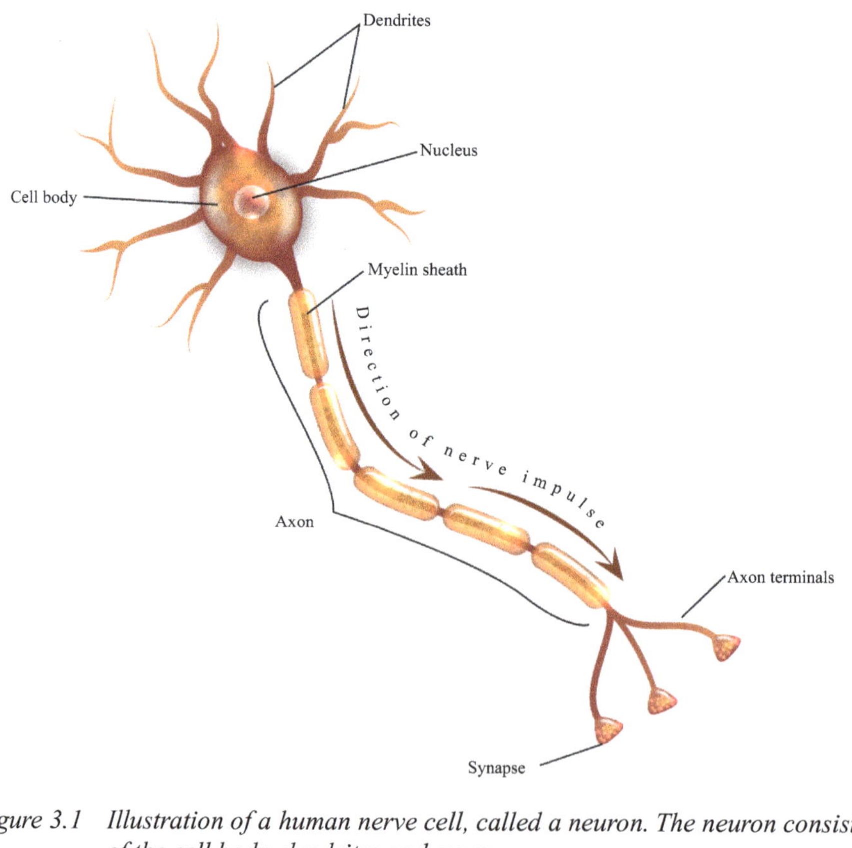

Figure 3.1 Illustration of a human nerve cell, called a neuron. The neuron consists of the cell body, dendrites and axons

(excitatory) or decrease (inhibitory) the activation potential between the neurons. This neuroplasticity is considered as a means of learning in the brain [8]. Only if a certain threshold of the incoming action potential is reached, the cell is said to "fire" and transmits an output signal along its axons to connected neurons (all-or-none law). Since each output signal is discrete, continuous values can be encoded in the firing rate of the neuron. The biological neuron is described by complex electrochemical processes. Even though its reaction speed is slow, neurons are very energy efficient, highly parallel and allow for computation of arbitrary functions. Artificial neural networks often do not have a direct biological match but share similar concepts and certainly inspired mathematical models for information processing as shown in the next section.

3.2 Perceptron

In analogy to the biological counterpart described earlier, an artificial neuron has to be modeled. It contains

1. connecting links, corresponding to the synapses;
2. a linear combiner to sum the input signals; and
3. a nonlinear activation function that decides if the neuron is firing or not.

A basic model for such an artificial neuron is the *perceptron* shown in Figure 3.2. It contains $m + 1$ input values $x_i \in \mathbb{R}$, a *summing output* $v \in \mathbb{R}$ and the *activation output* value $y \in \mathbb{R}$.

The original perceptron [9] contained a threshold such that the neuron, just like the biological one, would only fire, if the weighted input signal is larger than the

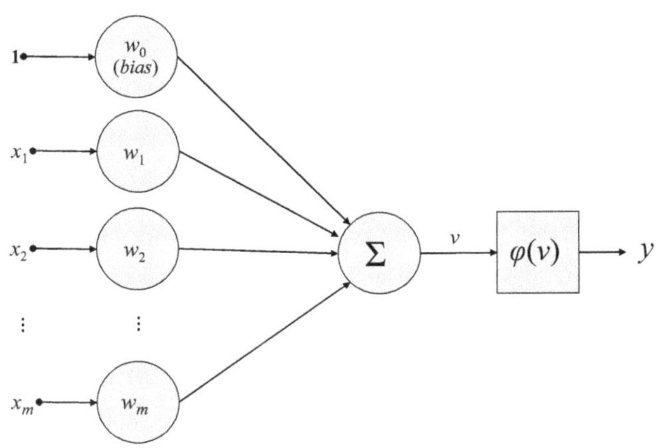

Figure 3.2 Perceptron model of an artificial neuron

threshold t, e.g., $\sum x_i \cdot w_i \geq t$. We can define a bias $b = -t$ and bring the term to the left side of the equation to obtain:

$$y = \begin{cases} 0 & \text{if} \sum x_i \cdot w_i + b \leq 0 \\ 1 & \text{if} \sum x_i \cdot w_i + b > 0 \end{cases} \tag{3.1}$$

Some textbooks treat the bias separately. In this chapter, we will introduce a weight w_0 that has a constant input value $x_0 := 1$ to represent the bias as shown in the figure. The summing output v can be computed by multiplying the input values x_i with the associated weights w_i and expressed in vector-notation:

$$v = \sum_{i=0}^{n} x_i \cdot w_i = \mathbf{x}^T \cdot \mathbf{w} \tag{3.2}$$

3.3 Sigmoid perceptron

The summing output serves as parameter for the nonlinear *activation function* $\varphi(v)$. Equation (3.1) is an example for a stepwise activation, e.g., the neuron only fires if a certain activation potential is reached.

The more common *sigmoid perceptron* uses the logistic function $y = \sigma(v)$ from (3.3) that produces a continuous output signal y between 0 and 1 as shown in Figure 3.6. Since the derivative of the activation function is required during training, it should be steady and easy to compute as shown in (3.4). Other activation functions are possible and explained in Section 3.7.1.

$$\varphi(v) = \sigma(v) = \frac{1}{1 + e^{-v}} \tag{3.3}$$

$$\varphi(v)' = \sigma'(v) = \sigma(v)(1 - \sigma(v)) = -ln\left(\frac{1}{v} - 1\right) \tag{3.4}$$

3.4 Multilayer perceptron

With the sigmoid perceptron as basic building blocks from Figure 3.2, we can build more complex hierarchies of artificial neurons. Figure 3.3 shows a multilayer perceptron (MLP) architecture. The network has three input signals $\mathbf{x} \in \mathbb{R} \times \mathbb{R} \times \mathbb{R}$, described here as *input layer*. The input variable is not mapped onto a neuron but just represents the values as real numbers. The network has two output neurons that are arranged in the *output layer*. Input and output layers are the *visible layers*, since they are directly connected to the environment. All layers inbetween are called *hidden layers* and may contain an arbitrary number of neurons. In this case, there are two hidden layers with four neurons. Layers, in which all neurons are connected to all neurons in the previous layer, are called *fully connected*. For short, the topology of this network can be denoted as $\{3, 4, 4, 2\}$. This topology is a *feedforward* neural network,

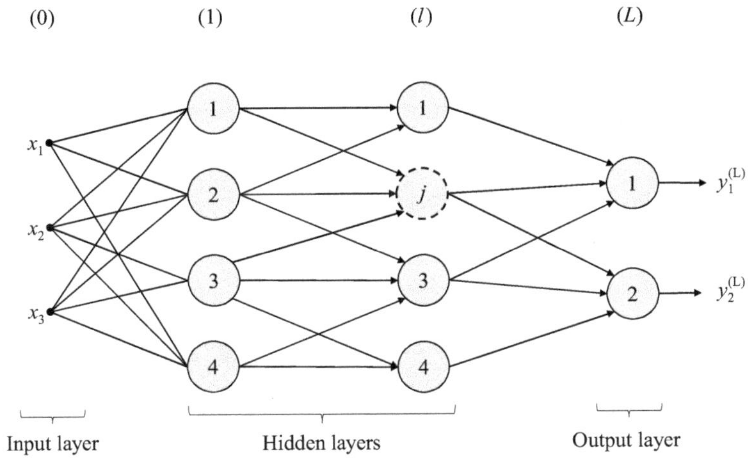

Figure 3.3 Multilayer-perceptron

since the information is propagated from the input to the output layer, without any loops.

In order to mathematically describe the neural network, a precise notation and indexing scheme is required. We use the subscript j and superscript l to address the jth neuron ($j = 1, 2, ..., m_l$) in the lth layer ($l = 1, 2, ..., L$) consisting of m_l neurons. The summing output is hence described as $v_j^{(l)}$ and the output activation as $y_j^{(l)} = \varphi(v_j^{(l)})$. The weights indices $w_{ji}^{(l)}$ may seem to be a bit counter-intuitive at first, but it will simplify notation later. The first index j denotes the neuron the weight belongs to at layer l, the second index i denotes the output neuron on the previous layer $l - 1$ the weight is connected to. The first layer $l = 0$ is connected to the input values $y^{(0)} = x_i$. The weight $w_{j0}^{(l)}$ captures the bias term in our notation and requires a constant output value $y_0^{(l-1)} = \varphi(v_0^{(l-1)}) = 1$ to be introduced for each layer. With these definitions, the summing output $v_j^{(l)}$ at layer l is computed by

$$v_j^{(l)} = \sum_{i=0}^{m_{l-1}} w_{ji}^{(l)} \cdot y_i^{(l-1)}. \tag{3.5}$$

Finally, we can now introduce a vector notation for the weights in layer l

$$\mathbf{w}^{(l)} = (w_{10}^{(l)}, w_{11}^{(l)}, ..., w_{m_l,m_{l-1}}^{(l)})^T \tag{3.6}$$

and obtain a vector containing all the weights by stacking

$$\mathbf{w} = (\mathbf{w}^{(1)}, \mathbf{w}^{(2)}, ..., \mathbf{w}^{(L)})^T \tag{3.7}$$

3.5 Gradient descent

The MLP illustrated in the previous section can be used to approximate an arbitrary function [10] given by a labeled training data set $\mathcal{T}$ consisting of N data samples $\mathbf{x}(n)$ and the desired output $\mathbf{d}(n)$.

$$\mathcal{T} = \{\mathbf{x}(n), \mathbf{d}(n)\}; \; n = 1, 2, \ldots, N \tag{3.8}$$

with the $m + 1$-dimensional input vector for training datum n

$$\mathbf{x}(n) = (1, x_1(n), x_2(n), \ldots, x_m(n))^T. \tag{3.9}$$

In order to determine the weights for the neural network that best approximate the given training data, a training algorithm for the network is required. A *cost function* $\mathcal{E} = \mathcal{E}(\mathbf{x}, \mathbf{d}, \mathbf{y}^{(L)})$ is required that numerically assesses how well the network has reconstructed the desired output $\mathbf{d}(n)$ for input $\mathbf{x}(n)$. A commonly used cost function is the mean-squared-error (MSE). For an unbiased Gaussian-distributed random variable, it corresponds to the minimum-variance estimate

$$\mathcal{E}_{\text{MSE}} = \frac{1}{2 \cdot N} \sum \|\mathbf{y}^{(L)} - \mathbf{d}\|^2. \tag{3.10}$$

The cost function should be zero if the network output $\mathbf{y}^{(L)}$ perfectly matches the desired training data $\mathbf{d}$, increase with deviations and be continuously differentiable. We can think of the neural network with its chained layers of summing junctions and activation functions as a (quite complex) but continuously differentiable multivariate mathematical function of the input data and the neuron weights and biases. The output of the network, and hence the cost function, can be approximated by a first-order Taylor expansion. The change in the cost function can locally be approximated by the total differential (for sufficiently small $\Delta\mathbf{w}$) as shown in Figure 3.4:

$$\Delta\mathcal{E} \approx \frac{\partial\mathcal{E}}{\partial w_{10}^{(1)}} \cdot \Delta w_{10}^{(1)} + \frac{\partial\mathcal{E}}{\partial w_{11}^{(1)}} \cdot \Delta w_{11}^{(1)} + \cdots + \frac{\partial\mathcal{E}}{\partial w_{ji}^{(L)}} \Delta w_{ji}^{(L)}. \tag{3.11}$$

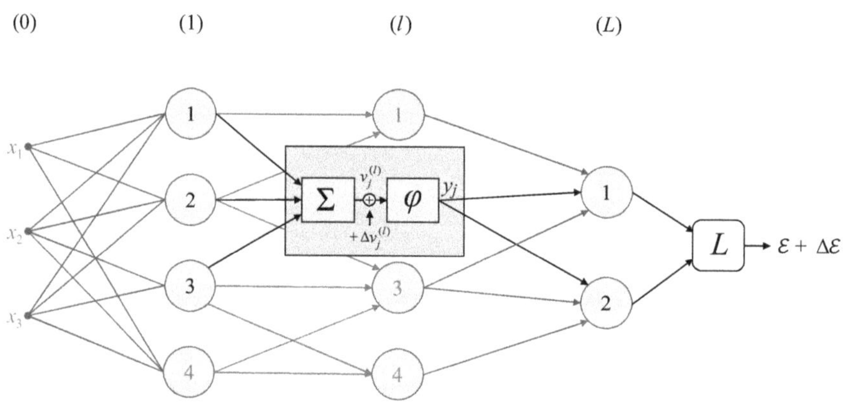

Figure 3.4 Effect of a local perturbation in the summing output to the cost function

Using the nabla-operator $\nabla\mathscr{E} = \left(\partial\mathscr{E}/\partial w_{01}^{(1)}, \partial\mathscr{E}/\partial w_{11}^{(1)}, ..., \partial\mathscr{E}/\partial w_{ji}^{(L)}\right)^T$ for the gradient yields

$$\Delta\mathscr{E} \approx \nabla\mathscr{E} \cdot \Delta\mathbf{w}. \tag{3.12}$$

An *iterative descent* algorithm is guaranteed to converge, if starting from an initial guess $\mathbf{w}(0)$, the cost function $\mathscr{E}(\mathbf{w}(n))$ is reduced in each iteration step

$$\mathscr{E}(\mathbf{w}(n+1)) < \mathscr{E}(\mathbf{w}(n)). \tag{3.13}$$

In the *method of steepest descent*, the weight vector is decreased in the opposite direction to the gradient vector

$$\Delta\mathbf{w} = -\eta \cdot \nabla\mathscr{E} \tag{3.14}$$

where η is a positive constant called the *learning rate*. If it is chosen low enough, so that (3.12) is still valid, this guarantees the reduction of the cost function, as can be seen by inserting (3.14) into (3.12)

$$\Delta\mathscr{E} \approx -\eta\nabla\mathscr{E} \cdot \nabla\mathscr{E} = -\eta\|\nabla\mathscr{E}\|^2 \leq 0 \tag{3.15}$$

Optimization and control of the learning rate is hence an important hyperparameter that must be carefully selected to make the algorithm converge quickly. For an iterative implementation, (3.14) can be rewritten as *gradient descent* weight update rule from step n to $n+1$

$$\mathbf{w}(n+1) = \mathbf{w}(n) - \eta \cdot \nabla\mathscr{E} \tag{3.16}$$

3.6 Backpropagation

For the gradient descent algorithm to be applied, we need to know the partial derivatives of the cost function with respect to the neuron weights $\partial\mathscr{E}/\partial w_{ji}^{(l)}$. To derive an equation for it, let us investigate how a (small) perturbation $\Delta v_j^{(l)}$ to the input of the activation function $\varphi = \varphi(v_j^{(l)}) = y_j^{(l)}$ at the jth neuron ($j = 1, 2, ..., m_l$) at layer l affects the output cost function $\mathscr{E}$. Assuming the network can be represented by a differentiable function and all other parameters remain unchanged, we can do a linear approximation of the increase $\Delta\mathscr{E}$ of the cost function by the total derivative $\Delta\mathscr{E} \approx (\partial\mathscr{E}/\partial v_j^{(l)}) \cdot \Delta v_j^{(l)}$. We define the *local gradient* as $\delta_j^{(l)} = (\partial\mathscr{E}/\partial v_j^{(l)})$.

To compute the local gradient for the jth neuron in the output layer L, we use its definition:

$$\delta_j^{(L)} = \frac{\partial\mathscr{E}}{\partial v_j^{(L)}}. \tag{3.17}$$

We can extend this expression by formally applying the (multivariate) chain rule from differential calculus with respect to the output activations $y_k^{(L)}$ of all m_L neurons in the output layer

$$\delta_j^{(L)} = \sum_{k=1}^{m_L} \frac{\partial \mathscr{E}}{\partial y_k^{(L)}} \cdot \frac{\partial y_k^{(L)}}{\partial v_j^{(L)}}. \tag{3.18}$$

Since we are investigating the influence of a single neuron, we assume that only the output activation $y_j^{(L)}$ of the jth neuron is changed, all other partial derivatives $k \neq j$ are zero.

$$\delta_j^{(L)} = \frac{\partial \mathscr{E}}{\partial y_j^{(L)}} \cdot \frac{\partial y_j^{(L)}}{\partial v_j^{(L)}} \tag{3.19}$$

With the definition of the activation function $y_j^{(L)} = \varphi(v_j^{(L)})$, it follows

$$\delta_j^{(L)} = \frac{\partial \mathscr{E}}{\partial y_j^{(L)}} \cdot \varphi'(v_j^{(L)}) \tag{3.20}$$

Then, we can derive an update rule from one layer to the next. We again start with the definition of the local gradient of the jth neuron at an arbitrary layer (l)

$$\delta_j^{(l)} = \frac{\partial \mathscr{E}}{\partial v_j^{(l)}}. \tag{3.21}$$

The chain rule can be used again to expand this term with respect to summing outputs $v_k^{(l+1)}$ of the m_{l+1} neurons in the next layer

$$\delta_j^{(l)} = \sum_{k=1}^{m_{l+1}} \frac{\partial \mathscr{E}}{\partial v_k^{(l+1)}} \cdot \frac{\partial v_k^{(l+1)}}{\partial v_j^{(l)}}. \tag{3.22}$$

Replacing the first term with the definition of $\delta_k^{(l+1)}$ we get

$$\delta_j^{(l)} = \sum_{k=1}^{m_{l+1}} \delta_k^{(l+1)} \cdot \frac{\partial v_k^{(l+1)}}{\partial v_j^{(l)}}. \tag{3.23}$$

Using the definition of the summing output $v_k^{(l+1)} = \sum_{j=0}^{m_l} w_{kj}^{(l+1)} \cdot y_j^{(l)}$ and $y_j^{(l)} = \varphi(v_j^{(l)})$, the required derivative can be computed as

$$\frac{\partial v_k^{(l+1)}}{\partial v_j^{(l)}} = \frac{\partial \sum_{j=0}^{m_l} w_{kj}^{(l+1)} \cdot y_j^{(l)}}{\partial v_j^{(l)}} = \frac{\partial \sum_{j=0}^{m_l} w_{kj}^{(l+1)} \cdot \varphi(v_j^{(l)})}{\partial v_j^{(l)}} = w_{kj}^{(l+1)} \cdot \varphi'(v_j^{(l)}) \tag{3.24}$$

and (3.23) can be rewritten as

$$\delta_j^{(l)} = \sum_{k=1}^{m_{l+1}} \delta_k^{(l+1)} \cdot w_{kj}^{(l+1)} \cdot \varphi'(v_j^{(l)}) = \varphi'(v_j^{(l)}) \sum_{k=1}^{m_{l+1}} \delta_k^{(l+1)} \cdot w_{kj}^{(l+1)}. \tag{3.25}$$

Finally, we can derive the cost with respect to the weights. We expand the desired expression using the chain rule:

$$\frac{\partial \mathscr{E}}{\partial w_{ji}^{(l)}} = \frac{\partial \mathscr{E}}{\partial v_j^{(l)}} \cdot \frac{\partial v_j^{(l)}}{\partial w_{ji}^{(l)}} \tag{3.26}$$

Using the definition of the summing output $v_j^{(l)} = \sum_{i=0}^{m_{l-1}} w_{ji}^{(l)} \cdot y_i^{(l-1)}$ again, we can compute the required derivative

$$\frac{\partial v_j^{(l)}}{\partial w_{ji}^{(l)}} = \frac{\partial \sum_{i=0}^{m_{l-1}} w_{ji}^{(l)} \cdot y_i^{(l-1)}}{\partial w_{ji}^{(l)}} = y_i^{(l-1)}. \tag{3.27}$$

We can then rewrite (3.26) with the definition of the local gradient and obtain the desired expression for the derivative of the cost function with respect to the weights $w_{ji}^{(l)}$ of the jth neuron at layer l, given its local gradient $\delta_j^{(l)}$ and all output activations $y_i^{(l-1)}$ from the previous layer

$$\frac{\partial \mathscr{E}}{\partial w_{ji}^{(l)}} = \delta_j^{(l)} \cdot y_i^{(l-1)}. \tag{3.28}$$

In our notation, these equations are also implicitly used to compute the derivatives of the biases $w_{j0}^{(l)}$ by remembering that $y_0^{(l-1)} = 1$. With (3.20), (3.25) and (3.28), the derivatives of the cost function can be computed and the neuron weights can be adjusted according to the gradient descent algorithm.

Figure 3.5 summarizes the *backpropagation* training algorithm that combines the computation of the gradient and the gradient descent method for weight adjustment. The timespan of N iterations in which the whole training set is presented one time is called an *epoch*. Note that the desired output of the training samples depends on the chosen activation function, the values of $d(n)$ depend on the range of the output layer activation function $\varphi(v_j^{(L)})$. The random weight initialization should be chosen such that the standard deviation of $v_j^{(l)}$ lies at the transition of the linear and the saturated parts of the activation function. If neuron j is not connected to all neurons in the following layer, the sum over k in the calculation of the local gradient runs only over the connected neurons in layer $l + 1$.

As stopping criteria, a fixed number of training epochs can be set. In the *early stopping* strategy, track is kept of the classification accuracy against the validation data set after each epoch, until either a desired accuracy has been reached, or the accuracy starts to decline again after reaching a maximum. The latter is a sign of *overfitting*, e.g., the network has "memorized" certain characteristics of the training data but is unable to generalize from it. Overfitting often occurs if there is an insufficient number of training data sets. If the stopping criterion is not met after N iterations, the complete training set should be presented again in a different order. In fact, it is common to introduce a third data set, called the *validation data set*, that is used to tweak and assess certain *hyperparameters* (such as the learning rate) or improvement techniques presented in the next section.

Summary of the backpropagation training algorithm

1. Define a training set $\mathscr{T}$, which consists of N training samples with the corresponding desired outputs

$$\mathscr{T} = \{\mathbf{x}(n), \mathbf{d}(n); n = 1, 2, \ldots, N\}.$$

2. Random initialization of the network weights $w_{ji}^{(l)}$.
Set $y_i^{(0)} = x_i, n = 1, y_0^{(l-1)} = 1, l = 1, 2, ..., L$.
3. Forward propagation of training sample $\mathbf{x}(n)$

$$v_j^{(l)} = \sum_{i=0}^{m_{l-1}} w_{ji}^{(l)}(n) y_i^{(l-1)}(n)$$

$$y_j^{(l)}(n) = \varphi_j^{(l)}\left(v_j^{(l)}\right).$$

4. Calculate the local gradient for all neurons. Equations (3.20) and (3.25).

$$\delta_j^{(l)}(n) = \begin{cases} \frac{\partial \mathscr{E}}{\partial y_j^{(L)}} \cdot \varphi_j'^{(L)}(v_j^{(L)}(n)) & \text{if} \quad l = L \\ \varphi_j'^{(l)}(v_j^{(l)}(n)) \sum_{k=1}^{m_{l+1}} \delta_k^{(l+1)}(n) w_{kj}^{(l+1)}(n) & \text{else} \end{cases}.$$

5. Compute gradient vector of weights according to (3.28)

$$\frac{\partial \mathscr{E}}{\partial w_{ji}^{(l)}} = \delta_j^{(l)} \cdot y_i^{(l-1)}$$

6. Weight update according to (3.16)

$$w_{ji}^{(l)}(n+1) = w_{ji}^{(l)}(n) - \eta \frac{\partial \mathscr{E}}{\partial w_{ji}^{(l)}}.$$

7. Iterate $n \rightarrow n+1$ and repeat steps 3–6. until stopping critarion is met.

Figure 3.5 Summary of the backpropagation algorithm

3.7 Improvements

In this section, we will illustrate some modifications to the approach presented earlier that are commonly used to make the backpropagation algorithm learn fast or generalize better from the data.

3.7.1 Activation functions

In Section 3.3, the logistic function was introduced as a differentiable activation function $\varphi(v)$. In this subsection, we will illustrate two other popular alternatives. A summary of activation functions and their derivatives is shown in Figure 3.6.

3.7.1.1 Hyperbolic tangent

The output of the logistic activation function tends to be zero, for a negative input. This may cause neurons to saturate during training. An activation function that alleviates

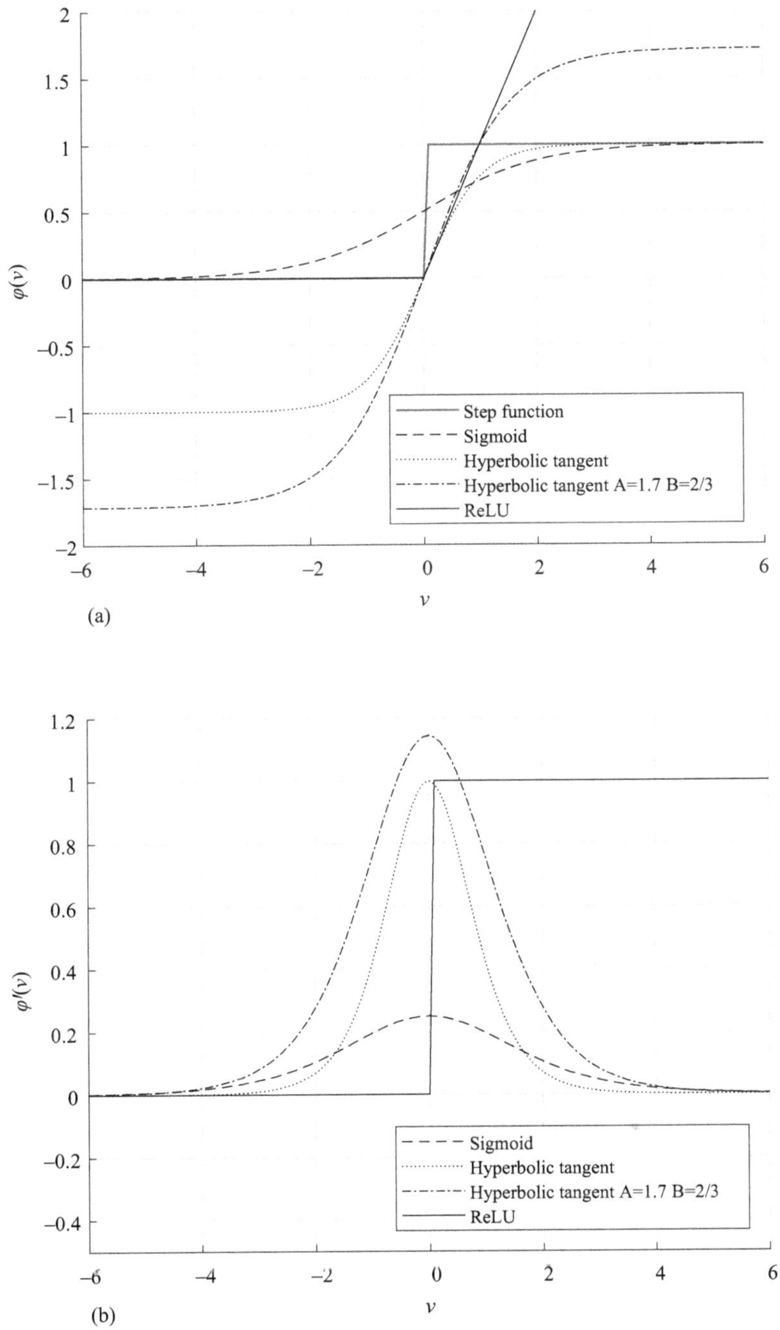

Figure 3.6 Comparison of different activation functions and derivatives: (a) activation functions $\varphi(v)$ and (b) derivatives of activation functions $\varphi'(v)$

this problem and is popular for classification is the hyperbolic tangent function. It is also sigmoid (s-shaped) and produces an output range from −1 to 1.

$$\varphi(v) = A \cdot (\tanh(b \cdot v)) \qquad (A, b) > 0, \tag{3.29}$$

and the derivative

$$\varphi'(v) = Ab(1 - \tanh^2(b \cdot v)) \tag{3.30}$$

3.7.1.2 Rectified linear unit

The rectified linear unit (ReLU) is probably most used in convolutional neural nets for image recognition. It has an output range from 0 to inf. From the definition in (3.31) it becomes clear why this function is called ReLU; for a positive input, it shows a linear behavior and negative inputs are mapped to 0, thus the signal gets rectified.

The ReLU activation function is defined as

$$\varphi(v) = max(0, v) \tag{3.31}$$

and the derivative

$$\varphi'(v) = \begin{cases} 0 & v < 0 \\ 1 & v > 0 \end{cases} \tag{3.32}$$

3.7.2 Momentum

The use of *momentum* in the weight change can stabilize and accelerate the training. In this variation, a certain amount of the weight change of the previous iteration is added to the current iteration by

$$\Delta w_{ji}^{(l)}(n) = \alpha \Delta w_{ji}^{(l)}(n-1) + \eta \delta_j^{(l)}(n) y_i^{(l-1)}(n), \tag{3.33}$$

where α is the *momentum constant.* With this variation, the training should be accelerated in steep areas and stabilized in oscillatory areas of the cost function.

3.7.3 Stochastic gradient descent

The stochastic gradient descent method is related to a change of the iterative behavior of the algorithm. So far each input sample was used independently to create a weight update, i.e., an iteration step of the backpropagation. The alternative to this is the so-called *batch update*, where the gradient of the cost function is calculated for K iterations and the weight update is performed using the mean value of these iterations

$$w_{ji}^{(l)}(n+1) = w_{ji}^{(l)}(n) + \frac{\eta}{K} \sum_{k=1}^{K} \delta_j^{(l)}(k) y_i^{(l-1)}(k). \tag{3.34}$$

In this case, the iteration index n refers to the iteration of the weight update and the index k is running over samples that have not been presented up to this weight update in the current epoch. Typical values for the batch size K vary between 10 and 50.

3.7.4 Cross-entropy cost function

In gradient descent, the learning rate per neuron depends on the gradient. For the mean-square error cost function (3.10), this yields $\partial\mathscr{E}/\partial w \sim (d-y)\cdot\varphi'(v)$. For a sigmoid neuron with an output close to one, the derivative $\varphi'(v)$ is very flat and learning is slow. An alternative cost function is the *CE* defined as

$$\mathscr{E}(n) = -\sum_{j=1}^{m_L} y_j(n)\ln(d_j(n)) + \big(1-y_j(n)\big)\ln(1-d_j(k)). \tag{3.35}$$

Its gradient does not depend on $\varphi'(v)$ anymore and is therefore generally preferred. This cost function can only be applied to output values between zero and one, which means that the activation function must be adapted to this interval. It is applicable especially to classification functions, where the output values must sum up to one and can thus be interpreted as likelihood (so-called *softmax* output layers). The cost function can be simplified in this case to

$$\mathscr{E}(n) = -\sum_{j=1}^{m_L} y_j(n)\ln(d_j(n)), \tag{3.36}$$

since the minimum of the first summand in (3.35) implies a minimum of the second summand.

3.7.5 Regularization

The idea behind a regularization technique known as *weight decay* is that a network with less and smaller weights should give a better generalization, and the risk of overfitting to the training data should be reduced. A regularization term $\sum_i w_i^2$ is added to the cost function to sum all the weights

$$\mathscr{E} = \mathscr{E}_0 + \frac{\varepsilon}{2n}\cdot\sum_i w_i^2. \tag{3.37}$$

All other things equal, the regularization term prefers a network with smaller weights. The regularization parameter ε controls the impact of the regularization versus the original cast function. There is no mathematical proof that this technique always works, but it was reported several times that this method is successful. Another reason why weight decay is so popular is its simplicity; after each iteration, all weights are simply "decayed" according to

$$w_{ji}^{(l)new}(n) = w_{ji}^{(l)old}(n)(1-\varepsilon) \qquad \text{with} \quad 0 < \varepsilon < 1. \tag{3.38}$$

With this decay, the irrelevant weights should decay to zero, while the important weights are kept by the algorithm.

3.8 The short history of deep learning

Even though the previous sections have laid the foundation for deep learning, a real breakthrough was the introduction of a network called *CNN* in [11]. This network was the first topology that consists of four or five layers and could thus be considered as "deep." If standard MLPs are built of more than two layers, a problem called the *vanishing gradient* arises. This means that the local gradient that is determined by (3.25) will diffuse while propagating through many neurons and layers. CNNs have mainly locally connected neurons, which helps the gradient to propagate based on the hypothesis that there is less influence from other neurons [12]. More details on CNNs will be given in Section 3.9.

The next big step in deep learning was the introduction of a layer-wise pretraining of the networks and finalize the fully connected network with some finetuning [13]. With this method, also fully connected MLPs with several layers are possible and as an example the so-called *AE* is presented in Section 3.10. Parallel to the introduction of the layer-wise pretraining, the use of graphical processor units (GPUs) with several cores for calculations to train deep learning architectures became popular.

The breakthrough of deep learning was in 2012 when a CNN was used in a competition to classify the ImageNet data set, which consists of 1.2 million images in 1,000 classes. The description of this network can be found in [14], it uses recent training techniques like GPU processing, rectified linear units as activation function and a regularization technique called dropout. The latter two, ReLU and dropout, will be shortly described in Section 3.11. Since 2012, deep learning techniques dominated the mentioned ImageNet competition, and many major companies, like Facebook, Google, IBM and Microsoft, invested significantly in research to thrive deep learning. Some applications that appeared since this time are the speech recognition systems from Apple and Google, the triumph of AlphaGo against the world champion in the board game Go and the face recognition system of Facebook to just name a few.

The application of deep learning became much easier during the last years, since there are many open source toolboxes to implement these kinds of networks. The most prominent one is probably Google's *TensorFlow*, but there are also others like the *Cognitive Tool Kit* from Microsoft or *Caffe* from UC Berkley. Many of these toolboxes can be integrated in *Keras* that is an open source Python toolbox that provides rapid prototyping and an easy way to build deep networks very fast.

3.9 Convolutional neural networks

CNNs are a special form of MLPs, which are designed specifically to recognize two-dimensional shapes with a high degree of invariance to translation, scaling, skewing and other forms of distortion [2]. This kind of network is inspired by the visual system of the brain and was for many years the only type of deep neural network that could efficiently be trained due to its structure using the technique of *weight sharing* [12].

3.9.1 Structure of a convolutional neural network

The basic structure of a network used for target classification in radar images is shown in Figure 3.7.

The first three layers of this network are an alternation of *convolutional* and *subsampling layers*. With these layers, a feature extraction stage is created, where each neuron in the first convolutional layer takes its inputs from a *local receptive field* in the input image. The neurons of the first layer are separated into *feature maps*, each of them corresponding to one local spatial feature that is learned during the training. Because all neurons in one feature map use the same weights, and the local receptive field of neighboring neurons is shifted only by one pixel in one direction, the values of each feature map correspond to the result of the two-dimensional

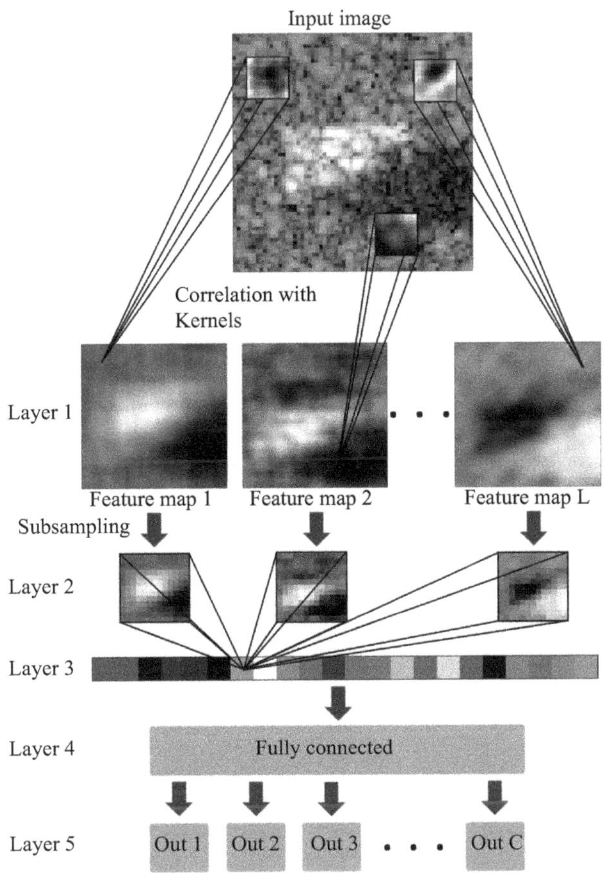

Figure 3.7 Structure of the used convolutional neural network. ©2017 IEEE. Reprinted, with permission, from Reference [15]

cross-correlation of the input image with the learned *kernel* K of each particular feature map F

$$F(i,j) = (I \times K)(i,j) = \sum_m \sum_n I(i+m, j+n)K(m,n). \tag{3.39}$$

The term "convolutional networks" is used, though actually a cross-correlation is performed here. It is also possible to implement a CNN with an actual convolution operation, which would lead to a flipped kernel. But in the opinion of the author, an implementation using cross-correlation is more intuitive than an implementation with convolution.

To show the benefit of the weight sharing and the local connections, a simple calculation for a single feature map of the network in Figure 3.7 is done. The size of the input image is 64 × 64 pixels and one correlation kernel is 13 × 13 pixels. Since only the output values of the correlation where image and kernel fully overlap are used, the size of the feature map in the first layer is 52 × 52 pixels. For comparison, the number of connection weights are calculated for three different connection schemes (the bias terms are neglected in this example for simplicity):

- A fully connected network. Each of the 2, 704 neurons in the feature map is connected to all of the 4, 096 input values, which gives roughly 11 million connection weights.
- A locally connected network with independent weights. Every neuron in the feature map is connected to a 13 × 13 pixels area in the input image, which gives around half a million connection weights.
- A CNN feature map connection scheme. All neurons in the feature map use the same weights, so that only 169 weights of the 13 × 13 receptive field must be determined for each feature map.

This example shows one of the reasons for the success of CNNs, it possesses a small number of parameter for a huge number of connections. Due to this reason, the so-called *Vapnik–Chervonenkis dimension* (VC-dimension) of a CNN is quite small and thus the probability for *overfitting* is small compared to fully connected neural networks of comparable size [1].*

In the input image of Figure 3.7, one target is visible in the center of the image. The correlation with the different kernels is visualized for three examples. The learned kernels are depicted inside the black squares on the input image and the result of the correlation can be seen in the feature maps of the first layer.

The second layer of the network is a subsampling layer and performs a reduction of the dimension by a factor of four. With this reduction, the exact position of the feature

*The VC-dimension is a measure for the complexity of a classifier, it is applicable to all classifier in the field of machine learning and is not limited to the field of deep learning. It gives the number of input samples that could be arbitrary separated by a binary classifier, which means that a classifier with a high VC-dimension could adapt very good to the distribution of the input data, instead of generalizing to a maybe simpler but more appropriate model for the test data. This "remembering" of the input data is also called overfitting, i.e., the classifier is tailored to the input data and not to the underlying distribution of the overall problem.

becomes less important and it reduces the sensitivity to other forms of distortion [2]. The subsampling is done by taking the maximum value of an area of 4 × 4 pixels, multiplying it with a weight w_j and adding a trainable bias b_j. This method is called *max-pooling*, other ways to perform the subsampling are averaging the chosen area or taking the median value.

The third layer is a convolutional layer again and relates the features found in the image to each other. This layer is trained to find patterns of features, which can be separated by the subsequent layers and discriminate the different classes. This alternation of different convolutional layers shows the main concept of deep learning; data is not classified directly, but different steps of feature extraction are performed. The first layer, which is applied to the input data, looks for abstract features and the subsequent layers combine these features to more concrete ones, and the last two layers perform the actual classification. This structure of feature extraction, "from abstract to class-dependent," is the main element in deep learning. Deeper networks with more layers might extract even more relevant features for discrimination, though there is still no tool to define an appropriate depth in advance depending on the problem.

The last two layers of the network form the decision part of the system are fully connected layers. The last layer consists of as many neurons as classes that have to be separated. The classification is done by assigning the corresponding class of the neuron with the highest output value.

3.9.2 Training of a convolutional neural network

The training of the network in Figure 3.7 is performed by a variation of the back-propagation algorithm coined *stochastic diagonal Levenberg–Marquardt algorithm,*[†] which is the first presented training algorithm for CNNs [11,16]. Because of the weight sharing, some changes in the notation compared to the previous sections are necessary. Since several connections use the same weight, these weights are grouped into the parameter vector $\boldsymbol{\alpha}$ with the elements α_k. Connection sets V_k are defined as the index triples (m, n, l) such that the connection between neuron m and n in layer l share the parameter α_k, i.e.,

$$w_{mn}^{(l)} = \alpha_k \qquad \forall\,(m, n, l) \in V_k. \tag{3.40}$$

The core of the actual learning algorithm is the same update rule as in (3.16) for the LMS algorithm, the parameters are updated according to the current gradient of the loss function at iteration n

$$\alpha_k(n+1) = \alpha_k(n) - \eta_k(n)\frac{\partial \mathscr{E}(n)}{\partial \alpha_k(n)}, \tag{3.41}$$

[†]The term stochastic is referring to the fact that the gradient is estimated based on a single snapshot or as the average of some snapshots of the available training samples. Thus the gradient is only an estimate of the true gradient.

where the learning rate η is no longer fixed but dependent on the parameter itself and iteration number. The calculation of the learning rate is done by

$$\eta_k(n) = \frac{\gamma(n)}{\mu + h_k(n)} \tag{3.42}$$

with the constant μ and a parameter $\gamma(n)$ that prevent the step size from becoming too large when the estimate of the second derivative $h_k(n)$, the Hessian, of the loss function $\mathcal{E}(n)$ with respect to α_k is small. For the calculation of $h_k(n)$, the Gauss–Newton approximation is used and guarantees a nonnegative estimate [16]. The use of the Hessian turns this variation into a second-order method, but the purpose is quite intuitive because a large Hessian indicates a changing gradient around this point. For this reason, the magnitude of the second derivative is used to control the learning rate of the algorithm. The parameter γ is marked here as dependent on the iteration but is fixed over several epochs of the training. Because of the weight sharing, the first and the second partial derivative of the loss function are the sums of partial derivatives with respect to the connections that actually share the specific parameter α_k

$$\frac{\partial \mathcal{E}(n)}{\partial \alpha_k(n)} = \sum_{(m,n,l) \in V_l} \frac{\partial \mathcal{E}(n)}{\partial w_{mn}^{(l)}(n)}, \tag{3.43}$$

$$\frac{\partial^2 \mathcal{E}(n)}{\partial \alpha_k^2(n)} = \sum_{(m,n,l) \in V_l} \sum_{(i,j,p) \in V_l} \frac{\partial^2 \mathcal{E}(n)}{\partial w_{mn}^{(l)}(n) \partial w_{ij}^{(p)}(n)}. \tag{3.44}$$

Further details of the algorithm and the approximations that are done to compute the derivatives can be found in [16].

A modern alternative to this second-order method is the so-called *Adam* algorithm that is nowadays used very often and some comparative investigations showed that this algorithm is always among the best results, for example, in [17]. A major difference to the previous algorithm is that it is a first-order method, i.e., only the gradient is used. Nevertheless, it is still adaptive, since the first and second moments of the gradient estimates are used to determine the learning rate. This estimation gave the name to the method, since the name is derived from *adaptive moment estimation*. The details of the algorithm and the derivation can be found in [18], here only the steps and the key points of the explanation are given.

This algorithm uses the batch update mentioned at the end of Section 3.6 and a fixed learning rate η with two hyperparameters β_1 and β_2. With these hyperparameters biased estimates of the first and second raw moments, the mean $m(n)$ and the uncentered variance $v(n)$, are calculated by

$$m(n) = \beta_1 m(n-1) + (1 - \beta_1) \frac{1}{K} \sum_{k=1}^{K} \frac{\partial \mathcal{E}(k)}{\partial w(k)} \tag{3.45}$$

and

$$v(n) = \beta_2 v(n-1) + (1 - \beta_1) \left(\frac{1}{K} \sum_{k=1}^{K} \frac{\partial \mathcal{E}(k)}{\partial w(k)} \right)^2. \tag{3.46}$$

Both $m(n)$ and $v(n)$ are initialized as a vector of zeros, which leads to a biased estimation toward zero. In the next step, these biased estimations are corrected to an unbiased estimation by

$$\hat{m}(n) = \frac{m(n)}{1 - \beta_1^n} \tag{3.47}$$

and

$$\hat{v}(n) = \frac{v(n)}{1 - \beta_2^n}. \tag{3.48}$$

With these values and the learning rate η, the final formula for the actual weight update is given by

$$w(n) = w(n-1) - \eta \frac{\hat{m}(n)}{\sqrt{\hat{v}(n)} + \varepsilon}, \tag{3.49}$$

where ε is a very small constant. In the original papers, default values for the tested machine learning problems are given by $\eta = 0.001$, $\beta_1 = 0.9$, $\beta_2 = 0.999$ and $\varepsilon = 10^{-8}$.

The scaling parameter of the learning rate, i.e., the ratio of mean to variance, is interpreted in [18] as signal-to-noise-ratio of the gradient. If this value is small, which corresponds to a large variance, the estimated direction of the gradient might differ from the true direction and the step size is reduced. If the value is high, i.e., the estimated gradient is always pointing in the same direction, this value is more trustworthy and thus the step size is increased.

Both of the methods here are extensions of the general backpropagation algorithm and can thus be used for any neural network, also for the AEs presented in the next section.

3.10 Autoencoders

In Section 3.9, CNNs were introduced to classify content of images and one elemental feature of these networks is the separation in feature extraction and classification part. Using this separation, the input data is transferred into an internal representation and the developer of the classifier does not have to design hand-crafted features that ensure accurate classification. This technique, which automatically extracts features from input data, is called *representation learning* and an *AE* is the most prominent example for this technique.

An AE is a combination of an *encoder*, which is represented by the function $f(\cdot)$, and a *decoder* realized by the function $g(\cdot)$. The encoder converts the input data into the internal representation $\mathbf{h}$, which is called *code*, and the decoder tries to recover the input data from this code. The output of the decoder is the *reconstruction* $\mathbf{r}$ of the input data. From these basic functions, it can be seen that the goal of an AE is to create a copy of the input data with the same meaningful latent space representation. The basic structure of an AE is shown in Figure 3.8.

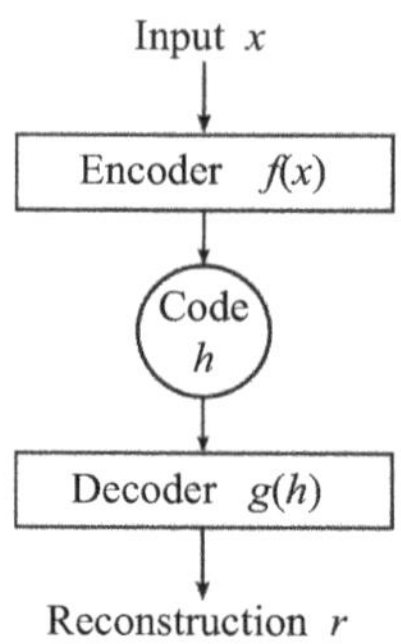

Figure 3.8 Basic structure of an autoencoder. ©2019 IEEE. Reprinted, with permission, From Reference [19]

The most commonly used forms for the encoder and decoder are of the same type as the perceptron in (3.2)

$$\mathbf{h} = f(\mathbf{x}) = \varphi_f(\mathbf{W}_f\mathbf{x} + \mathbf{b}_f) \tag{3.50}$$

$$\mathbf{r} = g(\mathbf{h}) = \varphi_g(\mathbf{W}_g\mathbf{h} + \mathbf{b}_g) \tag{3.51}$$

but in principal, any neural network structure can be used.

The outputs in (3.50) and (3.51) are vectors, since they represent the output values of all neurons in the encoder and decoder. For this reason, the weights $\mathbf{w}_j$ of the j neurons in these layers are combined as the rows of the weight matrix $\mathbf{W}$, which is multiplied with the corresponding input and the bias vector is added. The output of this affine mapping could be applied to a nonlinear function $\varphi(\cdot)$, which could also be omitted depending on the application.

One possible cost function that is often used is the squared difference between the input signal and the reconstruction

$$\mathscr{E}(n) = \frac{1}{2}\|\mathbf{x}(n) - \mathbf{r}(n)\|_2^2 = \frac{1}{2}\|\mathbf{x}(n) - g(f(\mathbf{x}(n)))\|_2^2. \tag{3.52}$$

This kind of cost function is similar to (3.10). It measures the instantaneous error energy between the desired output, which corresponds in this case to the input signal, and the actual output of the network at iteration n. Since the desired output of an AE is determined by the input signal, no labeling of the data is necessary. Due to this reason, the training of AEs is called *unsupervised training*, while the training with labeled data is called *supervised training*.

One possible application of AEs is data compression, which is presented in the following section.

3.10.1 Data compression with autoencoders

In some way, one might say that AEs are built for data compression. If the dimension of the code that is generated by the AE is smaller than the dimension of the input signal $\mathbf{x}$,

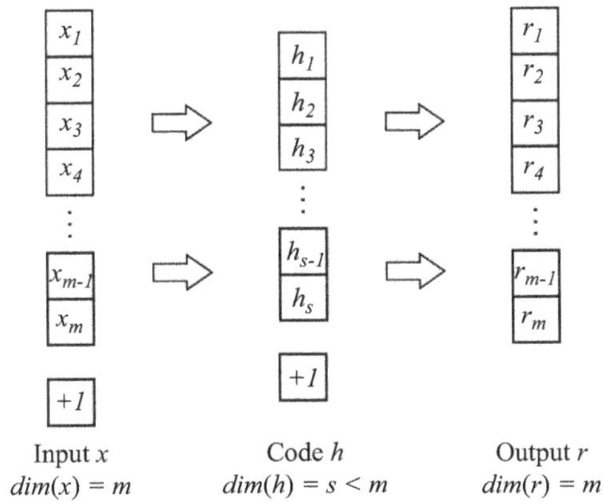

Figure 3.9 Structure of a two-layer undercomplete autoencoder

an intrinsic data compression is realized. These kinds of AE are called *undercomplete AEs* and the basic structure of a two-layer AE is shown in Figure 3.9. In this basic example, the input signal has dimension m and the code $\mathbf{h}$ that is produced in the first layer has dimension s, which is smaller than m. The reconstruction $\mathbf{r}$ has of course again the same dimension m as the input signal.

If linear decoder and encoder functions, i.e., no nonlinear activation function $\varphi(\cdot)$ and no bias, are used together with the squared reconstruction in (3.52), the subspace that is learned by the AE is the same as for *principal component analysis* (PCA). This subspace is spanned by the weight vectors of the neurons in the encoding layer, which are summarized in the weight matrix $\mathbf{W}$. These weight vectors are not constrained to form an orthonormal basis of the subspace, but the span is the same as for PCA [20].

To have an advantage over standard signal processing techniques like PCA, the nonlinearities in the neural network are used. Experience of the last years shows that this is a powerful tool to achieve good compression and automatic feature extraction [20]. A technique that is used quite often to produce more accurate autoencoders with higher compression ratios is stacking several undercomplete AEs, which is shown in Figure 3.10.

One way to train these *stacked AEs* is to split the layers into corresponding encoder–decoder pairs and pretrain these pairs before combining them into a deep AE. For the example in Figure 3.10, these would be the pairs consisting of layers one and six, layers two and five and layers three and four. Each of these pairs shows the appropriate dimension for training. Initially, layer one is trained as the encoder on the training data and layer six is trained as the decoder applied to the data in the first layer. In the next step, layer two is trained as the encoder applied to the values of layer one and layer five is trained to decode these values again, and so on.

To get an idea of the capacity of AEs, an example using the MNIST data set of handwritten digits is shown here. The MNIST data set is a public data set and is freely available from [21]. In this example, a one-dimensional code is created by a basic

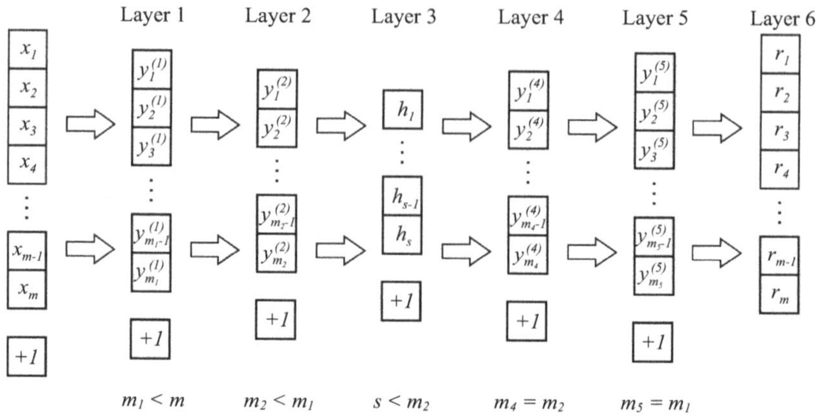

Figure 3.10 Example of a six-layer stacked undercomplete autoencoder

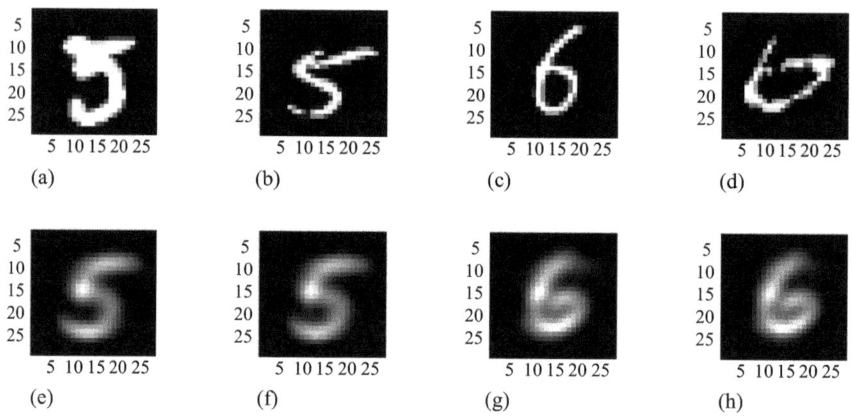

Figure 3.11 Comparison of input and reconstruction of MNIST examples with an autoencoder using a one element code: (a) input 1, (b) input 2, (c) input 3, (d) input 4, (e) output 1, (f) output 2, (g) output 3 and (h) output 4

two-layer AE with a sigmoid activation function in the code layer. As training data, the available training images of the digits 5 and 6 are used, i.e., two different classes. The training itself is unsupervised, i.e., the network is not aware of the digit in the image. As mentioned earlier, the aim of the AE is to reconstruct the input image at the output, but as an intermediate step, the image is compressed to a single value. Examples of the reconstruction of both digits are shown in Figure 3.11.

The output images of Figure 3.11 show that the reconstruction of digit 5 is the same for different input images. The same effect appears for digit 6, which is already

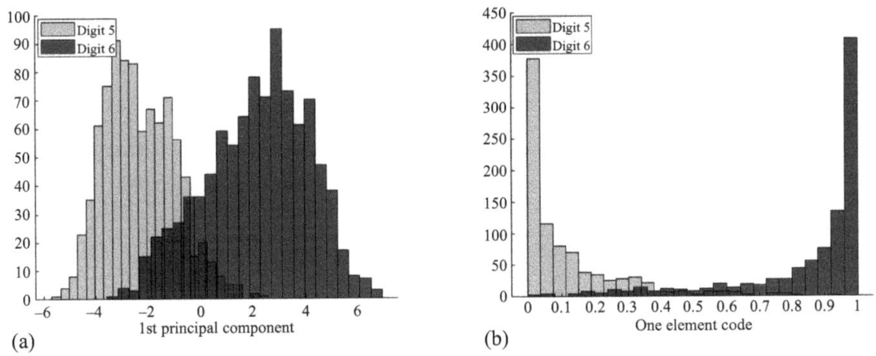

Figure 3.12 Histograms of one element code for digits 5 *and* 6 *of the MNIST test data set: (a) PCA and (b) autoencoder*

some sort of classification without knowledge of the input classes. It seems that the AE has learned average digits that minimize the reconstruction error over the complete data set. It should be mentioned that the images in Figure 3.11 are from the independent test set and have not been presented to the AE during training.

To compare the code separation with a standard signal processing approach, the PCA is calculated, and the histogram of the first principal component of the test data is shown in Figure 3.12(a). This histogram shows a clearly visible overlapping area for the two digits. The PCA was calculated using only the data of digits 5 and 6. The histogram of the one element code in Figure 3.12(b) shows a clear separation of the two digits, since most of the images of the two digits are pushed to the saturation of the activation function. Although there is an overlap for this code as well, the number of overlapping examples is reduced.

Further information on stacked AEs and data compression using these networks can be found, for example, in [13].

3.10.2 Other applications of autoencoders

In this section, some other possible applications of AEs taken from [3] are presented shortly. The training of the AEs presented here can be very different from the standard backpropagation and is not presented. For details on the training algorithms of the AEs in this section, the reader is referred to [3] and the references therein.

In the last section, the underrepresented AE was presented and its ability to extract the important features for data representation. In this section, the code of the AE has a higher dimension than the input data, thus it is an *overcomplete AE*. If these AEs are implemented with a linear activation function, i.e., $\varphi(x) = x$, they will most likely learn the identity and no useful information would be extracted and even if a nonlinear activation function is used, the code will probably not very useful. To achieve a representation that is useful for the user, the AE must be *regularized* to give a code with certain properties. To achieve this, a new general loss function is defined as the sum of the former loss function, for example, the MSE between input and reconstruction as in (3.52), and a regularization term $\Omega(\mathbf{h})$:

$$\mathscr{E}(\mathbf{x}, g(f(\mathbf{x}))) + \Omega(\mathbf{h}). \tag{3.53}$$

One of the most used regularization terms that is used in signal processing is a *sparsity constraint*

$$\Omega(\mathbf{h}) = \lambda \|\mathbf{h}\|_1 = \lambda \sum_i |h_i|, \tag{3.54}$$

which uses the common ℓ_p norm

$$\|\mathbf{x}\|_p = \left(\sum_i |x_i|^p\right)^{1/p} \tag{3.55}$$

with $p = 1$ as the convex relaxation of the ℓ_0 norm. The ℓ_0 norm is the limit of (3.55) for $p \to 0$ in the sense

$$\|\mathbf{x}\|_0 = \lim_{p\to 0} \|\mathbf{x}\|_p^p = \lim_{p\to 0} \sum_i |x_i|^p \tag{3.56}$$

and is strictly speaking not a norm, but it gives the number of nonzero elements of a vector [22]. An AE trained with the constraint in (3.54) is normally called *sparse AE* since this constraint will force the code to have only a few nonzero or significant entries for each example. The code that is produced in this way gives often useful features for classification as a byproduct.

A problem of the regularizer in (3.54) is that it cannot be implemented directly in the cost function, since the derivative with respect to the weights cannot be calculated. Therefore, another implementation must be found and the most common one is the *Kullback–Leibler* (KL)-divergence

$$\mathrm{KL}(q\|\hat{q}_j) = q \log\left(\frac{q}{\hat{q}_j}\right) + (1 - q) \log\left(\frac{1 - q}{1 - \hat{q}_j}\right) \tag{3.57}$$

between the average activation $\hat{q}_j$ of the code neuron j and the desired activation level q [4]. If, for example, an average activation of 0.05 is desired for each neuron, which means that in average 5% of the neurons are active in the code, the KL-divergence is added to the cost function for regularization and penalizes deviations from this goal. The resulting cost function is

$$\mathscr{E}(\mathbf{x}, g(f(\mathbf{x}))) + \lambda \mathrm{KL}(q\|\hat{q}_j). \tag{3.58}$$

Another example for regularized AEs is the so-called *contractive AE*. This AE is designed to create a code that does not change when the input signal shows small changes. To reach this, the gradient of the code with respect to the input signal is used as regularization term. This means that the regularization term is now also dependent on the input data $\mathbf{x}$:

$$\mathscr{E}(\mathbf{x}, g(f(\mathbf{x}))) + \Omega(\mathbf{h}, \mathbf{x}). \tag{3.59}$$

The regularization term is given in this specific case as

$$\Omega(\mathbf{h}, \mathbf{x}) = \lambda \sum_i \|\nabla_x h_i\|_2^2. \tag{3.60}$$

This means that changes in **h** with a changing **x** will produce high costs and thus the network is trying to minimize the variation within the code with changing input data.

A related AE structure is the *denoising AE*, which uses input data $\tilde{\mathbf{x}}$ that has been corrupted with an additional noise term **n**

$$\tilde{\mathbf{x}} = \mathbf{x} + \mathbf{n}. \tag{3.61}$$

This AE can be trained with the standard backpropagation using the MSE error function from (3.52) with the difference that the reconstruction is calculated using the noisy input data

$$\mathscr{E}(n) = \frac{1}{2}\|\mathbf{x}(n) - \mathbf{r}(n)\|_2^2 = \frac{1}{2}\|\mathbf{x}(n) - g(f(\tilde{\mathbf{x}}(n)))\|_2^2. \tag{3.62}$$

With this training method, an AE can, e.g., be trained to remove a specific form of noise from images. The training data is created by adding this specific noise to known images and use the original images without any noise as the reference for the reconstruction error. In this case, there is an intended difference between the input and the output of the AE.

3.11 Advanced training techniques and further applications of deep learning

Over the last years, the methods of deep learning networks have changed compared to the standard methods that were presented in this chapter so far. The purpose of this section is to fill this gap a bit and provide an overview of recent methods but not analyze the details.

A technique that was recently introduced is *dropout* [23]. This technique is based on ensemble methods, i.e., many classifiers are applied to the same data and the results are fused in some way. The basic idea of dropout is to randomly remove some neurons from the network at each training iteration. To test the network and the actual classification, all neurons of the network are used and the output layer gives the fused results of all classifiers. The only change that must be done after the training is the adaption of the weights to the dropping probability. If the neurons in a layer l are kept with probability p, then the weights of this layer are multiplied with the value p after the training. This regularization technique should give robust results and a network that shows good generalization capabilities. A general example of dropout with MLPs is shown in Figure 3.13; in each iteration, the crossed neurons are removed from the network. In this way, a new network is trained in each iteration. A drawback of this method is the increased training time, but the results achieved with this method show that it is a powerful method.

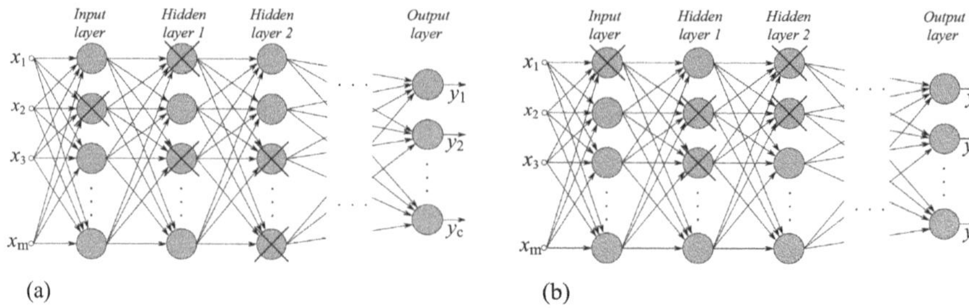

Figure 3.13 Example for the first iterations of training with dropout: (a) iteration 1 and (b) iteration 2

A problem that often arises in any machine learning system is the limited amount of training data. One possible solution is the use of *transfer learning*, which means that networks are trained on data that is different from the actual problem but shows some similarities. For example, the task at hand is to classify optical images of cows and horses to their categories, but the amount of labeled training data is very limited. The idea is now to use optical images from, for example, cats and dogs, or any other available data set with a huge amount of labeled data. After the training with the "wrong" data, the network is finalized with the small amount of labeled data from the original problem. This technique uses the different levels of abstraction in deep learning, very abstract features in upper layers close to the input can be shared for several applications, only the deeper layers, which are very problem specific, should be changed according to the problem [3].

An example for the application of deep learning is *speech recognition*. In [24], different research groups presented their experience and results using different deep learning networks for speech recognition. The algorithms are based on the use of hidden Markov models (HMM), which are built for the recorded speech. Deep learning methods are used to determine how well these HMM states fit a short frame of recorded speech to extract the spoken words.

In language processing, *recurrent neural networks* are also often used [25]. These networks posses a feedback loop, so that a time-delayed copy of the output of neurons is used as an input again. With these networks, interpretation of sentences is possible and the semantic meaning of the text can be extracted by the network [26]. These networks are also used for automatic translation of text [27].

Since the number of deep learning applications in image classification is large, only [28] is shown here exemplary. In this work, scientists from Google present a CNN with 11 layers to extract the address number from Google Street View imagery. With this data, Google is able to pinpoint an address to the physical house with very high accuracy.

These are just some applications and recent developments in deep learning. Nowadays, it is almost impossible to keep up with all the developments in deep learning, but with the material from this chapter it should be possible to understand the main points and categorize it to the fields presented here.

References

[1] Bishop CM. *Neural Networks for Pattern Recognition.* Oxford University Press, New York, NY, USA; 1995.

[2] Haykin S. *Neural Networks and Learning Machines.* 3rd ed. Upper Saddle River, NJ: Pearson Education Inc.; 2009.

[3] Goodfellow I, Bengio Y, Courville A, *et al. Deep Learning.* vol. 1. MIT Press, Cambridge, MA; 2016.

[4] Ng A. Sparse Autoencoder. CS294A Lecture Notes. 2011.

[5] Nielsen MA. *Neural Networks and Deep Learning.* Determination Press, San Francisco, CA; 2015.

[6] McCulloch WS and Pitts W. A logical calculus of the ideas immanent in nervous activity. *The Bulletin of Mathematical Biophysics.* 1943;5(4):115–133.

[7] Biran O and Cotton C. Explanation and justification in machine learning: a survey. In: *Proceedings of the Twenty-Ninth International Joint Conference on Artificial Intelligence,* Copyright © 2020 International Joint Conferences on Artificial Intelligence; 2020.

[8] Hebb DO. The organization of behavior; a neuropsychological theory. *A Wiley Book in Clinical Psychology.* J. Wiley; Chapman & Hall, London, 1949; p. 62–78.

[9] Rosenblatt F. The perceptron: a probabilistic model for information storage and organization in the brain. *Psychological Review.* 1958;65(6):386.

[10] Hornik K, Stinchcombe M, and White H. Universal approximation of an unknown mapping and its derivatives using multilayer feedforward networks. *Neural Networks.* 1990;3(5):551–560.

[11] LeCun Y. Generalization and Network Design Strategies [Technical Report]. 1989 June; (CRG-TR-89-4).

[12] Bengio Y. Learning deep architectures for AI. Foundations and Trends® in Machine Learning. 2009;2(1):1–127.

[13] Hinton GE and Salakhutdinov RR. Reducing the dimensionality of data with neural networks. *Science.* 2006;313(5786):504–507.

[14] Krizhevsky A, Sutskever I, and Hinton GE. Imagenet classification with deep convolutional neural networks. In: *Advances in Neural Information Processing Systems,* The MIT Press, Cambridge, MA; 2012. p. 1097–1105.

[15] Wagner S, Barth K, and Brüggenwirth S. A deep learning SAR ATR system using regularization and prioritized classes. In: *2017 IEEE Radar Conference (RadarConf),* IEEE, New York, NY, USA; 2017. p. 0772–0777.

[16] LeCun Y, Bottou L, Bengio Y, *et al.* Gradient-based learning applied to document recognition. *Proceedings of the IEEE.* 1998;86(11):2278–2324.

[17] Dong G, Liao G, Liu H, *et al.* A review of the autoencoder and its variants: a comparative perspective from target recognition in synthetic-aperture radar images. *IEEE Geoscience and Remote Sensing Magazine.* 2018;6(3):44–68.

[18] Kingma DP and Ba J. *Adam: A Method for Stochastic Optimization.* arXiv preprint arXiv:14126980. 2014.

[19] Wagner S and Johannes W. Target detection using autoencoders in a radar surveillance system. In: *IEEE International Radar Conference*, IEEE, New York, NY, USA; 2019.

[20] Bengio Y, Courville A, and Vincent P. Representation learning: a review and new perspectives. *IEEE Transactions on Pattern Analysis and Machine Intelligence.* 2013;35(8):1798–1828.

[21] LeCun Y, Cortes C, and Burges CJC. The MNIST Database of Handwritten Digits. 1998. Available from: http://yann.lecun.com/exdb/mnist/.

[22] Bruckstein AM, Donoho DL, and Elad M. From sparse solutions of systems of equations to sparse modeling of signals and images. *SIAM Review.* 2009;51(1):34–81.

[23] Srivastava N, Hinton G, Krizhevsky A, *et al.* Dropout: a simple way to prevent neural networks from overfitting. *The Journal of Machine Learning Research.* 2014;15(1):1929–1958.

[24] Hinton G, Deng L, Yu D, *et al.* Deep neural networks for acoustic modeling in speech recognition: the shared views of four research groups. *IEEE Signal Processing Magazine.* 2012;29(6):82–97.

[25] LeCun Y, Bengio Y, and Hinton G. Deep learning. *Nature.* 2015;521(7553): 436.

[26] Cho K, van Merrienboer B, Gülçehre Ç, *et al.* Learning phrase representations using RNN encoder-decoder for statistical machine translation. In: *Conference on Empirical Methods in Natural Language Processing (EMNLP)*, ACM, NY, USA; 2014.

[27] Bahdanau D, Cho K, and Bengio Y. Neural machine translation by jointly learning to align and translate. In: *Proceedings of the International Conference on Learning Representations 2015*, San Diego, USA; 2015.

[28] Goodfellow IJ, Bulatov Y, Ibarz J, *et al. Multi-Digit Number Recognition From Street View Imagery Using Deep Convolutional Neural Networks.* arXiv preprint arXiv:13126082. 2013.

Part II

Special topics

Chapter 4

Radar data representation for classification of activities of daily living

Baris Erol[1] *and Moeness G. Amin*[2]

4.1 Introduction

In this chapter, we address the problem of data representation and its impact on human motion classification using radar. In examining the motion classifier performance, it has become apparent that some human motion articulations are more distinguishable in some data domain representations than others [1,2]. The potential effect of data representation on motion classification performance calls for devising proper strategies and new approaches of how to best manipulate or preprocess the data for the sake of achieving most desirable results. We discuss domain integration and suitability using a single range-Doppler radar sensor [3,4]. No assumptions are made regarding the received radar backscattering data adhering to any specific models or signal structure [5–7].

Sensing technologies and data analytics are considered powerful tools in efficient indoor monitoring of human activities. Monitoring of activities of daily living (ADL) [8,9] can identify large variants in activity means and patterns over time, which may indicate alerting changes in routines and lifestyle as well as the state of physical, cognitive, and psychological health of the person. Human daily activities can be categorized into translation and in-place motions [10,11]. Whereas the former mainly describes crawling and gait articulations, the latter is primarily associated with motions that do not exhibit considerable changes in position. In-place motions include sitting, standing, kneeling, and bending; each is performed without any stride or noticeable range swath. Translation and in-place motions by themselves can be used as barometers to quantify the person's daily energy exertion and point to active and passive periods in a day.

Apart from translation vs. in-place motion characterization, many of ADL are considered normal, such as walking sitting and standing, while others are infrequent and far in between and, in this sense, may be viewed as abnormal, such as falling and crawling. However, normality and abnormality are tied in with age-groups. For example, falling is not common among teenagers, it is a likely occurrence in elderly

[1]Siemens Corporate Technology, Princeton, NJ, USA
[2]Center for Advanced Communications, Villanova University, Villanova, PA, USA

population. In fact, falls are considered as the leading cause of fatal and nonfatal injuries for people aged 65 and over [1,12] –[17]. To this end, monitoring human indoor activities is more urgent for the vulnerable population, including the elderly and disabled. In these situations, the principal task is to detect and classify motion abnormalities as well as anomalies in activity patterns that may be linked to deteriorating health conditions.

ADL and fall detection systems have been recently developed and are continuously being improved for enhancing the quality of life for all ages, and more specifically for the elderly. Radio-frequency (RF)-based sensors work in all lighting and weather conditions. They are safe, nonintrusive, preserve privacy, and provide reliable home automation and security. In addition to ADL, RF-based gesture recognition using hands and arms is also an important contactless technology for human–machine interface [18]-[21]. Adding to the indoor applications, RF-based vital sign monitoring has vast medical use, as respiration and heartbeats are essential diagnostic barometers for many health problems [22–25]. More recently, RF sensors have also been proposed for gait analysis and recognition for rehabilitation and timely diagnosis of many neurological, orthopedic, and medical conditions [26–28] Changes in gait patterns can be precursors of falls.

Various contributions to radar recognition of ADL have used both predefined and automatically learned types of features. Typically, two-dimensional (2D) joint-variable domains, i.e., range-map and time–frequency (TF) domains are used to reveal detailed and salient discriminatory properties of each motion articulation. This requires the use of range-Doppler radar and processing the signal backscattering to provide moving target information along the three variables of fast-time, slow-time, and Doppler frequency. The ADL feature extraction techniques can be speech inspired [29,30], derived from principal component analysis (PCA) [30], based on specific kinematics, or drawn from transform domains [31]. Examples of transform-based features include features derived from the cadence velocity diagram, such as pseudo-Zernike moments [32]. Recent human motion classification efforts have applied deep neural networks [33–35].

It is well established that, unlike the one-dimensional (1D) representation, such as time and frequency, the 2D data representations can expose intricate properties of the target complex motion, specifically the time-dependency of target velocity, i.e., the signal local frequency behavior. Each 2D motion data representation provides distinct and valuable information that might not be present or difficult to glean in another 2D domain. In essence, some motions are more distinguishable in range map while others differ distinctively in the TF domain[1]. The latter is offered by the spectrograms [2,36] and carries the micro-Doppler signature of the target. The dependency of classification performance on the data domain puts in question the use of the same 1D or 2D domain for feature extraction without vetting its suitability across all motion classes.

Rather than debating or attempting to associate 2D data domains to the different motions, a three-dimensional (3D) joint-variable signal representation can be adopted irrespective of the type of motions considered. The use of 3D representation in lieu of 2D representation has the same motivation as that of preferring 2D over

1D representations. The goal is to exploit the hidden dependency and correlations among increased radar signal variables that, in 3D, are slow-time, fast-time, and frequency. We use radar data cube (RDC) constituted by these three variables and pursue multilinear dimensionality reduction methods using tensor analysis [37].

In this chapter, we employ PCA for multifold illustrations of the influence of data representation on classification performance. PCA is used as the feature extraction method followed by the nearest neighbor classifier to compare the suitability of 2D domain representations for different classes of motions. PCA is an unsupervised dimensionality reduction method proposed to solve problems related to high dimensionality. It reduces the dimensionality of an input dataset consisting of correlated variables into an output set of linearly uncorrelated variables, referred to as principal components. The output set, with reduced dimensionality, attempts to capture the variation present in the input data. PCA can be seen as a smart compression technique; however, since the principal components are uncorrelated, it can also be categorized as an automatic feature learning method, which is proven effective, offering high fall classification rates. Conventional PCA-based human motion classification methods deal with one joint-variable representation of the radar data, whether it is TF, range map, or their fusion. In each case, the 2D data representation is vectorized and the result is used to determine the principal components. Generalized PCA, on the other hand, does not require a vectorization step in the beginning and seeks to find two projection subspaces that operate on different modes of the 2D joint-variable representation. Variations of this approach have been used in radar applications, more specifically for activity recognition and gait abnormality detection.

Moving to more generalized treatment, we apply multilinear PCA (MPCA) to the RDC, rather than lower dimension processing. A comparison between the baseline performances of predefined physical features, 1D-PCA, 2D-PCA, a 12-layer convolutional neural network (CNN) (all operate on TF domain), and MPCA is provided to show the merits of using RDC. We use a database that contains multiple aspect angles, a large variance within class samples due to performing each motion with different speeds, and collecting data from different locations and subjects.

The chapter is organized as follows. In Section 4.2, the radar signal model, as well as discussions on domain suitability for feature extraction, is presented. In Section 4.3, multilinear subspace learning methods are introduced. In Section 4.4, the design and optimization considerations of multilinear subspace learning methods are discussed in terms of iterative projections, initialization, termination criteria, convergence, and number of projections. In Section 4.5, we introduce a boosting scheme of MPCA by combining it with two supervised methods followed by decision fusion. Section 4.6 presents a comparison of specific subspace learning methods under different RDC dimensions. Finally, conclusions are provided in Section 4.7.

4.2 Radar signal model and domain suitability

Frequency-modulated continuous wave (FMCW) radars are in compact form and can provide range, time, and Doppler information. The radar system used in this work is SDRKIT 2500B, which is developed by Ancortek, Inc. Operating parameters of this radar system are transmitting frequency 25 GHz, sampling frequency 1 kHz, and bandwidth of 2 GHz which provides 0.075 m range resolution. The radar can switch between continuous wave and FMCW modes and operates in a monostatic configuration that consists of one transmitting and one receiving antennas.

For an FMCW radar, the backscattering signal from a target located at a distance R can be expressed as

$$s_{\mathrm{rx}}(t) = A_{\mathrm{rx}} \cos\left(2\pi j(f_0(t-\tau) + \frac{r}{2}(t-\tau)^2) + \phi_{\mathrm{rx}}\right), \tag{4.1}$$

where τ is the round trip time delay, f_0 is the carrier frequency, r is the chirp rate, ϕ_{rx} is the phase of the received signal, and A_{rx} is the amplitude of the received signal computed from the radar range equation as

$$A_{\mathrm{rx}} = \frac{G\lambda\sqrt{P\sigma}}{(4\pi)^{1.5}R^2\sqrt{L_s}\sqrt{L_a}}, \tag{4.2}$$

where G is the antenna gain, P is the transmitter power, σ is the target radar cross section (RCS), and L_s and L_a represent the system and atmospheric losses, respectively.

The received signal is then demodulated by the I/Q demodulator, providing the in-phase and quadrature-phase components of the complex baseband signal, expressed as

$$s(t) = I(t) + jQ(t) = A \exp \psi(t), \tag{4.3}$$

where ψ is the phase of the signal.

The received I/Q signal from FMCW radar can be transformed into a 2D matrix representation where the columns and rows, respectively, represent fast-time and slow-time variables. The range map is computed by column-wise discrete Fourier transform. Some examples of the range maps for different human motion activities are depicted in Figure 4.1(a)–(e). In Figure 4.1(a), the human subject falls toward the radar and, as a result, we observe a high range spread that is determined by the subject's height. Sitting, bending, and kneeling (Figure 4.1(b), (d), and (e)) have small range spreads due to the limited translational motion of the body in range. In the case of a walking, subject starts walking toward the radar around 3.5 m away and stops around 1 m, as it can be clearly seen in Figure 4.1(c).

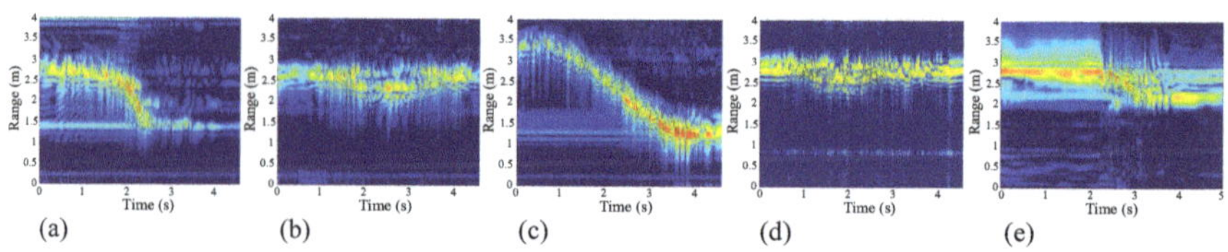

Figure 4.1 Range maps of different motions: (a) falling, (b) bending, (c) walking, (d) kneeling, (e) sitting

The Doppler information of the human motion activities are often discriminative and descriptive of the nature of the signal, its source, and the mechanism of its generation. Raw slow-time data, x(n) where n is the slow-time, can be computed by agglomeration of the range map along its range axis (columns). The resulting data from the human (individual being monitored) is nonstationary with time-varying frequencies, which are associated with velocity, acceleration, and higher order terms of different human body parts. Spectrograms are the simplest and most commonly used TF distribution. It is the energetic form of the short-time Fourier transform that is obtained by splitting the time domain signal into many overlapping or disjoint consecutive segments, and then taking the fast Fourier transform of each segment. This provides the signal's local frequency behavior. The spectrogram is mathematically defined as

$$\mathrm{S}(n,k) = \left| \sum_{m=0}^{N-1} \mathrm{h}(m)\mathrm{x}(n-m)e^{-j2\pi km/N} \right|^2, \tag{4.4}$$

where h(m) is a window function that can affect both the time and frequency resolutions. The TF representation of five different motions is provided in Figure 4.2(a)–(e). Note that in the walking motion (Figure 4.2(c)), the torso exhibits a sinusoidal pattern with the strongest return. The periodic motion of the limbs causes frequency oscillations modulated around the main Doppler shift. The motion of the legs causes the highest frequency oscillation, followed by that of the arms, which appear to be more distinct. Sitting spectrogram, provided in Figure 4.2(e), appears to be more dense along the frequency axis due to the motion of the upper body as a whole. Interestingly, in bending, positive and negative frequency components are present within the same time interval (Figure 4.2(b)). As the upper body translates closer to the radar, the lower body (human back) moves away from the radar. Kneeling motion has the same kinematics as bending with one significant difference; motion of the knee. Kneeling is performed as in tying the shoelace. Therefore, subjects put their one knee on the ground while tying the shoelace. This effect can be clearly seen in Figure 4.1(d). Fall spectrogram, shown in Figure 4.2(a), depicts sudden drop in frequency in the shape of a positive "hump." In falling, the human subject puts the knee first on the ground to slow down the fall and this effect can be clearly seen around 1.8 s in the TF domain.

Different motions manifest themselves differently depending on the domain used. Since there is no one single domain that provides distinctive signatures for all motion types, an effective human motion recognition approach should undertake multiple

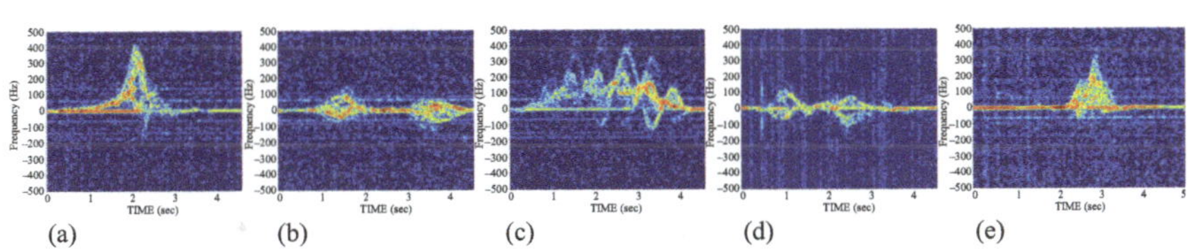

Figure 4.2 Time–frequency representations of different motions: (a) falling, (b) bending, (c) walking, (d) kneeling, (e) sitting

data domain representations. Deciding on a proper domain for a given motion would depend on the resemblance/distinction between the respective motion signatures and other motion signatures in the same domain. The spectrograms for different motions assume different behaviors and characteristics over time and frequency. The similarity of motion spectrograms can be measured using different metrics. One possible metric is the Pearson correlation coefficient defined by

$$\rho = \frac{\sum_{i=1}^{N} (x_i - \overline{x})(y_i - \overline{y})}{\sqrt{\sum_{i=1}^{N} (x_i - \overline{x})^2} \sqrt{\sum_{i=1}^{N} (y_i - \overline{y})^2}} \tag{4.5}$$

where x_i and y_i are the ith pixels of the two images X and Y, and the bar signifies average value. Another possible similarity metric is the canonical correlation between two motion spectrogram subspaces. The motion subspace is established by vectorizing all spectrograms for the same motion, putting them in a matrix, and then applying singular value decomposition (SVD) to capture the principal eigenvalues. If $\boldsymbol{V}$ and $\boldsymbol{U}$ correspond the dominant eigenmatrices of two motion classes, then the canonical correlation is given by the maximum value of the SVD decomposition

$$\boldsymbol{U}^H \boldsymbol{V} = \boldsymbol{P} \Lambda \boldsymbol{Q} \tag{4.6}$$

The canonical correlation is the cosine of the canonical angle. So, the maximum canonical correlation corresponds to minimum angle. The two abovementioned similarity measures can also be applied to the range map. To demonstrate different levels of motion similarities, we consider five motions: walking, sitting, standing, falling, and bending. Figure 4.3 shows both the canonical and Pearson correlation

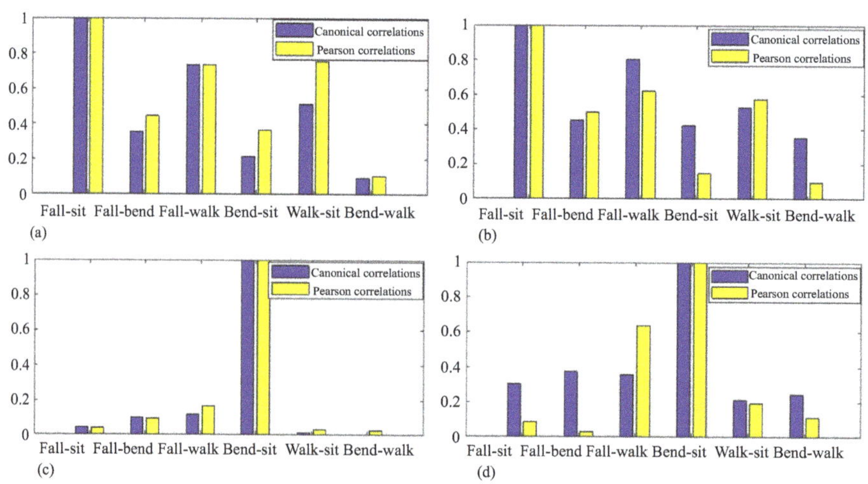

Figure 4.3 Canonical and Pearson correlation coefficients for a pair of subspaces when (a) and (c) one and (b) and (d) five principal components are used: (a) and (b) spectrograms; (c) and (d) range maps. Correlation coefficient values are normalized in the range [0, 1]

coefficients in different domains when using the strongest principal component of each motion class. It is evident that high correlation coefficients properly describe similar motions, i.e., falling and sitting in the TF domain, and bending and sitting in the range map. This suggests grouping the different motions into subsets and applying a decision-tree based-classification approach utilizing domain sensitivities, which was discussed in [2].

Compared to 1D single-variable domain, the 2D joint-variable representations are more effective in human motion classification. Each 2D motion data representation provides distinct and valuable information that might not be present in other 2D domains. Acquiring all the information that is needed for better classification often requires a parallel processing of 2D domains, which might result in increased computational complexity and loss of information present in the interrelation between the domains. Therefore, we utilize a 3D joint-variable signal representation to exploit the underlying dependency and correlations among the three radar signal variables: range, slow-time, and Doppler frequency. RD representation includes the effects of both target velocity and range [38]–[40]. Typically, a single RD frame can be obtained by applying the Fourier transform for each range bin over a nonoverlapping coherent processing interval (CPI) of slow-time on the range map. In this chapter, the CPI is determined as 32 pulses. We create a tensor structure, called RDC, by stacking consecutive RD frames. Visualization of the RDC is usually done by a video sequence of RD frames. However, in [41], RDC is visualized by creating a surface that has the same intensity value within the slices of data cube. This can be accomplished by isosurface method that is a 3D extension of an analog isoline. Volumetric representations of the RDC for falling, bending, walking, kneeling, and sitting are presented in Figure 4.4(a)–(e), respectively. Note that RDC gathers fast-time, slow-time, and Doppler frequency information in a single domain, aiming at increased detection and reduced false alarms.

Our RDC extraction produce starts with a preprocessing approach called extended CLEAN (eCLEAN) [42,43]. This methods aims at suppressing unwanted distortions or noise effects while enhancing the natural structural integrity of the data. The eCLEAN algorithm directly operates on the individual RD frames. The original CLEAN algorithm strives to find the highest peaks in an image, which correspond to a real target location. At each step of the algorithm, the maximum peak is extracted, and a portion of the point spread function centered at that peak is subtracted until some

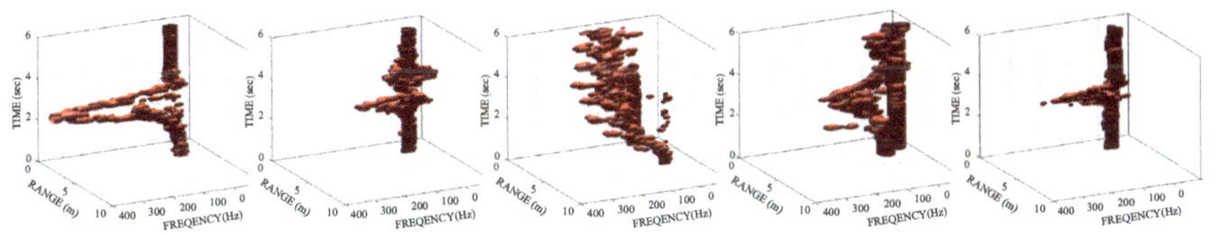

(a) (b) (c) (d) (e)

Figure 4.4 Multidimensional radar data cube (RDC) representation of different motions: (a) falling, (b) bending, (c) walking, (d) kneeling, (e) sitting

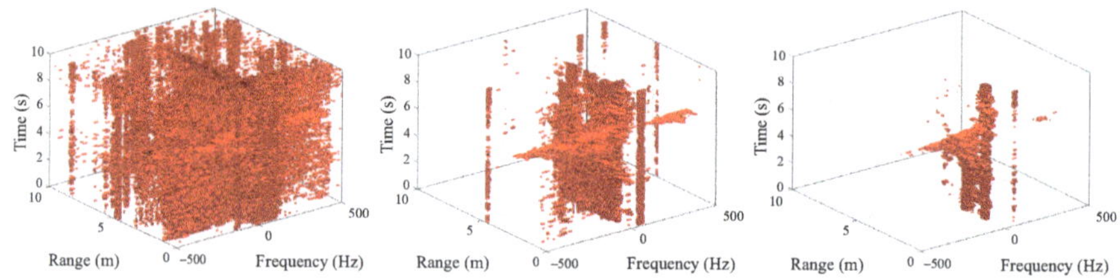

Figure 4.5 Visualization of a noisy falling RDC data (a) without any filtering, (b) after basic thresholding, and (c) after proposed eCLEAN

threshold is met [41]. The challenge in the original CLEAN algorithm is to determine the number of peaks in the RD frame. In our scheme, the number of points that are needed to be removed are automatically determined prior to extraction of peaks using a simple and efficient histogram-based method. The output of the algorithm is depicted in Figure 4.5(c) when a noisy falling RDC data is given in Figure 4.5(a).

4.3 Multilinear subspace learning

In dimensionality reduction, the simplest approach contains a preprocessing step where the different domain data (spectrograms or range maps) are reshaped into vectors. This operation is called vectorization or flattening, which entails a high processing cost in terms of increased computational time and memory access. For example, a typical RDC of size ($64\times64\times64$) would translate into a vector with a size of ($262{,}144\times1\times1$) upon flattening. Moreover, it is well known that the reshaping operation breaks the natural structure and correlation in the original data [44].

In order to overcome the limitations and problems with flattening, some approaches have been recently introduced where the images are represented as matrices (second-order tensors) [45]. The image covariance matrix can then be constructed. Spatial correlation of the image pixels within a localized neighborhood is used to create a less restrictive PCA. However, these approaches cannot be applied to higher other tensors (three or higher). Since our RDC structure has three dimensions (slow-time, range, and frequency), we employ the MPCA proposed in [46]. In addition, we apply multilinear discriminant analysis (MLDA) and compare its performance with that of MPCA [47].

4.3.1 Multilinear algebra basics and notations

This section briefly introduces basic multilinear notations that are commonly used in multilinear subspace learning algorithms. Vectors are denoted by lowercase symbols, such as p; the normal uppercase symbols represent matrices, e.g., $\boldsymbol{A}$. An Nth-order tensor is denoted as $\mathcal{X} \in \mathbb{R}^{I_1 \times I_2 \times \cdots \times I_N}$. The implemented algorithms often require reshaping of the data which, in the case of tensors, is called unfolding, or matricization.

Unfolding $\mathcal{X}$ along the n-mode is denoted as $\mathcal{X}^{(n)} \in \mathbb{R}^{I_n \times (I_1 \times \cdots \times I_{n-1} \times I_{n+1} \times \cdots \times I_N)}$. Finally, the n-mode product of a tensor $\mathcal{X}$ by a matrix $\boldsymbol{A}$ is defined by $\mathcal{Y} = \mathcal{X} \times_n \boldsymbol{A}$.

4.3.2 Multilinear principal component analysis (MPCA)

The MPCA is an unsupervised subspace learning algorithm for higher order tensors targeting the same objective function as in 1D-PCA: variation maximization while reducing the dimensions along multiple modes. Assume a set of training tensor samples ($m = 1, 2, \ldots, M$) given as $\mathcal{X}_m \in \mathbb{R}^{I_1 \times I_2 \times \cdots \times I_N}$. The objective is to find a tensor-to-tensor projection (TTP) matrix subspace $\tilde{\boldsymbol{U}}^{(n)} \in \mathbb{R}^{I_N \times P_n}$ that projects the original tensors into low-dimensional tensors $\mathcal{Y}_m \in \mathbb{R}^{P_1 \times P_2 \times \cdots \times P_N}$ (with $P_n \leq I_n$). To project an RDC sample, we utilize three projection matrices, $\boldsymbol{U}^{(1)} \in \mathbb{R}^{I_1 \times P_1}$, $\boldsymbol{U}^{(2)} \in \mathbb{R}^{I_2 \times P_2}$, and $\boldsymbol{U}^{(3)} \in \mathbb{R}^{I_3 \times P_3}$, as

$$\mathcal{Y}_m = \mathcal{X}_m \times_{(1)} \boldsymbol{U}^{(1)^T} \times_{(2)} \boldsymbol{U}^{(2)^T} \times_{(3)} \boldsymbol{U}^{(3)^T}. \tag{4.7}$$

Note that $\mathcal{Y}_m$ captures most of the variation observed in the original RDC sample. RDC-based dimensionality reduction procedure is illustrated in Figure 4.6. It depicts the TTP of an RDC (shown in individual range-Doppler slices) with a size of (64×64×64) to a smaller size of tensor with a size of ($P_1 \times P_2 \times P_3$). Mathematically, the RDC-based objective function can be expressed as the determination of the three projection matrices that maximize the total tensor scatter:

$$\{\tilde{\boldsymbol{U}}^{(n)}\} = \arg\max_{\boldsymbol{U}^{(n)}} \sum_{m=1}^{M} \|\mathcal{Y}_m - \tilde{\mathcal{Y}}\|_F^2, \tag{4.8}$$

where the mean of the projected tensor features is defined as $\tilde{\mathcal{Y}} = (1/M)\sum_{m=1}^{M} \mathcal{Y}_m$. After few iterations, the objective function converges to an eigenvalue decomposition problem and can be solved by using Lagrangian optimization. The maxima of the objective function are obtained if $\{\tilde{\boldsymbol{U}}^{(n)}\}$ consists of the P_n eigenvectors of the

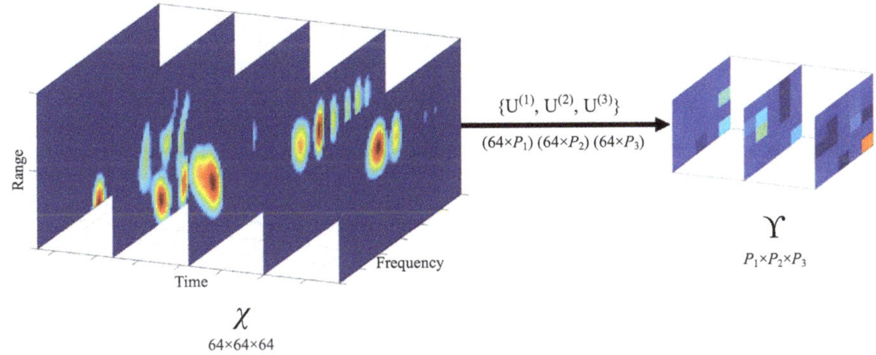

Figure 4.6 Illustration of the RDC-based dimensionality reduction procedure through tensor-to-tensor projections

covariance matrices (along each mode) corresponding to the largest P_n eigenvalues [44]. Since the computation of the approximation is multilinear, iterative optimization of (4.8) can be solved through a sequence of linear subproblems using alternating least squares (ALS), whereby the least squares of the cost function is optimized for one mode at a time, while keeping the other mode subspace matrices fixed. Finally, the feature tensor is obtained by projecting the original tensor using optimized subspace, $\tilde{\mathrm{U}}^{(n)}$, as

$$\tilde{\mathcal{Y}}_m = \mathcal{X}_m \times_{(1)} \tilde{\mathrm{U}}^{(1)^T} \times_{(2)} \tilde{\mathrm{U}}^{(2)^T} \times_{(3)} \tilde{\mathrm{U}}^{(3)^T}. \tag{4.9}$$

Then, feature vector of the mth training tensor can be constructed as $f_m = \mathrm{vec}(\mathcal{Y}_m)$, $\in \mathbb{R}^{1\times(P_1\times P_2\times P_3)}$, and vec(·) is the matrix vectorization operator. This framework is followed for each of the samples in the database and stored in a matrix representation. Note that the dimensionality of the projected tensor subspace, P_n, is assumed to be known or predetermined. The effect of the P_n is shown in Section 4.4 in terms of classification accuracy and computational complexity.

4.3.3 Multilinear discriminant analysis

MLDA is a supervised algorithm aiming at maximizing the between-class scatter while minimizing the within-class scatter in the projected subspace. In this work, we use the discriminant analysis with tensor representation (DATER) algorithm to reduce the dimensionality of the RDC in a supervised fashion [47]. Similar to MPCA, DATER aims to solve the optimization problem through TTPs $\{\mathrm{U}^{(n)} \in \mathbb{R}^{I_N\times P_n}\}$ by projecting the original RDC to a low-dimensional tensor while maximizing a discriminative objective criterion. Since MLDA is a supervised method, class label for the mth RDC sample is defined as c_m, and there are C classes in total. In this specific application, we have 5 classes and total of 827 training samples. The objective function of DATER can be expressed as

$$\{\tilde{\mathrm{U}}^n\} = \arg\max_{\{\mathrm{U}^n\}} \frac{\sum_{c=1}^{C} M_c\|\tilde{\mathcal{Y}}_c - \tilde{\mathcal{Y}}\|_F^2}{\sum_{m=1}^{M} \|\mathcal{Y}_m - \tilde{\mathcal{Y}}_{c_m}\|_F^2}, \tag{4.10}$$

where $\tilde{\mathcal{Y}}$ and $\tilde{\mathcal{Y}}_c$ are defined as the overall and the class means, respectively.

The abovementioned optimization problem can be treated as an eigenvalue problem similar to MPCA. The objective function is maximized only if $\{\tilde{\mathrm{U}}^n\}$ consists of the P_n eigenvectors of the between-class and within-class scatter matrices associated with the P_n largest eigenvalues. Feature extraction can be performed by following the same steps as in (4.7).

4.4 Optimization considerations for multidimensional methods

4.4.1 Iterative projections

TTP-based projections usually have N sets of parameters for N projection matrices, one in each mode. These N-mode optimization problems cannot be solved simultaneously, except for the case with vectorization. Therefore, the solution of one mode generally depends on the remaining mode projection matrices. To overcome this problem, an iterative algorithm, inspired by ALS, is employed for multilinear subspace learning, referred to as alternating partial projections (APP) [46]. This algorithm does not cover the whole optimization space; hence, it can be categorized as a suboptimal solution. The APP basically splits the underlying optimization problem into N conditional problems that can be solved sequentially, one by one. In each mode n, it solves the defined conditional subproblem through n-mode partial multilinear projections utilizing the projection matrices obtained from the other modes.

4.4.2 Initialization

Due to the iterative nature of the APP solutions, the projection matrices have to be initialized at the beginning of the algorithm. Then, with each iteration, initialization subspaces are updated. Random initialization is one of the most popular approaches where each element of the n-mode projection matrices is constructed from a zero-mean uniform or Gaussian distribution. Another method for initialization is called full-projection truncation [44], which requires the utilization of some of the training data.

Since, we have a small database (827 samples), compared with those in image recognition (approximately 1.5 million samples), we utilize a diversified simulation database proposed for transfer learning of spectrograms. This database is generated from 55 MOCAP measurements of 5 different test subjects using a Kinect RGB-D camera. The Kinect sensor is used as a markerless system for capturing the time-varying coordinate information of human joints needed for radar simulations. The radar return from the human body can be computed as the sum of the reflected signals. For a human target, the RCS is represented by a sphere for the head, and ellipsoid for the torso and limbs. According to these changes, simulation can be computed by following the steps in the radar signal mode by only adding the individual K received signals.

In this work, four different activities were recorded: walking, running, falling, and sitting. Simulation parameters are matched with the real radar system as center frequency of 25 GHz, sampling frequency of 1 kHz, and bandwidth of 2 GHz. Three transformations are applied to the underlying skeletal model tracked by the Kinect sensor:

1. scaling in size, to capture the effect of different heights;
2. scaling in time, to capture different motion speeds; and
3. random variation of the parameters in a Fourier series model of the trajectory of different points on the body to emulate the effect of different walking styles.

A more detailed description of the diversification methodology and its validation can be found in [48]. Using this approach, original 55 MOCAP measurements were used to generate simulations of 32,000 RDCs. An example for simulated falling RDC is illustrated in Figure 4.7. Consistent with the arguments made earlier, the highest

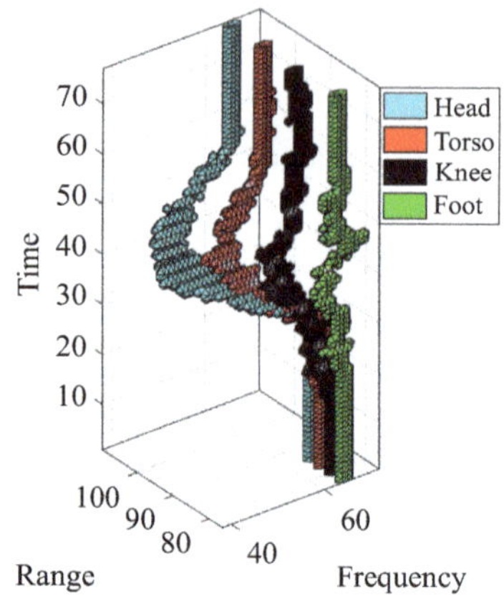

Figure 4.7 Simulated falling RDC from motion capture data for individual joint trajectories

Doppler return for falling is obtained from the upper body (head), followed by torso, and lower body. The representation is also similar to a real radar RDC shown in Figure 4.4(a).

In image databases, initialization plays an important role in both computational cost, speed of convergence, and performance, albeit this role is both task and domain dependent. In the underlying problem, the diversified initialization achieves a much faster convergence compared to random initialization, as depicted in Figure 4.8, while gaining an increase in the overall classification performance. The convergence termination criteria are met within three iterations with the proposed initialization method, whereas for random initialization, it nearly takes 20 iterations to achieve the same value. Clearly, these results prove the merit of using the diversified database to initialize the subspace learning methods.

In this work, the optimized subspace matrices obtained from diversified database are used in two different ways [49]: (1) **bottleneck analysis:** diversified projection matrices are directly utilized to extract the features from the real radar data without any fine-tuning (no training performed on real radar data) and (2) **fine-tuning:** diversified projection matrices are fine-tuned with the real radar data.

4.4.3 Termination criteria and convergence

In subspace learning, there are two commonly used termination criteria. In this work, we examine the tensor distance between the original samples and the reconstructed approximations. The approximation of the mth RDC sample can be expressed as

$$\tilde{\mathcal{X}}_m = \tilde{\mathcal{Y}}_m \times_{(1)} \tilde{\mathrm{U}}^{(1)} \times_{(2)} \tilde{\mathrm{U}}^{(2)} \times_{(3)} \tilde{\mathrm{U}}^{(3)}. \tag{4.11}$$

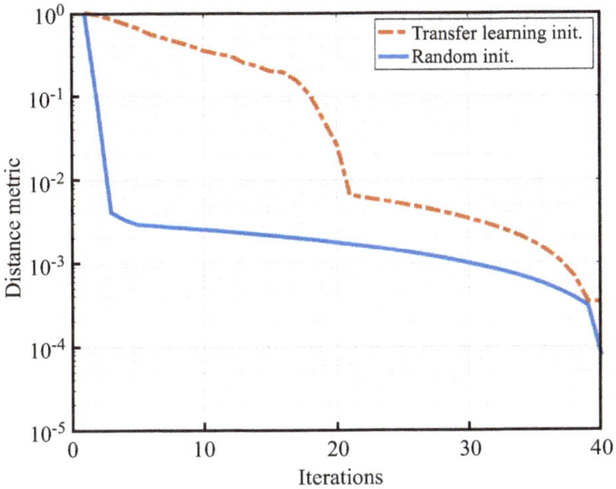

Figure 4.8 Convergence of diversified and random initialization for MPCA

Then, the termination distance criteria can be defined as

$$d = \frac{1}{M}\sum_{m=1}^{M} \|\mathcal{X}_m - \tilde{\mathcal{X}}_m\|_F^2. \tag{4.12}$$

This distance is computed in each iteration and when it drops below some user-defined hyper-parameter threshold, the algorithm breaks and outputs the optimized projection matrices. However, for computational considerations and in order to avoid any infinite loops, we also set a maximum number of iterations in case the algorithm would not convergence. Note that this iteration is only for controlling the convergence of the algorithm and different than the APP iterations.

Agglomeration along different modes in RDC will result in individual domains, such as TF (note that TF domain is obtained without performing any TF techniques), range map, and time-integrated RD map as depicted in Figure 4.9(a). To evaluate and validate our termination criteria, we also analyze the reconstructed RDC (with only five components) and perform the agglomeration on the reconstructed RDC, as shown in Figure 4.9(b). It is clear from the figure, the main discriminative properties of these domains are still present after the reconstruction only with five principal components in each mode.

4.4.4 Number of projections

Prior to solving the multilinear subspace projections, the number of projections (P_1, P_2, P_3) needs to be determined for different modes. For the TTP-based method, the search space could become very large, depending on the size of the tensor $(\prod_{n=1}^{3} I_n)$.

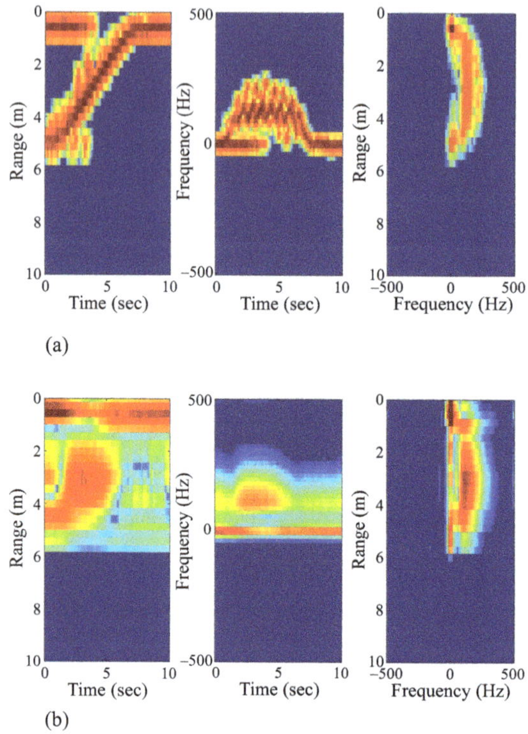

Figure 4.9 Visualization of agglomeration operator along different modes for a walking sample: (a) original and (b) reconstructed

For the RDC activity recognition task, there are 262,144 possible subspace dimensions. There are several ways to determine the number of projections used in the projection matrices.

1. *Q metric:* This metric is a suboptimal, simplified dimension determination procedure that requires no iteration or extensive testing [44]. We start with defining the ratio as

$$Q^{(n)} = \frac{\sum_{i_n=1}^{P_n} \lambda_{i_n}^{(n)}}{\sum_{i_n=1}^{I_n} \lambda_{i_n}^{(n)}}, \tag{4.13}$$

   where $\lambda_{i_n}^{(n)}$ is the i_nth full-projection n-mode eigenvalue. In the Q-based method, the first P_n eigenvectors are kept in n-mode so that it ensures the user-defined threshold parameter. Q metric is an extension of the traditional dimension selection in PCA. In essence, this method discards the least significant eigenvectors in each mode and does not require any classification.

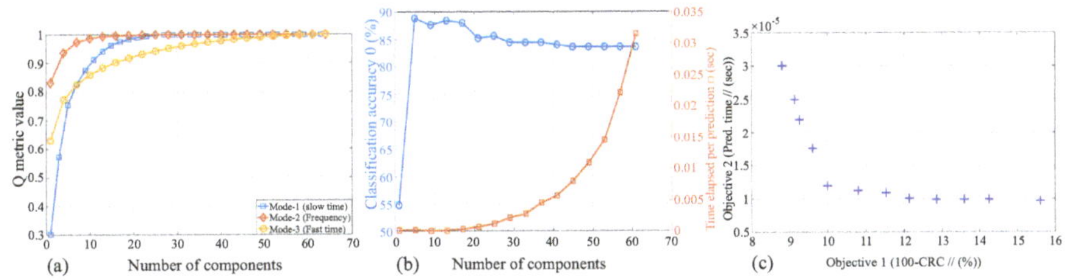

Figure 4.10 Determination of the number of projections: (a) Q metric along different modes, (b) brute-force classification with time elapsed per prediction, and (c) GA-based multiobjective parameter selection—Pareto plot –

The $Q^{(1)}$, $Q^{(2)}$, and $Q^{(3)}$ ratios are provided in Figure 4.10(a) for different number of components. For mode-2, 6 components are enough to capture 90% of the variation, whereas for mode-1 and mode-3, it nearly takes 15 and 25 components to exceed 90%, respectively.

2. *Brute-force selection:* In brute-force selection, we directly compute the classification performance for different number of projections. However, this method does not allow coverage of the entire feature space because the same number of projections is used for each mode. The classification accuracies of MPCA are provided in Figure 4.10(b) for different numbers of projections. Moreover, the run-time complexity (time elapsed per prediction) increases in this process owing to the higher dimensionality search. This specific phenomenon is indicated in juxtapositions in the same plot (Figure 4.10(b)) where the prediction time follows nearly a linear trajectory for increased number of projections.
3. *Genetic algorithms (GA)-based multiobjective parameter selection: Q* metric does not guarantee any classification improvement, and brute-force selection does not span the whole projection space. Therefore, we take a more heuristic and supervised approach where we attempt to find the optimized number of projections in each mode, while maximizing the classification performance together with minimizing the prediction time per sample. This forms a multi-objective optimization (MOO) problem and can be formulated by considering a decision-maker that seeks to optimize K objective functions. Without loss of generality and overfitting, all objective functions need to be the minimization type. Given an n-dimensional (in our case $n = 3$, (P_1, P_2, P_3)), decision variable vector $\mathrm{x} = x_1, \ldots, x_n$ in the solution space. The main idea is to find a vector $\tilde{\mathrm{x}}$ that minimizes the given set of objective functions. The general solution space is generally restricted by a series of constraint, such as the bounds on the decision variables (size of the RDC). To describe the terminology of MOO, several explanations have to be made at this point. A feasible solution is said to be Pareto optimal if it is not dominated by any other solution. A Pareto optimal solution cannot be improved with respect to any objective without worsening at least one

other objective. The set of all feasible non-dominated solutions in solution space is referred as the Pareto optimal set, and for a given Pareto optimal set, the corresponding objective function values in the objective space are called Pareto front. A more detailed explanation about MOO can be found in [50].

GA can be categorized as a population-based optimization approach, and because of their properties, they are well suited for MOO problems. GA is inspired by the evolutionist theory where in nature, weak and unfit species within their environment are faced with extinction by natural selection. The strong ones have greater opportunity to pass their genes to future generations via reproductions [51]. Fitness functions of the GA are determined as the classification accuracy and the computation time attained by the minimum distance classifier (MDC). We employed the NSGA-II that is one of the most popular MOO algorithms with three special characteristics: fast non-dominated sorting approach, fast crowded distance estimation procedure, and simple crowded comparison operator. Resulting Pareto plot can be seen in Figure 4.10(c), where the two objective functions are plotted: classification error and prediction time per sample.

4.5 Boosting the MPCA

Boosting is an ensemble meta-algorithm for reducing bias and variance in supervised learning. It is stemmed by the question of whether a set of weak learners can be boosted into an accurate strong learner. In this work, the boosting terminology used in a different sense, where two supervised methods are attached to the features extracted by MPCA. The idea is to extend the feature extraction procedure by selecting more discriminative features from the f_m (Equation (4.7)). Therefore, we added two supervised methods, LDA and shallow neural network (SNN) and fuse the achieved soft outputs in decision level. The results were also computed where LDA and SNN used separately without any decision fusion at the end; however, the achieved results were not satisfactory.

Since the core matrices extracted from the RDCs are already vectorized, assume that S_B and S_W are to be the between-class scatter and within-class scatter based on the f_m, respectively. Then, the corresponding LDA projection matrix can be expressed as $\mathrm{U}_{\mathrm{LDA}}$ where it consists of the generalized eigenvectors associated with the largest generalized eigenvalues of the generalized eigenvalue problem. Finally, the feature vector can be extracted as $z_m = \mathrm{U}_{\mathrm{LDA}}^T f_m$ and given as an input to the classifier.

In this work, an SNN is also implemented in the boosting scheme. Weight-elimination property of the neural network is included by the backpropagation algorithm to place a penalty on the MPCA features with redundant information in the fitted neural network model. Weight and biases of neurons are learned such that the cost function J is minimized. Given a set of training MPCA features and corresponding class labels, the cost function can be expressed as

$$J(W, b) = \left[\frac{1}{m}\sum_{i=1}^{m}\left(\frac{1}{2}\|h_{W,b}(f_m) - c_m\|^2\right)\right], \tag{4.14}$$

where W and b are the weights and biases of the network. $h_{W,b}$ is defined as the nonlinear hypothesis that provides the classification. The cost function represents the error between the input data and the neural net output. Backpropagation exploits these error values to compute the gradient of the cost function with respect to the weights and biases in the neural network. Afterward, the gradient is fed to the scaled conjugate gradient descent method, which in turn uses it to update the weights and biases, in an attempt to minimize the cost function. Finally, the output of the fitted neural network is given as an input to the softmax regression classifier, which is a generalization of logistic regression when there are multiple classes.

In the final step of the scheme, the soft classification labels obtained from LDA and SNN are fused using the L-place operators minimum, maximum, average, and product as the aggregation rule F as

$$c_m^F = F(d^1_{(m,i)}, d^2_{(m,i)}), \tag{4.15}$$

where $d^1_{(m,i)}$ and $d^2_{(m,i)}$ are the soft outputs (each class) of the LDA and SNN, respectively. An illustration of the boosting scheme is shown in Figure 4.11.

4.6 Experimental results

Two datasets were collected from two different indoor environments. The first location is the Radar Imaging Lab (RIL), which is a semi-controlled laboratory environment, located at Villanova University, CEER Building. Back and rear walls of the RIL covered with pyramid radiant-absorbent material to eliminate unwanted reflections and avoid any risk of measurement errors and ambiguities. In the RIL experiments, the radar was placed on a table raised 3.2 ft. above the ground and pointing directly the back wall.

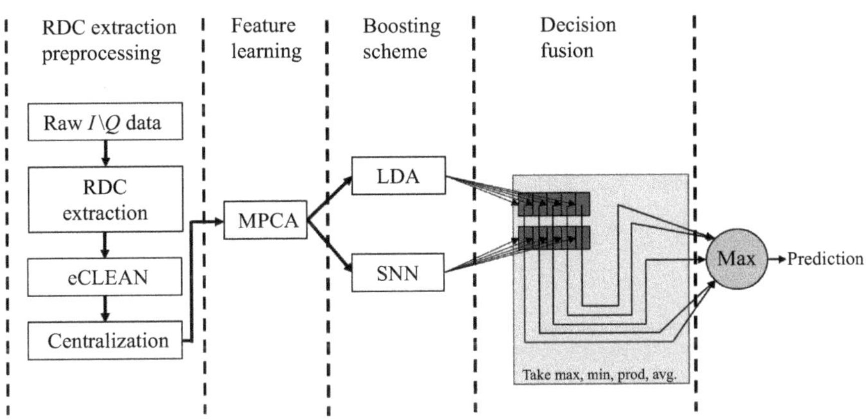

Figure 4.11 RDC-based MPCA boosting scheme with LDA, SNN, and decision fusion

The second dataset was acquired in the Center for Advanced Communications conference room located at Villanova University, Tolentine Building. This location is selected to mimic an uncontrolled environment similar to a senior residence apartment where office furniture such as television, plants, tables, bookshelves, and chairs were used. The radar was again placed on a table with a height of 3.2 ft. A Kinect was also located next to the radar in both configurations to record the ground truth optical videos of the motions.

Experiments were performed by 14 human subjects. A total of 827 signals were collected from the test subjects posed heights ranging from 5.1 to 6.3 ft., weights ranging from 119 to 220 lb, and included 12 male and 2 female subjects. Five different motions were considered in the experiments: falling (191 samples), sitting (213 samples), bending (203 samples), kneeling (108 samples), and walking (112 samples). Motions were performed for five different directions: 0°, 30°, 45°, 60°, and 90°. Each subject performed the experiments with three different speeds: slow (enacting as elderly people), medium, and fast.

We start the simulations by a comparison with 1D-PCA, predefined physical features, 2D-PCA, a 12-layer CNN, and RDC-based MPCA as depicted in Table 4.1. Note that the 1D-PCA, pre-defined features, 2D-PCA, and CNN are applied on the spectrograms. The average classification accuracies for 1D-PCA, predefined features, 2D-PCA, CNN, and MPCA are determined to be 65.32%, 73.65%, 83.10%, 84.54%, and 91.4%, respectively. The MPCA produces the lowest number of missed detections and highest fall detection rate. In essence, MPCA performance is drastically higher than that achieved with existing commonly used algorithms, demonstrating the importance of considering RDC in the application at hand. Then, we move to compare four different RDC-based methods: MPCA with MDC, MLDA with MDC, boosted-bottleneck performance of the simulation projection matrices on real data (BMPCA-Bottleneck), and boosted-fine-tuned-MPCA (BMPCA-Fine-tuned).

The classification accuracies are provided in Table 4.2 for different dimensions of the RDC. As we increase the dimension, the performances of the RDC MPCA-MDC, BMPCA-Bottleneck, and BMPCA-Fine-tuned also improve. The MLDA-MDC provides its best performance as 87.22% when the dimensions are set to $(16 \times 16 \times 16)$. BMPCA-Bottleneck features provide a good classification performance of 93.6%. Note that this result is achieved only using the simulation projection subspaces, proving the importance and merits of the initialization step. Finally, the nearly perfect classification accuracy belongs to the BMPCA-Fine-tuned at a rate of 97.2%.

Table 4.1 Note that only MPCA-MDC operates on the RDC.

	1D-PCA-MDC features-SVM	**Physical**	**2D-PCA-MDC**	**CNN**	**MPCA-MDC**
Test accuracy (%)	65.32	73.65	83.10	84.54	91.4

Table 4.2 Classification performance (on test data) comparison for different dimensions of the RDC between MPCA-MDC, MLDA-MDC, BMPCA-Bottleneck, and BMPCA-Fine-tuned in terms of accuracy

RDC dimension	MPCA-MDC (%)	MLDA-MDC (%)	BMPCA (bottleneck) (%)	BMPCA (fine-tuned) (%)
16×16×16	90.56	87.22	89.14	94.1
32×32×32	90.94	82.58	92.56	95.14
64×64×64	91.4	65.4	93.6	97.2

Table 4.3 Confusion matrix of BMPCA-Fine-tuned features on the test data (TA: 97.20%)

Activity class	Falling	Sitting	Bending	Kneeling	Walking
Falling	98.7	0	0	0	1.3
Sitting	2.3	94.1	3.6	0	0
Bending	1.3	2.4	95.1	1.2	0
Kneeling	0	0	0	100	0
Walking	2	0	0	0	98

The confusion matrix performance for the test set given by BMPCA-Fine-tuned is shown in Table 4.3. The primary source of misclassification is observed to be between the classes of sitting and bending, with about 3.6% confusion. This is not surprising as these two classes are most similar, and because our database contains different speeds of motions.

4.7 Conclusion

In this chapter, we considered a RDC-based multilinear subspace method for human activity recognition. Utilization of RDC offers an effective way to combine motion information from individual domains to capture cross-correlations and interdependency. The subspace method considered benefits from a single representation utilizing the entwined relationship between fast-time, slow-time, and Doppler frequency and their corresponding joint-variable domains. The proposed RDC-based approach starts with a preprocessing step called eCLEAN and attempts to suppress unwanted distortions and noise effects in the data. By employing MDC, we demonstrated that the multidimensional PCA method outperforms those based on predefined features, 1D- and 2D-PCA features, and a 12-layer CNN. We showed that improved performance RDC-based MPCA under diverse data repository can be achieved by using different optimization techniques. We employed a transfer learning initialization and GA-based optimization method to find the optimum number of projections. Further, we introduced a boosting scheme with two supervised methods to enhance the classification performance attained by the RDC-based MPCA. The proposed approach provided a high classification accuracy of 97.2% for a five-class activity recognition problem.

References

[1] M. G. Amin, *Radar for Indoor Monitoring: Detection, Classification, and Assessment*, 1st ed. CRC Press, 2017. [Online]. Available: https://www.crcpress.com/Radar-for-Indoor-Monitoring-Detection-Classification-and-Assessment/Amin/p/book/9781138746091.

[2] B. Jokanović and M. Amin, "Suitability of Data Representation Domains in Expressing Human Motion Radar Signals," *IEEE Geoscience and Remote Sensing Letters*, vol. 14, no. 12, pp. 2370–2374, 2017.

[3] D. C. Schleher, *English MTI and Pulsed Doppler Radar With MATLAB*, 2nd ed. Boston: Artech House, 2009.

[4] "Introduction to Radar Systems | MIT Lincoln Laboratory." [Online]. Available: https://www.ll.mit.edu/outreach/introduction-radar-systems.

[5] M. G. Amin, "Time-Frequency Spectrum Analysis and Estimation for Nonstationary Random-Processes," 1992. [Online]. Available: https://www.semanticscholar.org/paper/Time-Frequency-Spectrum-Analysis-and-Estimation-for-Amin/b428dcc25c1fb70fce005f0a0085cd8b5dbe17fd.

[6] P. Setlur, M. Amin, and F. Ahmad, "Analysis of micro-Doppler signals using linear FM basis decomposition," in *Radar Sensor Technology X*, vol. 6210. International Society for Optics and Photonics, May 2006, p. 62100M. [Online]. Available: https://www.spiedigitallibrary.org/conference-proceedings-of-spie/6210/62100M/Analysis-of-micro-Doppler-signals-using-linear-FM-basis-decomposition/10.1117/12.669003.short.

[7] L. Cirillo, A. Zoubir, and M. Amin, "Parameter Estimation for Locally Linear FM Signals Using a Time-Frequency Hough Transform," *IEEE Transactions on Signal Processing*, vol. 56, no. 9, pp. 4162–4175, 2008.

[8] Y. Kim and H. Ling, "Human Activity Classification Based on Micro-Doppler Signatures Using a Support Vector Machine," *IEEE Transactions on Geoscience and Remote Sensing*, vol. 47, no. 5, pp. 1328–1337, 2009.

[9] B. Tekeli, S. Z. Gurbuz, and M. Yuksel, "Information-Theoretic Feature Selection for Human Micro-Doppler Signature Classification," *IEEE Transactions on Geoscience and Remote Sensing*, vol. 54, no. 5, pp. 2749–2762, 2016.

[10] M. G. Amin and R. G. Guendel, "Radar Human Motion Recognition Using Motion States and Two-Way Classifications," *arXiv:1911.03512 [cs, eess]*, 2019. [Online]. Available: http://arxiv.org/abs/1911.03512.

[11] M. G. Amin, "Micro-Doppler classification of activities of daily living incorporating human ethogram," in *Radar Sensor Technology XXIV*, vol. 11408.

International Society for Optics and Photonics, Apr. 2020, p. 1140808. [Online]. Available: https://www.spiedigitallibrary.org/conference-proceedings-of-spie/11408/1140808/Micro-Doppler-classification-of-activities-of-daily-living-incorporating-human/10.1117/12.2558322.short.

[12] C. Li, V. M. Lubecke, O. Boric-Lubecke, and J. Lin, "A Review on Recent Advances in Doppler Radar Sensors for Noncontact Healthcare Monitoring," *IEEE Transactions on Microwave Theory and Techniques*, vol. 61, no. 5, pp. 2046–2060, 2013.

[13] G. Bergen, M. R. Stevens, and E. R. Burns, "Falls and Fall Injuries Among Adults Aged 65 Years – United States, 2014," *MMWR. Morbidity and Mortality Weekly Report*, vol. 65, no. 37, pp. 993–998, 2016.

[14] M. G. Amin, Y. D. Zhang, F. Ahmad, and K. D. Ho, "Radar Signal Processing for Elderly Fall Detection: The Future for In-Home Monitoring," *IEEE Signal Processing Magazine*, vol. 33, no. 2, pp. 71–80, 2016.

[15] B. Y. Su, K. C. Ho, M. J. Rantz, and M. Skubic, "Doppler Radar Fall Activity Detection Using the Wavelet Transform," *IEEE Transactions on Biomedical Engineering*, vol. 62, no. 3, pp. 865–875, 2015.

[16] R. M. Narayanan and M. Zenaldin, "Radar Micro-Doppler Signatures of Various Human Activities," *Sonar Navigation IET Radar*, vol. 9, no. 9, pp. 1205–1215, 2015.

[17] Q. Wu, Y. D. Zhang, W. Tao, and M. G. Amin, "Radar-Based Fall Detection Based on Doppler Time–Frequency Signatures for Assisted Living," *Sonar Navigation IET Radar*, vol. 9, no. 2, pp. 164–172, 2015.

[18] S. Wang, J. Song, J. Lien, I. Poupyrev, and O. Hilliges, "Interacting with Soli: Exploring Fine-Grained Dynamic Gesture Recognition in the Radio-Frequency Spectrum," in *Proceedings of the 29th Annual Symposium on User Interface Software and Technology*, ser. UIST '16. Tokyo, Japan: Association for Computing Machinery, Oct. 2016, pp. 851–860. [Online]. Available: https://doi.org/10.1145/2984511.2984565.

[19] S. Skaria, A. Al-Hourani, M. Lech, and R. J. Evans, "Hand-Gesture Recognition Using Two-Antenna Doppler Radar With Deep Convolutional Neural Networks," *IEEE Sensors Journal*, vol. 19, no. 8, pp. 3041–3048, 2019.

[20] S. Zhang, G. Li, M. Ritchie, F. Fioranelli, and H. Griffiths, "Dynamic hand gesture classification based on radar micro-Doppler signatures," in *2016 CIE International Conference on Radar (RADAR)*, Chinese Institute of Electronics (CIE), Guangzhou, China, Oct. 2016, pp. 1–4.

[21] Z. Chen, G. Li, F. Fioranelli, and H. Griffiths, "Dynamic hand gesture classification based on multistatic radar micro-Doppler signatures using convolutional neural network," in *2019 IEEE Radar Conference (RadarConf)*, IEEE, Boston, MA, USA. Apr. 2019, pp. 1–5, iSSN: 2375-5318.

[22] J. Lin, "Noninvasive Microwave Measurement of Respiration," *Proceedings of the IEEE*, vol. 63, no. 10, pp. 1530–1530, 1975.

[23] C. Li, J. Cummings, J. Lam, E. Graves, and W. Wu, "Radar Remote Monitoring of Vital Signs," *IEEE Microwave Magazine*, vol. 10, no. 1, pp. 47–56, 2009.

[24] K. Higashikaturagi, Y. Nakahata, I. Matsunami, and A. Kajiwara, "Non-invasive respiration monitoring sensor using UWB-IR," in *2008 IEEE International Conference on Ultra-Wideband*, IEEE, Hannover, Germany, vol. 1, Sep. 2008, pp. 101–104, iSSN: 2162-6596.

[25] T. Kellomaki, W. G. Whittow, J. Heikkinen, and L. Kettunen, "2.4 GHz plaster antennas for health monitoring," in *2009 3rd European Conference on Antennas and Propagation*, Berlin, Germany, Mar. 2009, pp. 211–215, iSSN: 2164-3342.

[26] A.-K. Seifert, M. G. Amin, and A. M. Zoubir, "Toward Unobtrusive In-Home Gait Analysis Based on Radar Micro-Doppler Signatures," *IEEE Transactions on Biomedical Engineering*, vol. 66, no. 9, pp. 2629–2640, 2019.

[27] S. Z. Gurbuz, C. Clemente, A. Balleri, and J. J. Soraghan, "Micro-Doppler-Based In-Home Aided and Unaided Walking Recognition With Multiple Radar and Sonar Systems," *Sonar Navigation IET Radar*, vol. 11, no. 1, pp. 107–115, 2017.

[28] A.-K. Seifert, L. Schäfer, M. G. Amin, and A. M. Zoubir, "Subspace classification of human gait using radar micro-Doppler signatures," in *2018 26th European Signal Processing Conference (EUSIPCO)*, Rome, Italy, Sep. 2018, pp. 311–315, iSSN: 2076-1465.

[29] D. Yessad, A. Amrouche, M. Debyeche, and M. Djeddou, "Micro-Doppler classification for ground surveillance radar using speech recognition tools," in *Progress in Pattern Recognition, Image Analysis, Computer Vision, and Applications*, ser. Lecture Notes in Computer Science, C. San Martin and S.-W. Kim, Eds. Berlin, Heidelberg: Springer, 2011, pp. 280–287.

[30] B. Erol and M. Amin, "Generalized PCA fusion for improved radar human motion recognition," in *2019 IEEE Radar Conference (RadarConf)*, IEEE, Boston, MA, USA, Apr. 2019, pp. 1–5, iSSN: 2375-5318.

[31] F. H. C. Tivive, S. L. Phung, and A. Bouzerdoum, "Classification of Micro-Doppler Signatures of Human Motions Using Log-Gabor Filters," *Sonar Navigation IET Radar*, vol. 9, no. 9, pp. 1188–1195, 2015.

[32] C. Clemente, L. Pallotta, A. De Maio, J. J. Soraghan, and A. Farina, "A Novel Algorithm for Radar Classification Based on Doppler Characteristics Exploiting Orthogonal Pseudo-Zernike Polynomials," *IEEE Transactions on Aerospace and Electronic Systems*, vol. 51, no. 1, pp. 417–430, 2015.

[33] S. Z. Gurbuz and M. G. Amin, "Radar-Based Human-Motion Recognition With Deep Learning: Promising Applications for Indoor Monitoring," *IEEE Signal Processing Magazine*, vol. 36, no. 4, pp. 16–28, 2019.

[34] M. S. Seyfioğlu, S. Z. Gürbüz, A. M. Özbayoğlu, and M. Yüksel, "Deep learning of micro-Doppler features for aided and unaided gait recognition," in *2017 IEEE Radar Conference (RadarConf)*, IEEE, Seattle, WA, USA, May 2017, pp. 1125–1130, iSSN: 2375-5318.

[35] B. Jokanović and M. Amin, "Fall Detection Using Deep Learning in Range-Doppler Radars," *IEEE Transactions on Aerospace and Electronic Systems*, vol. 54, no. 1, pp. 180–189, 2018.

[36] V. Chen, F. Li, S.-S. Ho, and H. Wechsler, "Micro-Doppler Effect in Radar: Phenomenon, Model, and Simulation Study," *IEEE Transactions on Aerospace and Electronic Systems*, vol. 42, no. 1, pp. 2–21, 2006.

[37] B. Erol and M. G. Amin, "Radar Data Cube Processing for Human Activity Recognition Using Multisubspace Learning," *IEEE Transactions on Aerospace and Electronic Systems*, vol. 55, no. 6, pp. 3617–3628, 2019.

[38] F. R. Dickey, M. Labitt, and F. M. Staudaher, "Development of Airborne Moving Target Radar for Long Range Surveillance," *IEEE Transactions on Aerospace and Electronic Systems*, vol. 27, no. 6, pp. 959–972, 1991.

[39] S. Fukushima, H. Yamada, H. Kobayashi, and Y. Yamaguchi, "Human motion detection and extraction by using FM-CW range-Doppler radar," in *2014 International Symposium on Antennas and Propagation Conference Proceedings*. Kaohsiung, Taiwan: IEEE, Dec. 2014, pp. 173–174. [Online]. Available: http://ieeexplore.ieee.org/document/7026586/.

[40] D. Tahmoush and J. Silvious, "Time-integrated range-Doppler maps for visualizing and classifying radar data," in *2011 IEEE RadarCon (RADAR)*, IEEE, Kansas City, MO, USA, May 2011, pp. 372–374.

[41] Y. He, P. Molchanov, T. Sakamoto, P. Aubry, F. L. Chevalier, and A. Yarovoy, "Range-Doppler Surface: A Tool to Analyse Human Target in Ultra-Wideband Radar," *Sonar Navigation IET Radar*, vol. 9, no. 9, pp. 1240–1250, 2015.

[42] J. Hogbom, "Aperture synthesis with a non-regular distribution of interferometer baselines," in *Astron. Astrophys. Suppl. Ser.*, vol. 15, 1974, pp. 417–426.

[43] J. Li and P. Stoica, "Efficient Mixed-Spectrum Estimation With Applications to Target Feature Extraction," *IEEE Transactions on Signal Processing*, vol. 44, no. 2, pp. 281–295, 1996.

[44] H. Lu, K. N. Plataniotis, and A. N. Venetsanopoulos, *Multilinear Subspace Learning: Dimensionality Reduction of Multidimensional Data*. Chapman and Hall/CRC Press, Boca Raton, FL, 2012.

[45] A. Cichocki, D. Mandic, L. D. Lathauwer, *et al.*, "Tensor Decompositions for Signal Processing Applications: From Two-Way to Multiway Component Analysis," *IEEE Signal Processing Magazine*, vol. 32, no. 2, pp. 145–163, 2015.

[46] H. Lu, K. N. Plataniotis, and A. N. Venetsanopoulos, "MPCA: Multilinear Principal Component Analysis of Tensor Objects," *IEEE Transactions on Neural Networks*, vol. 19, no. 1, pp. 18–39, 2008.

[47] S. Yan, D. Xu, Q. Yang, L. Zhang, X. Tang, and H. Zhang, "Multilinear Discriminant Analysis for Face Recognition," *IEEE Transactions on Image Processing*, vol. 16, no. 1, pp. 212–220, 2007.

[48] M. S. Seyfioglu, B. Erol, S. Z. Gurbuz, and M. G. Amin, "Diversified radar micro-Doppler simulations as training data for deep residual neural networks," in *2018 IEEE Radar Conference (RadarConf18)*, IEEE, Oklahoma City, OK, USA, Apr. 2018, pp. 0612–0617.

[49] S. J. Pan and Q. Yang, "A Survey on Transfer Learning," *IEEE Transactions on Knowledge and Data Engineering*, vol. 22, no. 10, pp. 1345–1359, 2010.
[50] K. Miettinen, *Nonlinear Multiobjective Optimization.* Springer Science & Business Media, Norwell, MA, 1999, google-Books-ID: ha_zLdNtXSMC.
[51] K. Deb, A. Pratap, S. Agarwal, and T. Meyarivan, "A Fast and Elitist Multi-objective Genetic Algorithm: NSGA-II," *IEEE Transactions on Evolutionary Computation*, vol. 6, no. 2, pp. 182–197, 2002.

Chapter 5
Challenges in training DNNs for classification of radar micro-Doppler signatures

Sevgi Z. Gurbuz[1], *Moeness G. Amin*[2], *Mehmet S. Seyfioğlu*[3], *and Baris Erol*[4]

One of the principal challenges in applying deep learning to classify radar data is how to train deep models when there is very little data available for training. Ideally, a training database should have a sufficient and equal number of samples in each class, which are also statistically consistent with the entire range of probable target profiles. In practice, the acquisition of such training data is quite difficult for radar applications. In airborne or spaceborne applications, the cost to fly or launch a platform is quite high, and even then acquisition of data spanning all possible scenarios and targets would be impossible. In ground-based radar applications for human activity recognition, the costs of the actual radar and its deployment are much less, but still there is a high human cost in data collection. Each activity repetition in front of the radar system costs time and money, this time in the form of compensation to participants. Even supposing a certain number of participants could be recruited, how well would this group represent the spectrum of heights, weights, walking styles, and speeds of the human population? Training data insufficiency becomes more daunting when dealing with minority or vulnerable populations, such as the elderly, children, or people with any sort of illness. Moreover, all human subject testing is regulated by strict ethical guidelines, and no data can be collected without first getting approval from an institutional review board.

This presents an interesting dilemma when considering deep learning for radar applications. On the one hand, in other fields, deep learning has outperformed state-of-the-art, conventionally supervised and unsupervised learning techniques by leveraging features learned directly from the data, *without using any explicit domain knowledge*. On the other, radar engineers will never have the availability of a database like ImageNet [1], which contains over a million images, for the purposes of training. Small training data support raises an important question of how can deep learning be effectively applied to radar data?

[1]Department of Electrical and Computer Engineering, The University of Alabama, Tuscaloosa, AL, USA
[2]Department of Electrical and Computer Engineering, and Center for Advanced Communications, Villanova, University, Villanova, PA, USA
[3]Department of Electrical and Computer Engineering, University of Washington, Seattle, WA, USA
[4]Siemens Corporate Technology, Princeton, NJ, USA

In this chapter, we explore the motivation for constructing very deep neural networks (DNNs) and investigate the theoretical requirements for supporting such complex models. Techniques for training DNNs when little real data is available are discussed, such as DNN architectures less susceptible to overfitting and the use of transfer learning to pre-train on alternative data sources. Detailed techniques for synthetic data generation for micro-Doppler signature classification are presented, including the challenges faced when adapting techniques developed for computer vision to radar datasets.

5.1 Theory of training complex models

The foremost requirement on training DNNs is that our algorithm must perform well on *new, previously unseen inputs*—not just on the subset of data used to train the network. This is known as *generalization* and is typically evaluated by testing the algorithm on a different set of data that is *independent* from the training dataset. When training a machine learning model, we desire not just the training error to be minimized, but also the generalization (or test) error to be low.

5.1.1 Bias–variance trade-off

The conventional wisdom in machine learning recommends controlling a model's ability to fit a wide variety of functions (also known as its *capacity*) by balancing the *bias–variance trade-off* [2]. Suppose we wish to predict y based on the inputs x. Then, $y = f(x) + e$, where e is the error. The model, in this case, would be some estimate, $\hat{f}(x)$, of the true underlying function, $f(x)$. The *bias* (μ) is the difference (on average) between the prediction and true value:

$$\mu = E[y - \hat{f}(x)] \tag{5.1}$$

If the bias is high, the model is relatively uninfluenced by the training data and the resulting model is too simple to represent the data accurately. Both the training error and test error will be high, causing *underfitting*.

The *variance* (σ^2) is a reflection of the variability in the prediction and is defined as

$$\sigma^2 = E[(\hat{f}(x) - E[\hat{f}(x)])^2] \tag{5.2}$$

Models with high variance may perform well on training data but do not generalize well on test data, resulting in high test error, i.e., *overfitting*.

Our goal, however, is not to minimize just the bias or the variance but to determine an optimal estimator that is unbiased and with minimum variance. In this case, the natural optimality criterion [3] is the *mean square error* (MSE), i.e., the *risk* for a quadratic cost function, defined as

$$MSE(\hat{f}(x)) = E[(y - \hat{f}(x))^2] \Rightarrow \mu^2 + \sigma^2 + Error \tag{5.3}$$

Thus, minimization of the MSE results in a trade-off between bias and variance [4], as illustrated by the classical "U"-shaped curve depicted in Figure 5.1. The model

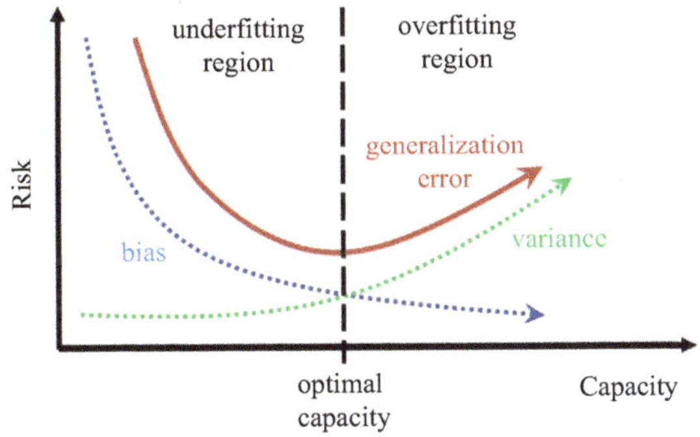

Figure 5.1 Risk curve illustrating bias–variance trade-off

training process aims at finding a "sweet-spot" between bias and variance yielding optimal capacity; namely, a model good enough to "understand" the general structure of data, while simple enough to prevent further memorization of noise or unrelated samples. This can be done in DNNs through a number of ways, for example, by modifying the architecture, by regularizing the loss function, or by using techniques such as early stopping.

5.1.2 Requirements on training sample support

Equally critical as identifying the optimal model capacity is ensuring that the model is trained with a sufficient number of training samples. Generally speaking, the greater the model capacity, the greater an amount of training data are required. Although there are various means for quantifying model capacity, one of the most well-known metrics is the *Vapnik–Chervonenkis (VC) dimension* [5]. The VC dimension provides an upper bound on the test error of a model:

$$Pr\left(e_{test} \leq e_{train} + \sqrt{\frac{1}{N}\left[VC\left(\log\left(\frac{2N}{VC}\right)+1\right) - \log\left(\frac{\eta}{4}\right)\right]}\right) = 1 - \eta \quad (5.4)$$

where e_{train} and e_{test} are the training and testing errors, respectively. VC is the VC dimension of the model, η is a number between $0 \leq \eta \leq 1$, and N is the size of the training data set. The previous inequality is valid when $VC \ll N$, otherwise the model may be overfitting. This leads to a criterion on training sample size in terms of the VC dimension as

$$N = \Theta\left(\frac{VC + \ln(1/P_{err})}{\varepsilon}\right) \quad (5.5)$$

where ε is the learning error and P_{err} is the error probability.

Let us define E as the number of connects (or edges) and V as the number of nodes in the network. Then, it can be shown [6] that the VC dimension of a DNN with general weights is at most $O(|E| \cdot \log(|E|))$ if the activation function is the *sign* function, and between the bounds $\Omega(|E|^2) \leq VC \leq O(|E|^2 \cdot |V|^2)$ if the activation function is a sigmoid function. Alternatively, if the weights are drawn from a finite family of real numbers, then for both *sign* and sigmoid activation functions, the VC dimension is at most $O(|E|)$. For a DNN with hundreds of nodes and thousands of connections, billions of training samples will be required. It is therefore a somewhat surprising result that DNNs do not overfit and instead exhibit good generalization performance. There is a dichotomy between deep learning theory and practice, as discussed in more detail in the next section.

Some practical rules of thumb have been suggested as guidelines for estimating the size of the training database required. For example, it has been recommended that the amount of training data should be at least 10 times the number of dimensions of the model [2]. And, there should be at least 1,000 representative images per class. Regardless of the number, the generalization performance of a model tends to increase with the size of the training data set utilized. However, these gains are not unlimited. It has been observed that performance eventually does plateau. For example, in [7], the authors classified 100 million Flickr images and captions using a convolutional neural network (CNN). They reported that performance increased with growing training data set size, until about 50 million images. For radar data sets, however, the typical training data set size is only on the order of several thousand at most. Thus, there is a threshold beyond which the performance plateaus has not yet been discovered for RF datasets. On the contrary, the availability of sufficient training data remains a serious obstacle to the implementation of DNNs.

5.1.3 Machine learning theory versus practice: empirical studies

One consequence of the bias–variance trade-off is the widely held belief that a model with zero training error will overfit the data and have poor generalization. Since DNNs have complex architectures, we would expect, therefore, that they would suffer from problems of overfitting, with high variance and low bias. However, research [8–10] have empirically demonstrated that in DNNs the *opposite* is true—a result that contradicts our classical understanding of the bias–variance trade-off. DNNs *do have* very low training risk while *also providing very accurate predictions on new data*. Deep over-parameterized networks can achieve *zero training loss* when several weak conditions are met [11]. Moreover, even when the training data have high levels of noise, it has been shown that DNNs trained to interpolate the training data (also known as zero loss training) have exceedingly high generalization properties yielding near-optimal test results [8,12].

This apparent dichotomy is further revealed by empirical studies [10] that extend the U-curve beyond the point of high risk interpolation and show that DNNs exhibit a "double descent" risk curve, shown in Figure 5.2. When capacity (i.e., number of model parameters) is low, the classical trends are followed. There is indeed an optimal point that minimizes risk for modest capacities. But if the capacity of the network were to push high enough, beyond the interpolation threshold, by increasing the number of features or DNN size/depth, the risk will gradually decrease again to levels that in fact may be even lower than the optimal risk found for modest capacities. This is indeed what has been witnessed in the trend to construct very DNNs with over a 100 layers. So what is the catch? Well, the training of very deep neural networks

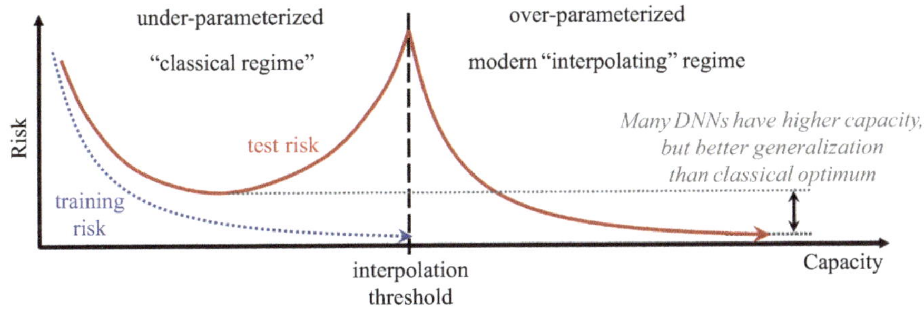

Figure 5.2 Empirically observed risk curves for a DNN [10]

becomes increasingly difficult with depth, even with mechanisms such as residual learning [13].

However, this also represents a great opportunity for deep learning solutions to challenges in radar applications: DNNs that have been applied so far have been greatly limited in depth, and, hence, capacity, due to the lack of sufficient training data. Through the application of transfer learning, domain adaptation, and synthetic target modeling, however, arbitrarily large training databases can be generated. This will pave the way for future breakthroughs and novel applications in RF sensing.

5.2 Training with small amounts of real data

To get an idea of the current state of the art in DNNs for micro-Doppler (μD) classification [14,15], consider the network depths and train/test sample sizes indicated for a sampling of recent papers, as shown in Table 5.1, which used real data only for training. It is typical for networks trained on real data alone that only several hundred samples are available for training. A notable exception is the work by Trommel [20], where over 17,000 real training samples were independently acquired. In contrast, the work by Cao [21] used data augmentation via image transformations to obtain 64,000 training samples. Although image transformations are not as effective as other training data synthesis techniques (see Section 5.5.2), the significance of sufficiently

Table 5.1 DNN training data size used in recent works

Application	Reference	No. of classes	Layers	No. of training data	No. of test data
Fall detection	Jokanović [16]	2	2 f.c.	200	208
Hand gesture	Kim [17]	10	3 conv. + 1 f.c.	450	50
Human activity	Kim [18]	7	2 conv. + 1 f.c.	756	252
Human activity	Seyfioğlu [19]	12	3 conv. + 2 f.c.	900	100
Human activity	Trommel [20]	4	3 conv. + 2 f.c.	17,580	5,776
Human ID	Cao [21]	20	AlexNet	64,000	1,600

large training datasets is underscored by its ability to train AlexNet, an 8-layer network with 5 convolutional (*conv.*) and 3 fully connected (*f.c.*) layers, for identification of 20 different people.

5.2.1 Unsupervised pretraining

Given the same amount of training data, the classification accuracy attainable and tendency of a DNN to overfit is dependent upon the structure of the network. To illustrate this, consider a 12-class human activity recognition problem, illustrated in Figure 5.3, for which on average 72 samples per class have been measured. CNNs represent the current state of the art in computer vision and have advantages in image processing due to their use of convolutional filtering, which allows the network to capture local spatial variations. Typically, the training process begins with a random assignment of network weights. However, the objective function of CNNs is highly non-convex, meaning that the parameter space of the model may contain many local minima. As a result, gradient-based optimization algorithms may converge to local minima, which are actually not optimal in a global sense [22].

An alternative way to train networks, which has been shown to yield greater accuracy with smaller amounts of training data, is *unsupervised pretraining*. An encoder–decoder structure, as used in autoencoders (AEs), can be used to greedily pretrain network weights [23] to learn an identity mapping in an unsupervised fashion. This approach has been shown to be effective in capturing the underlying variations of the data without any labels [24]. The benefits of both convolutional filtering and unsupervised pretraining are captured by the *convolutional AE* (CAE) architecture, pictured in Figure 5.4, which utilizes both an encoder–decoder structure *and* convolutional filtering.

The efficacy of unsupervised pretraining can be seen by comparing the performance of all three architectures on the 12-class activity recognition dataset [19]. Applying 10-fold cross-validation, the classification accuracy achieved is tabulated for different network configurations in Figure 5.5. Notice that for all DNNs, beyond a depth of three layers, the test accuracy begins to drop, indicating an increase in generalization error and model overfitting. At four layers, however, the CAE continues to give performance quite close to the maximum and at six layers only drops $\Delta = -4.1\%$ in comparison to a drop of $\Delta = -5.8\%$ by the AE and $\Delta = -5.2\%$ for the CNN. Nevertheless, all of the previous networks are *shallow* DNNs, offering a stark demonstration of training sample size limitations on model capacity.

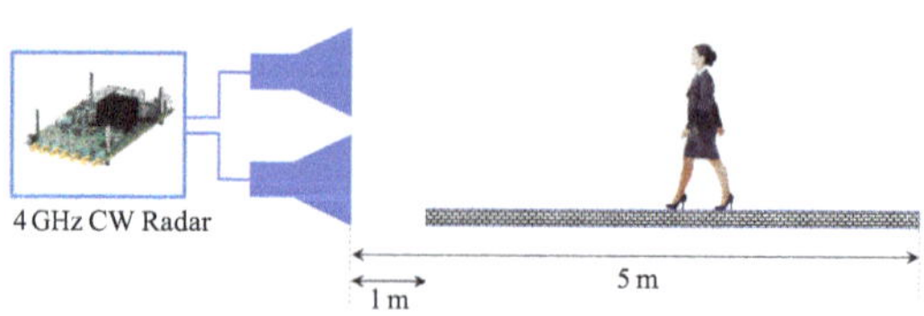

Class	# Samples	Class	# Samples
walking	71	crutches	74
jogging	72	sitting	50
crawling	74	walker	121
limping	104	creeping	56
cane usage	123	wheelchair	121
falling	53	falling off chair	53

Figure 5.3 Measured dataset for 12-class activity recognition problem with 4 GHz CW radar data

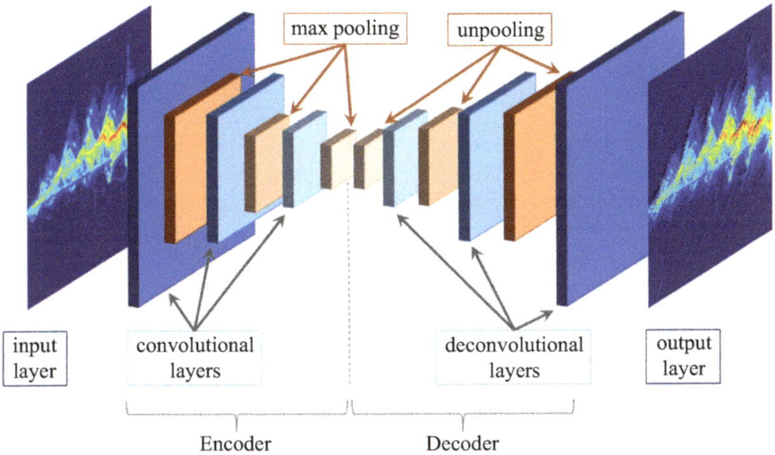

Figure 5.4 Architecture for convolutional autoencoder

5.3 Cross-frequency training using data from other radars

One possible alternative is to train on data acquired from a *different* radar system than that which acquired the test data. In this case, the transmit frequency, bandwidth, and waveform of the training radar may be different than that of the test radar—leading to differences in time–frequency representation, also known as the micro-Doppler signature, of the target. The transmit frequency affects how sensitive the radar will be to target motion. From physics, we know the Doppler shift (f_D) incurred in the radar return is proportional to velocity (v) and transmit frequency (f_t):

$$f_D = \frac{2v}{c} f_t = \frac{2v}{\lambda_t} \tag{5.6}$$

where c is the speed of light. Thus, for a higher transmit frequency, a greater Doppler shift frequency will be recorded, which improves the ability to classify activity classes with slight kinematic differences. Bandwidth improves range resolution, which aides in recognition using joint-domain representations, but it also increases the electronic noise floor. The scattering properties of targets also change as a function of frequency, resulting in different target radar cross section (RCS) and received signal power from both direct path and multipath returns. Finally, there may be differences due to the RF sensors themselves in terms of type of waveform transmitted (e.g., continuous-wave (CW), frequency modulated CW (FMCW), and impulse), sensor artifacts, antenna gain, and beam width, which all affect the measurement.

Consider the measurements shown in Figure 5.6 for the same person walking directly toward the radar, but for five different radar systems. The essential characteristics of walking are common to all the signatures, e.g., the slow oscillation, higher power reflection of the torso; the high amplitude but more rapid oscillations of the legs; the straight line at 0 Hz corresponding to the ground clutter. However, the overall

AUTOENCODER		
Depth	Width*	Acc. (%)
1	20	74.1
1	50	76
1	100	75.5
2	20–50	79.8
2	50–100	81.1
2	100–200	80.9
3	20–50–100	83.4
3	40–80–160	83.2
3	50–100–200	**84.1**
4	20–50–100–200	83.4
4	40–100–200–400	83.1
4	100–200–400–800	79.8
5	20–50–100–200–400	82.7
5	40–100–200–400–800	82.8
5	100–200–400–800–1600	79.5
6	20–50–100–200–400–800	80.1
6	40–100–200–400–800–1600	78.9
6	100–200–400–800–1600–3200	78.3

CONVOLUTIONAL NEURAL NETWORK		
Depth	Width*	Acc. (%)
1	5	80.2
1	30	81.4
1	100	81.5
2	5	82.1
2	30	86.2
2	100	87.1
3	5	88.4
3	30	**90.1**
3	100	90
4	5	87.6
4	30	89.9
4	100	89.5
5	5	87.2
5	30	88.1
5	100	88.4
6	5	84.9
6	30	85.5
6	100	86.8

*Width applies to each layer

CONVOLUTIONAL AUTOENCODER		
Depth	Width*	Acc. (%)
1	5	83.8
1	30	84.9
1	100	84.7
2	5	88.2
2	30	88.4
2	100	89
3	5	92.2
3	30	**94.2**
3	100	93.2
4	5	94
4	30	94.1
4	100	93.7
5	5	93.2
5	30	93.4
5	100	93.9
6	5	90.1
6	30	90.9
6	100	90.3

*Width applies to each layer

Figure 5.5 Classification accuracy for different combinations of network depth and width (number of neurons in a layer) [19]

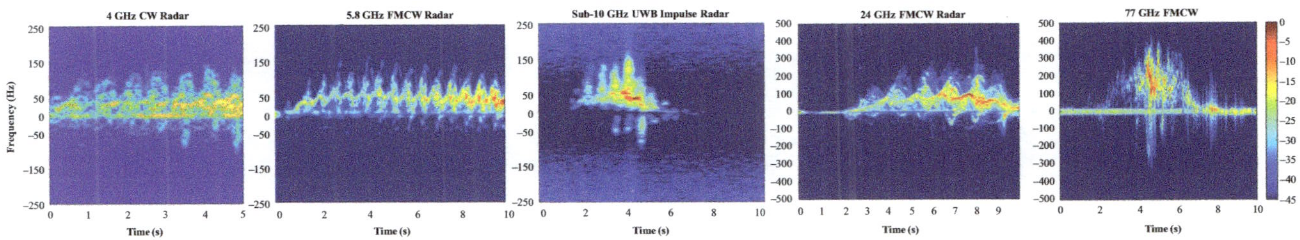

Figure 5.6 Comparison of micro-Doppler signatures of walking for different types and frequencies of radar

Doppler spread of the signature changes across frequency. While the peak frequency at 4 GHz is about 100 Hz, at 77 GHz it is 400 Hz. The texture of each signature as well as strength of the ground clutter is also variable across sensors.

The efficacy of cross-frequency training was tested by conducting an experiment involving simultaneous measurements by 5.8 and 24 GHz FMCW radars of the same person walking [25], as shown in Figure 5.7. The bandwidth of 5.8 GHz FMCW radar was set at 400 MHz, while that of the 24 GHz FMCW was set at 750 MHz. Both systems had a pulse repetition frequency of 1 kHz. A total of 1,009 independent recordings of 6 different activity classes were made for *each* radar, with the test subject walking along either a direct path, or slightly angled path at 15° or 30°. A three-layer CNN was trained with the 25 GHz radar recordings, while accuracy was assessed on the 5.8 GHz recordings. Note that ***no fine tuning*** was done after the initial training with the 25 GHz data. Results show that for certain classes, for example, ambulatory activities, such as walking or dragging one's foot, high classification accuracies are maintained. However, for other activities, where the differences between the signatures of each class are more slight, the classification accuracy plummets to as low as 53%. It is also significant to note that when the walking gait changes noticeably, such as with short steps, again over a 10% drop in accuracy is observed.

Thus, for fine-grain motion recognition applications, such as gesture recognition or biomedical gait analysis, the frequency-based scattering difference between RF sensors is significant enough that it precludes data from other RF sensors to provide adequate training. Moreover, this approach does not resolve the more fundamental issue that regardless of what sensor is acquiring the data, human time, cost, and effort are still required, which limits the overall availability of training data. Unlike computer vision applications, where there are numerous and massive datasets published for

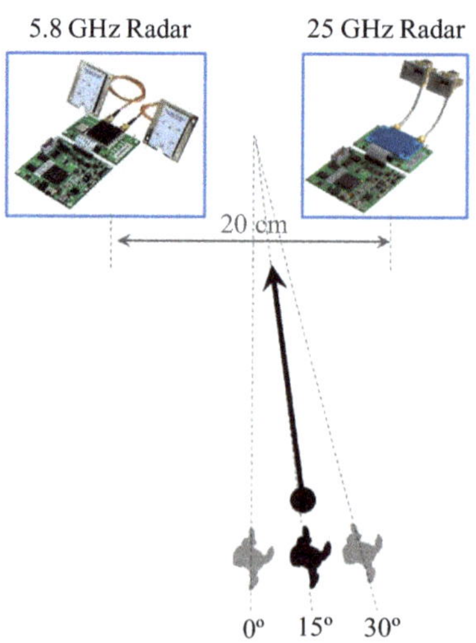

Cross-Frequency Training Experiment

Train on 25 GHz Data → Test on 5.8 GHz Data

True Class \ Predicted Class	1	2	3	4	5	6
1	**90**		1	1	2	3
2		**99**		1		
3			**66**		1	3
4		10	22	**53**	13	2
5			1	1	**96**	2
6	14			2	1	**83**

1: walking towards radar
2: walking away from radar
3: bowing forwards
4: picking up object
5: dragging foot
6: short steps

Figure 5.7 Classification accuracy for testing 5.8 GHz-data with a three-layer CNN trained on 25-GHz data [25]

research, hardly any databases of radar micro-Doppler signatures are readily available. Moreover, it is highly likely that the classes recorded in any "dataset of opportunity" will be different from those desired to be tested. This results in an *open set problem* [26] caused by mismatch of test classes with training classes and will limit the efficacy of such datasets for pretraining. The design of DNNs that can generate novelty outside trained model classes remains an active area of research [27,28].

Recent research [29] has proposed the use of adversarial learning (discussed in Section 5.6) for bridging sensor differences and growing the size of training database. Consider the cross-frequency training performance, shown in Figure 5.8, between a 77 GHz FMCW radar with 750 MHz bandwidth (660 samples), a 24 GHz FMCW radar with 1,500 MHz bandwidth (840 samples), and an ultra-wideband impulse radar operating about 10 GHz (660 samples). A CAE with 5 convolutional layers of 64 filters with sizes of 3 × 3 and 9 × 9 is used to classify signatures from each sensor. Notice that the gap between the classification accuracies attained under cross-frequency scenarios versus matched frequency scenarios is now as much as 70%. Consistency between the frequency of the test data and data used to fine tune the network is of utmost importance, as in that case the performance differences reduce to roughly 15%. When an auxiliary conditional generative adversarial network (ACGAN) is used to synthesize a total of 6,000 signatures, representing a 10-fold increasing in the amount of data used for pretraining, the performance degradation due to cross-frequency training is reduced, but this gain is at the cost of about a 5% degradation in training and testing on the same frequency (e.g., 77 GHz). Thus, simply utilizing more training data does not solve the problem of cross-frequency training losses. Application of domain adaptation networks, such as CycleGAN [30], is currently being considered as a mitigation strategy.

CAE with Training on Real Data only

Pre-train	Fine Tune	Test	Testing Accuracy
77 GHz	77 GHz	77 GHz	91.5%
		24 GHz	22.5%
		10 GHz	18.9%
	24 GHz	24 GHz	74.4%
	10 GHz	10 GHz	75.5%
24 GHz	24 GHz	24 GHz	91.2%
		77 GHz	28.5%
		10 GHz	40.5%
24 GHz	77 GHz	77 GHz	83.8%
	10 GHz	10 GHz	81.6%
10 GHz	10 GHz	10 GHz	91.8%
		77 GHz	24.1%
		24 GHz	18.8%
	77 GHz	77 GHz	80.0%
	24 GHz	24 GHz	79.1%

Human Activity Classes	
Walking towards the radar	Walking away from the radar
Picking up an object from the ground	Bending
Sitting on a chair	Kneeling
Crawling towards the radar	Limping with right leg stiff
Walking on both toes	Walking with short steps
Scissor gait	

CAE with Pre-Training on Synthetic Data

Pre-train	Fine Tune	Test	Testing Accuracy
Synthetic 77 GHz	77 GHz	77 GHz	85.40%
		24 GHz	8.85%
		10 GHz	7.20%
	24 GHz	24 GHz	77.15%
	10 GHz	10 GHz	80.80%

Figure 5.8 Cross-frequency training between 77 GHz FMCW, 24 GHz FMCW, and 10 GHz UWB impulse radar systems [29]

5.4 Transfer learning using pretrained networks

A widely used method that requires only small amounts of real data for testing and fine-tuning is *transfer learning*. The objective of transfer learning is to learn the weights of a network from a *source domain* and then transfer those weights to a model for another "related" problem, or *target domain*. As with unsupervised pretraining, transfer learning can be used for initializing the model weights to values that bring the model closer to a global optimum. A small amount of data from the target domain can then be applied to fine-tune the weights in a supervised fashion.

Transfer learning also allows us to leverage very DNNs designed for computer vision (source domain) to initialize networks for learning models of radar data (target domain). There are an abundance of image databases available, including ImageNet [31], which contains over 1 million images and can be used for pretraining. It has been popular to use DNNs that have achieved success in the ImageNet large-scale visualization challenge, such as AlexNet [32], VGGnet [33], GoogLeNet [34], and ResNet [13], for transfer learning. Transfer learning has been used in radar micro-Doppler applications, such as human activity recognition [11,35–37], gesture recognition [38], and human voice recognition [39].

The efficacy of transfer learning can be compared with other DNN architectures, such as CNNs and CAEs, using the 4 GHz 12-class experiment summarized in Figure 5.3. In particular, two popular CNN models, VGGNet and GoogleNet, are pretrained on ImageNet before being fine-tuned with a randomly selected subset of the measured micro-Doppler data. As both VGGnet and GoogLeNet require RGB inputs, we generated an RGB spectrogram by copying the same spectrogram onto each band. Of the total acquired measured data, 80% is used for training and fine-tuning, while the remaining 20% is reserved for testing. Ten-fold cross validation was applied by setting aside 20% of the training data for validation.

Results for the validation accuracy, as a function of epoch and test accuracy, as well as the amount of real data used during training, are shown in Figure 5.9. The greatest overall accuracy is achieved by the CAE, when all available training data (80% of total measurements) are utilized, yielding a result that surpasses transfer learning with VGGnet by 5%. The worst result is achieved by the three-layer CNN, with GoogLeNet only slightly better. This is an interesting result because on the radar micro-Doppler dataset, VGGnet outperforms GoogLeNet, even though in the ILVSC challenge, on ImageNet, GoogLeNet had given better performance than VGGnet. As such, it represents an example of how results from computer vision do not necessarily translate directly to RF datasets. Even more interesting, when the dependency of accuracy on an amount of real training samples is compared, we can see that the CAE maintains its superiority only above a certain threshold, or crossover point, at about 650 samples. Below this threshold, VGGnet consistently outperforms the other networks. For the case of the fewest training data samples, at about just 200 samples, VGGnet surpasses the three-layer CNN by over 20%, and the CAE by nearly 10%.

This brings us to the significant conclusion that if a moderate amount of data are available, say, on the order of a 1,000 samples, use of unsupervised pretraining

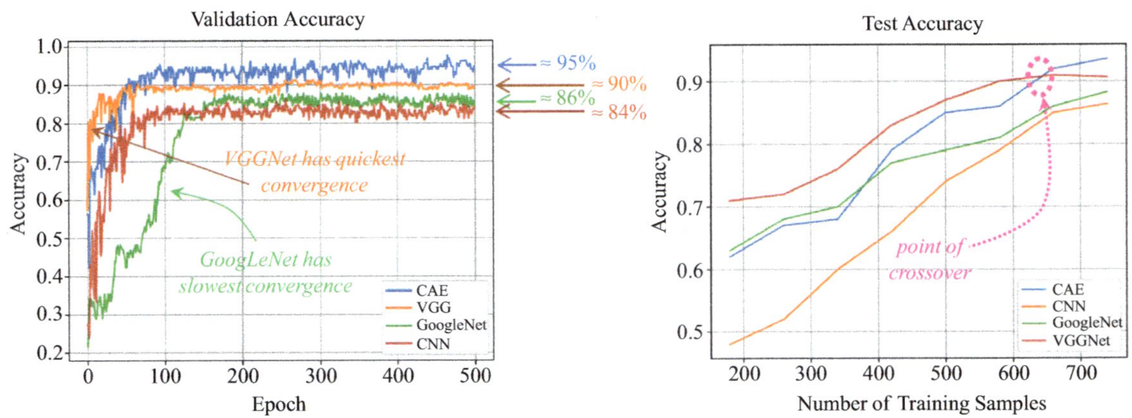

Figure 5.9 Comparison of transfer learning from networks pretraining on ImageNet with DNNs trained on real radar data [36]

(i.e., CAEs) is more advantageous. But, if truly not many samples are available, transfer learning can be applied to provide reasonable accuracy. However, the gains of transfer learning are limited: beyond 580 training samples, the performance attainable by VGGnet ***levels off***. This should not be surprising, since, after all, we are training on source data that have no common phenomenology with the target data, inherently limiting our ability to train our model.

5.5 Training with synthetic data from kinematic models

As shown in Section 5.1, DNNs with a high capacity, which is achievable with greater depth, are required to drive down the generalization error. To break the inherent limitations of transfer learning from optical imagery, the use of *synthetic data* has been proposed for pretraining.

5.5.1 Modeling radar return with MOCAP data

Accurately synthesizing the radar return is not a trivial task, as there will be scattering from every object or surface in the environment. Depending on the classification problem at hand, however, we will be primarily interested in just the scattering from the target of interest. All other scattering will be result in clutter, or unwanted reflections, in the signal. Thus, a general way of decomposing the received radar return (s_r) is

$$s_r(t) = s_t(t) + s_c(t) + w(t) \tag{5.7}$$

where $s_t(t)$ and $s_c(t)$ represent the target and clutter returns, respectively, and $w(t)$ is the noise. Although more intricate clutter models have been developed [40–42], oftentimes probabilistic models are used to represent clutter and noise, especially the Gaussian probability density function (PDF), which is mathematically one of the more tractable PDFs. In contrast, because targets are generally spatially well-defined, computational electromagnetics and finite element modeling of the electromagnetic scattering [43] can be used to generate a good, although complex, target model.

A simpler, practical approximation to the real target return can be accomplished by decomposing the return as the sum of scattering from point targets. Although this approach ignores multipath and applies geometric approximations to the true RCS of larger volumetric scatterers, for many targets it does provide a sufficient representation of the target response [44].

As a case in point, let us consider the problem of modeling the radar return from the human body. Mathematically, the radar return signal from K point targets is

$$s_t(t) = \sum_{i=1}^{K} a_{t,i} \exp\left\{-j\left[(2\pi f_0)t + \frac{4\pi}{\lambda} R_{t,i}\right]\right\} \tag{5.8}$$

where f_0 is the transmit frequency, λ is the wavelength, t is time, and $R_{t,i}$ is the time-varying range to each point target. The amplitude $a_{t,i}$ can be found from the radar range equation as

$$a_{t,i} = \frac{G\lambda\sqrt{P_t}\sqrt{\sigma_i}}{(4\pi)^{3/2}R_{t,i}^2\sqrt{L_s}\sqrt{l_a}} \tag{5.9}$$

Here, G is the antenna gain, P_i is the transmitter power, σ_i is RCS for each point target, and L_s and L_a represent the system and atmospheric losses, respectively. Typically the RCS for each body part is approximated with the RCS of standard geometric shapes, such as a sphere, cylinder or ellipsoid.

The time-varying range, $R_{t,i}$, can be represented using kinematic models, such as the Boulic walking model [45]. However, animation using motion capture (MOCAP) data provides a more accurate way of representing the entire range of human motion [46], not just walking, and is thus preferred for animation of a broad spectrum of motions. MOCAP data were first exploited to simulate human micro-Doppler signatures by Blasch and Majumder [47,48], who used the Poser^{TM} system to animate human figures, whose electromagnetic properties were modeled with the aide of the Xpatch software. Ram and Ling [49] used data from a MOCAP database to simulate not just human micro-Doppler signatures, but also that of animals, such as dogs and horses, and signatures in through-the-wall configurations [50]. One free source of MOCAP data, which has been widely used for micro-Doppler studies, is the Carnegie Mellon University (CMU) Motion Capture Library [51], which includes a wide range of activities acquired by placing infrared sensors at key joints on the body and recording their time-varying positions. A study [52] on human activity recognition with a 77 GHz radar showed that a k-nearest neighbors classifier exhibited only a 1% difference in accuracy when trained with synthetic signatures generated from CMU MOCAP data versus measured radar data. The main drawback of relying on databases, however, is that they contain a limited number of data samples with no control over the data collection duration, or the characteristics, speed, duration, and trajectory of the test subject.

A practical, low-cost solution for generating MOCAP-based simulated micro-Doppler signatures was proposed [53,54] by Erol and Gurbuz in 2014, which exploited the Kinect sensor for 3D skeleton tracking. While the Kinect-based system may not be as accurate as the MOCAP data acquired by CMU with a costly but high-quality markered system, use of the Kinect sensor greatly eased the process of collecting MOCAP data. The main drawback was that the Kinect still required each measurement to be individually acquired from test subject participation—***the human effort and costs for data collection are no different than if a radar had been operated instead of the Kinect.***

5.5.2 Diversification of synthetic micro-Doppler signatures

A common way to grow the amount of training data while also increasing diversity is with *data augmentation*, applied by using transformations to slightly modify the existing data samples. In image processing, for example, techniques such as translation, rotation, cropping, and flipping can be used to augment datasets. The problem with using such operations on micro-Doppler signatures, however, is that these tend to corrupt the data by generating data samples that map to kinematically impossible target behavior.

MOCAP data, however, provides a way of capturing the underlying target motion that can be exploited to augment the dataset in a way that is kinematically consistent with human motion. Rather than transform the micro-Doppler signature directly, transformations can be applied to the *skeletal model* animated by the Kinect MOCAP data. Scaling the dimensions of the skeleton can model different body sizes while scaling the time axis results in different translational speeds. Individual differences in gait can be emulated by adding perturbations to a parametric model of the time-varying joint trajectories. An overview of the diversification transformations applied to the skeletal model is given in Figure 5.10. Although there is no way to guarantee that the result of such skeletal transformations is fully compatible with the kinematic constraints of human motion, the extent of any discrepancies are greatly limited by the skeletal model. Thus, the diversified MOCAP data provide an alternative source of training data that better captures intra-class variations while providing a systematic way of growing the amount of statistically independent synthetic micro-Doppler signatures.

This proposed diversification methodology [37,55] was used to generate a training database of 32,000 synthetic micro-Doppler signatures from just 55 Kinect MOCAP measurements collected from 5 test subjects performing 7 different activities: walking, jogging, limping, walking with a cane, falling, and sitting. To demonstrate the efficacy of this diversified database for training, let us consider the classification of these seven classes as collected from the 4 GHz CW radar described in Figure 5.3. The diversified database was used to initialize the weights of a CNN, which was later fine-tuned with just 474 real radar measurements. The results comparing pretraining with diversified synthetic data versus training on real data only are shown in Figure 5.11, with configurations exhibiting overfitting highlighted in yellow. Notice that when only real data are used for training, the highest classification accuracy attained for a 90×120-sized input is 87% with a four-layer CNN. But, when the diversified synthetic data are used for pretraining, the performance yielded with a four-layer CNN increases to 94%, with the highest accuracy of 96% using a seven-layer CNN. Thus, we can see that the increased amount of training data, albeit synthetic, has permitted the construction of a deeper neural network with greater capacity and reduced generalization error. However, note that there is a drop in the training accuracy when diversified data training is performed on 11-layer and 15-layer networks. To remedy this problem, residual neural networks have been proposed [13].

5.5.3 Residual learning

Deep residual networks comprise building blocks, which rather than computing the original mapping of $y_l(x) := h(x_l)$ compute the residual mapping of

$$y_l = h(x_l) + F(x_l, W_l) \tag{5.10}$$

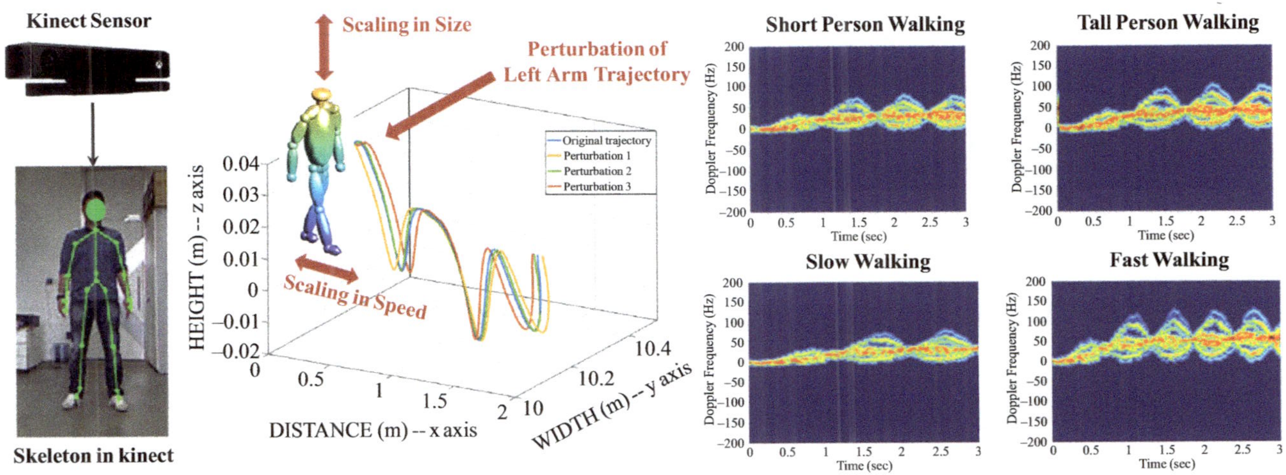

Figure 5.10 Diversification transformations used to augment initial Kinect MOCAP skeleton tracking for training dataset generation [37]

Convolutional Neural Network

7-Classes

	Trained with Measured Data Only						Trained with Diversified Data		
	90 x 120 Pixel Input			327 x 436 Pixel Input			90 x 120 Pixel Input		
No. of Conv. Layers	**VA Max@Epoch**	**VA @500**	**TA**	**VA Max@Epoch**	**VA @500**	**TA**	**Train Acc.**	**VA**	**TA**
2	86% @ 30	83%	84%	85% @ 30	79%	80%	99%	95%	93%
4	86% @ 65	86%	87%	85% @ 25	84%	83%	100%	94%	94%
7	81% @ 151	81%	77%	85% @ 40	84%	79%	99%	96%	96%
11	83% @ 132	79%	80%	82% @ 60	82%	81%	95%	90%	91%
15	80% @ 116	76%	74%	72%@ 13	64%	70%	88%	84%	85%
20	74% @ 97	68%	67%	65%@ 30	61%	61%	100%	78%	80%
25	64%@ 74	52%	49%	69% @ 21	43%	42%	100%	63%	64%
30	14%	14%	14%	14%	14%	14%	100%	14%	14%

Residual Neural Network

	7-Classes			11-Class Generalization		
No. of Conv. Layers	**Train Acc.**	**VA**	**TA**	**Train Acc.**	**VA**	**TA**
5	100%	96%	96%	100%	93%	91%
10	100%	95.6%	96%	100%	94.5%	95%
15	100%	98%	97%	100%	96%	96%
20	100%	95%	94%	100%	90%	91%
25	100%	78%	73%	100%	80.5%	76%
30	100%	14%	14%	100%	9%	9%

12-Class Generalization

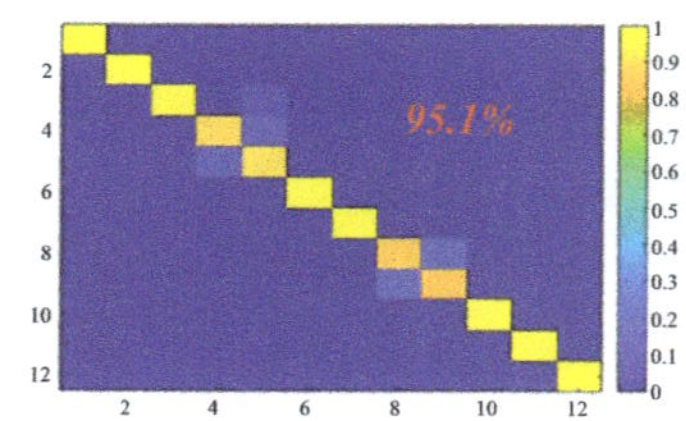

Open Set Pre-Training

Figure 5.11 Classification results for pretraining with diversified synthetic micro-Doppler signatures [37]

$$x_{l+1} = f(y_l) \tag{5.11}$$

where x_l is the input to the lth residual unit (RU), $W_l = W_{l,k|1\leq k\leq K}$ is a set of weights and biases for the lth RU, K is the number of layers in an RU, F is a residual function, $h(x_l) = x_l$ is an identity mapping that is performed by using a shortcut path, and f is an activation function (e.g., ReLU). Batch normalization layers are used to reduce the interval covariance shift and enable the use of larger learning rates [56]. In modified RUs [13], the activation function is placed before the final addition, as opposed to afterward, resulting in a more easily training network with improved generalization properties. RUs allow for the implementation of shortcut connections, i.e., identity mappings, which allows the network to effectively self-adapt its depth: if adding layers give no benefit, identity mappings can be used and prevent "too deep" a network from being implemented.

When a residual neural network is applied to the seven-class problem, it may be seen from Figure 5.11 that the irregularities in training accuracy have been eliminated, and the overall classification accuracy further increases to 97% with a 15-layer DNN. The architecture of this network, called DivNet-15, is depicted in Figure 5.12 and includes 7 RUs with a total of 32 layers (15 convolutional layers).

Another important benefit of utilizing diversified synthetic micro-Doppler signatures for pretraining is that it increases the intra-class variability as well as the ability of the network to generalize the classes outside of those used during pretraining. Figure 5.11 also shows results for applying residual learning for generalizing the 11-class and 12-class problems. In these cases, the pretraining is done over seven classes of synthetic diversified data, while the fine-tuning and testing of the network is done using 11 or 12 classes of real data. Thus, the test sets now include target classes not included in the training set, an open-set pretraining situation. In this case, high classification rates continue to be obtained, with the 95.1% accuracy reported for the 12-class problem being the best achieved on that dataset in comparison to other DNN architectures [57].

5.6 Training with synthetic data generated by adversarial neural networks

While the methods for diversifying MOCAP data to synthesize radar micro-Doppler signatures are very effective at representing the target return for the various types of walking styles and speeds over a range of heights, it does not model the clutter, $s_c(t)$, or noise terms, $w(t)$, in (5.7). For example, consider the micro-Doppler signatures from a 24 GHz radar collected in two different environmental conditions: (1) with the radar having direct line-of-sight (LOS) to the target and (2) with a plywood wall between the radar and target (through-the-wall radar (TWR)), as shown in Figure 5.13. The signatures for each case are given for eight different types of human activities in Figure 5.14. It may be observed that while signatures from both environments contain noise, the TWR configuration has higher power scattering outside of the regions of the target micro-Doppler. Both datasets also contain horizontal noise artifacts due to

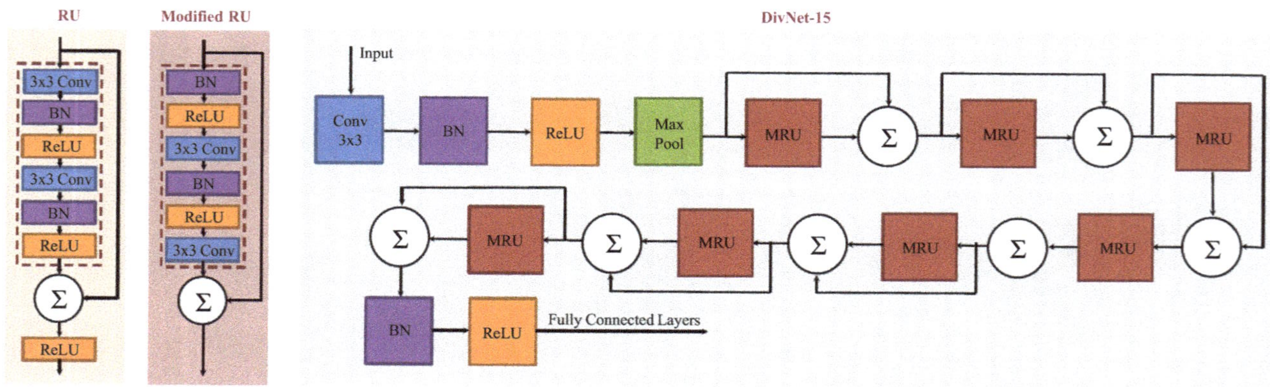

Figure 5.12 Architecture of residual unit (RU), modified residual unit (MRU), and DivNet-15 [37]

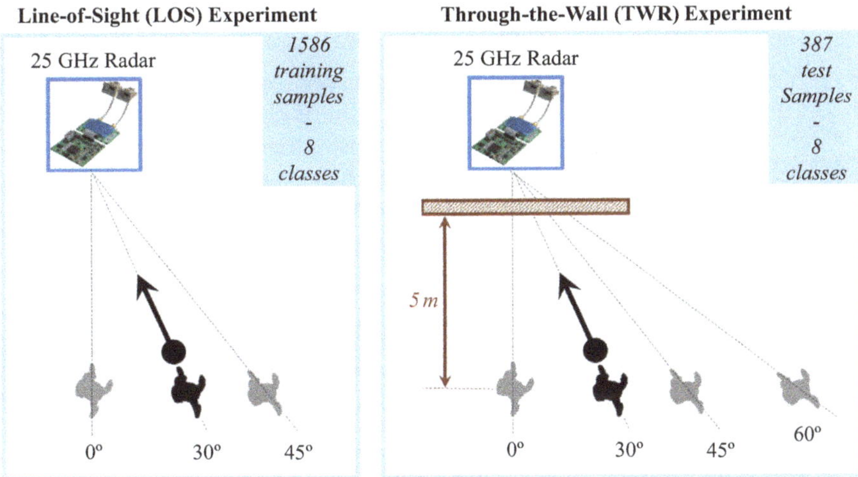

Figure 5.13 LOS and TWR experiment configurations

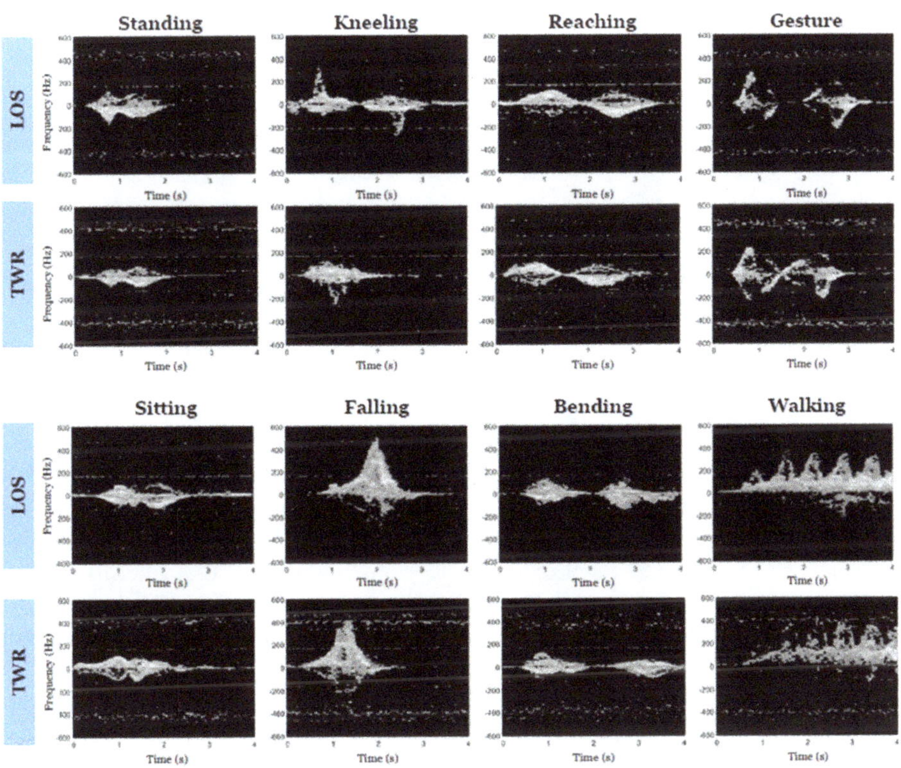

Figure 5.14 Example signatures collected from a 24 GHz radar in LOS and TWR configurations

the RF sensor itself. If synthetic data were generated using MOCAP-based models of target kinematics alone, there would still be significant mismatch between training and test data that would degrade test accuracy. This raises the question then of how best to simulate not just the target response but also the clutter or noise components.

Generative adversarial networks (GANs) have been proposed for synthesizing realistic images in a variety of applications [58], including synthetic aperture radar [59] and micro-Doppler classification [26,60]. The goal of generative models is to take training data of distribution p_{data} and learn a close estimate of it, p_{model}. In this way, generative models attempt to predict features given a certain label. Variational AEs (VAEs) [61,62] are one example of an *explicit* generative model that explicitly assumes a certain prior and likelihood distribution. In contrast, GANs [63,64] use implicit statistical models, providing new data points through simulation, rather than explicit definition of a distribution. The GAN itself is structured as a game-theoretic problem in which two players (the generator and the discriminator) compete in a zero-sum game. The generator seeks to generate samples that are of the same distribution as the input data. Meanwhile, the discriminator aims at classifying whether the resulting image is real or fake. GANs are trained by maximizing the probability that the discriminator makes a mistake. The loss function of adversarial networks can thus be expressed as

$$\min_G \max_D \mathrm{E}_{x \sim p_{data}} \log(D(x) + \mathrm{E}_{z \sim p_z}[\log(1 - D(G(z)))]), \tag{5.12}$$

where $G(z)$ is the synthetic image generated and $D(x)$ is the probability of the image being real. The discriminator is trained to maximize $D(x)$ for images with $x \sim p_{data}$, while the generator tries to produce images such that $D(G(z)) \sim p_{data}$.

5.6.1 Wasserstein GAN

The Wasserstein GAN (WGAN) is a popular variant of the GAN architecture, which employs the 1-Wasserstein distance, also known as the Earth-mover (EM) distance rather than alternative metrics, such as the Kullback–Leibler divergence or the Jenson–Shannon divergence, to quantify the distance between the model and target distributions [65]. The WGAN is advantageous because it provides for a more stable training process, with proven convergence of the loss function and is less sensitive to model architecture or hyper-parameter selection.

The results of applying a WGAN to synthesize radar micro-Doppler signatures for walking are shown in Figure 5.15. Most of synthetic micro-Doppler signatures generated exhibited features that may “look” similar to real micro-Doppler signatures but, in fact, map to impossible kinematic situations or physical occurrences inconsistent with the motion class. Consider the signatures that appear pointed downwards (e.g., samples 2, 3, 5, and 6). While the periodic peaks are characteristic of walking, the pattern in the negative frequency region (lower half of the image) indicates a person walking *away* from the radar, not toward it. In other cases, the signature may not have any peaks at all (e.g., sample 1), which contradicts the necessity for periodic leg swings during walking. In fact, the shapes and periodicities that appear in actual measurements of micro-Doppler for walking are constrained by the

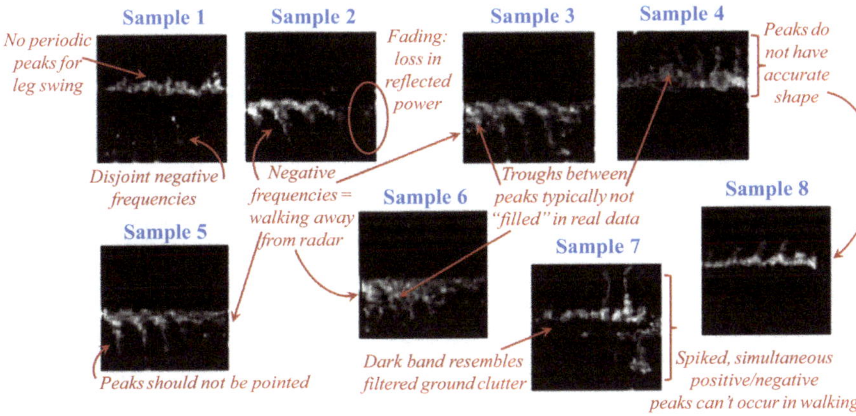

Figure 5.15 Sample walking signatures synthesized by a Wasserstein GAN

kinematics of the motion. But, the network optimization of the WGAN, a network originally proposed for image synthesis, has no information or constraint put upon it that relates to kinematics. Unfortunately, what visually may appear similar can be entirely impossible kinematically. Such discrepancies will lead to models being mistakenly trained, and, ultimately, cause a degradation in classification accuracy.

One way to improve the accuracy of the synthetic signatures generated is to allow the generative model to condition on external class labels, as implemented by conditional VAEs (CVAEs) or ACGANs.

5.6.2 Conditional variational autoencoder

CVAEs are an extension of the vanilla VAE, where the input observations modulate the prior on Gaussian latent variables that generate the outputs [66]. A vanilla VAE consists of an encoder and decoder, implemented as neural networks, with weights optimized according to a loss function. The encoder takes an image as input and outputs a stochastic latent space representation in a lower dimension. It is common for the encoder to output parameters for a Gaussian probability density, which can then be sampled to obtain noisy values of the latent representation. Then, the decoder takes the encoded latent representation as an input and outputs parameters to the probability distribution of the data.

CVAEs have a similar structure, with the exception of additional input branches, called conditions (e.g., external class labels), to both the encoder and decoder. Due to embedding of class labels, the encoder is conditioned on the micro-Doppler signature and corresponding class labels, whereas the decoder is conditioned on latent variables and class labels. The CVAE is typically trained to maximize the conditional log-likelihood. Sample outputs for an implementation of a CVAE [67] as applied to eight-class 24 GHz data (see Figure 5.14) are illustrated in Figure 5.16. In comparison to the signatures synthesized by the WGAN, the CVAE outputs appear to

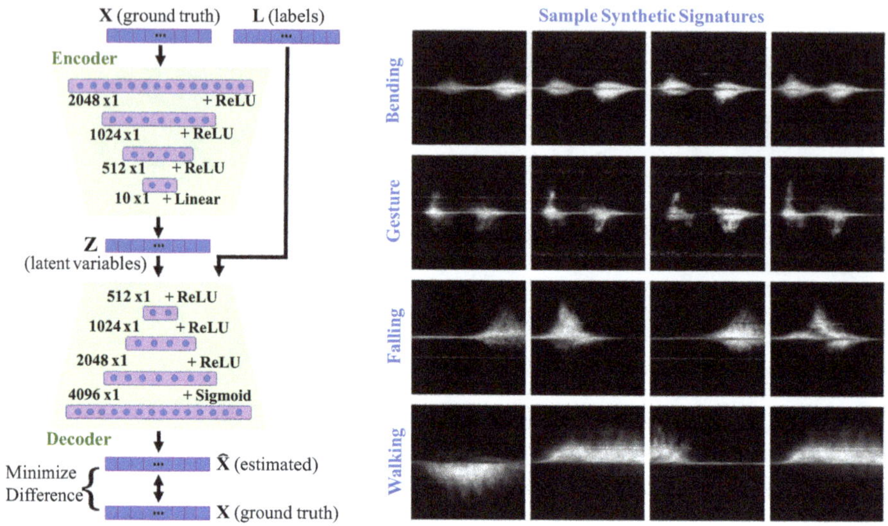

Figure 5.16 CVAE architecture and sample synthetic signatures

be somewhat blurred, although, especially in the case of the gesture, the horizontal, noise-like sensor artifacts have also been captured somewhat. The CVAE appears to have been able to capture the gross features of the data, conforming somewhat to the expected overall shape of the signature. However, details of the internal structure of the micromotions are not visible, indicating that the CVAE is not modeling fine-grain motions adequately.

5.6.3 Auxiliary conditional GAN

ACGANs attempt to improve the quality of the generated images by tasking the discriminator directly with reconstructing labels, as opposed to feeding it class information. This is done by modifying the discriminator to include an auxiliary decoder network, which outputs the class labels for the training data [68]. This modification results in the objective function having two parts: the log-likelihood of the correct source, L_s, and the log-likelihood of the correct class, L_Y. Thus, the discriminator is trained to maximize $L_s + L_Y$, while the generator is trained to maximize $L_Y - L_s$, where

$$L_s = E[\log p(s = real|x_{real})] + E[\log p(s = fake|x_{fake})], \tag{5.13}$$

$$L_Y = E[\log p(Y = y|x_{real})] + E[\log p(Y = y|x_{fake})], \tag{5.14}$$

and s are the generated images. The discriminator is trained in order to maximize the $L_s + L_Y$, while the generator is trained to maximize $L_s - L_Y$.

5.6.4 Analysis of kinematic fidelity and diversity

Let us now visually evaluate the ability of the ACGAN to synthesize micro-Doppler signatures. The specific ACGAN architecture implemented and a few examples of the resulting synthetic samples are shown in Figures 5.17 and 5.18, respectively. Notice that the ACGAN is able to both capture the target signature with greater clarity, sharpness, and richness, as well as some of the noise components. However, some of the similar issues with the accuracy of WGAN signatures remain an issue with the ACGAN-synthesized signatures: disjoint components of micro-Doppler, frequency components opposing the direction of motion, portions of the signature seemingly overflowing, and merging with noise components. There are also instances where the variations generated by the ACGAN map to the signatures of other motion classes, resulting cases where a signature resembling a "progressive fall" is generated under the *falling* class, or a progression of "walk–stop–walk" or walking away from, as opposed to toward, the radar is generated under the *walking* class.

Two essential qualities of a good training dataset are high fidelity with the radar return and diversity. The principal objective in generating synthetic data is to grow the size of the training database. But if the signatures do not accurately map the target classes, or if each of the samples is simply identical, degenerates copies of one

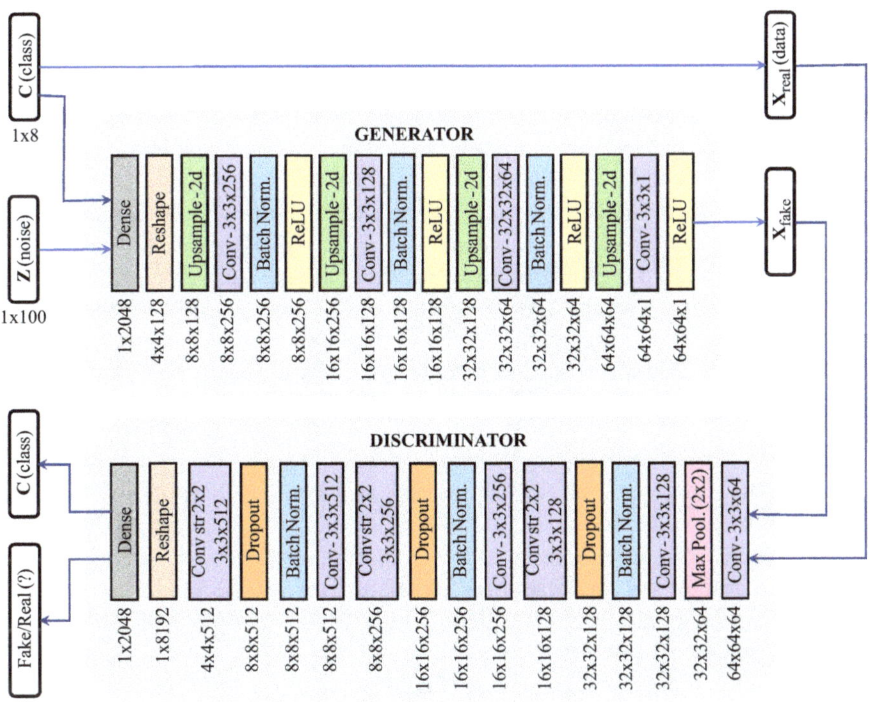

Figure 5.17 Implementation of ACGAN architecture [69]

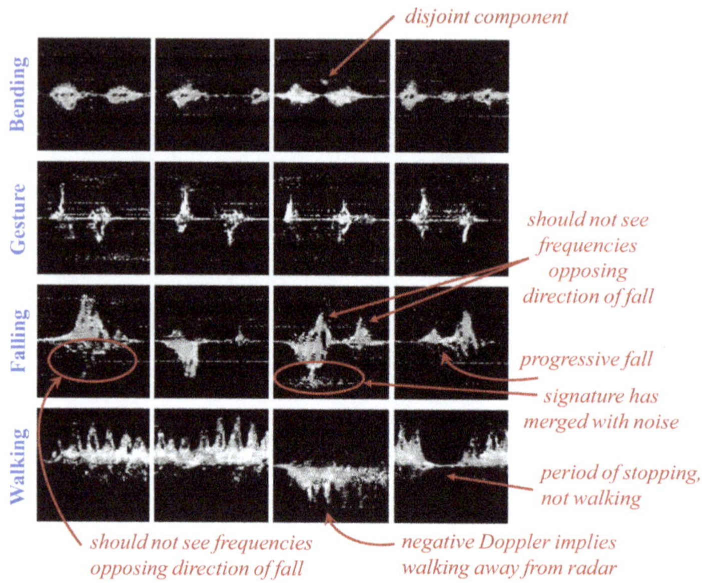

Figure 5.18 Sample signatures synthesized by a ACGAN

another, then the training database will not have the statistical properties desired for a model to accurately learn the underlying data representation. GANs, in particular, do have a tendency to collapse by converging to parameters that result in the generation of replicas.

First, let us examine the kinematic properties of the synthetic signatures for two particular activity classes in comparison to their real counterparts: walking and falling. These represent challenging cases for the ACGAN due to the great diversity within the real training samples as well as the greater richness of frequencies comprising the signature. The kinematic fidelity of the synthetic data can be evaluated by considering its consistency with certain properties that are known to be common to all samples within a class. For example, walking micro-Doppler exhibits the following properties:

1. Periodicity: the cyclic motion of the left and right legs results in a periodic micro-Doppler signature.
2. Leg velocity > torso velocity: walking requires the legs to move much faster than the average speed of the torso; thus, the maximum micro-Doppler frequency of the signature should be greater than the maximum frequency of the strongest return (given by the torso).
3. Frequency composition of signature should match direction of motion: if moving toward the radar, most of the micro-Doppler frequencies should be positive, with any negative frequency components having a very small Doppler frequency. The reverse will be true if moving away from the radar.

These properties can be converted into quantitative rules for evaluating the kinematic fidelity of synthetic signatures. For example, to test walking signatures, the upper and lower envelopes are extracted, from which periodicity can be assessed. The minimum and maximum values of the envelopes can also be used in conjunction with the Doppler frequency corresponding to the peak response (e.g., torso return) to assess the second and third rules. As an example, consider the 12 randomly selected synthetic walking signatures shown in Figure 5.19. The green labels indicate that the synthetic images passed all three kinematic rules, while orange indicates minor issues (i.e., only one or two rules failed), and red indicates that the image fails all rules. When these rules are applied to the 5,000 synthetic walking spectrograms generated by ACGAN, 15% of the signatures failed all three of the kinematic rules. In the case of falling, only the second and third rules are applicable—falling is an impulsive, not a period motion. When these rules are enforced on the synthetic falling samples, 10% of the generated images fail.

The diversity of the synthetic signatures generated can be assessed using a quantitative similarity measured, such as *MS-SSIM* [70]. The MS-SSIM attempts to weight different parts of the image according to its importance for human perception of similarity. Defined over the interval of 0–1, the MS-SSIM metric takes on higher values when two images are very similar. Mathematically, it is defined as

$$\mathrm{SSIM}(\mathrm{x}, \mathrm{y}) = [l_M(\mathrm{x}, \mathrm{y})]^{\alpha_M} \prod_{j=1}^{M} [c_j(\mathrm{x}, \mathrm{y})]^{\beta_j} [s_j(\mathrm{x}, \mathrm{y})]^{\gamma_j}, \tag{5.15}$$

where x and y are the compared images; α_M, β_j and γ_j are used to adjust the relative importance of different components; $l((\mathrm{x}, \mathrm{y}))$, $c((\mathrm{x}, \mathrm{y}))$, $s((\mathrm{x}, \mathrm{y}))$ are defined as the luminance, contrast, and structural comparison metrics, respectively; and M depicts the scale that will be used in the iterative filtering and downsampling.

The kinematic fidelity and diversity of the ACGAN-generated signatures for walking and falling are evaluated as a function of the amount of real training data provided in Figure 5.19. The diversity of the signatures rapidly improves until about 100 samples for both classes, at which point the MS-SSIM metric levels off at about 0.32. The fidelity of the signatures, however, continuously improves with the amount of training samples used. With a total of 350 training samples, or on average 175 samples per class, the percentage of signatures passing all the kinematic rules reaches 90%. Using the ACGAN, a total of 40,000 signatures for 8 classes were generated using 1,586 training samples. This results in a little under 200 samples per class being used for training. The correlation between kinematic fidelity and classification accuracy is evaluated next.

5.6.5 Boosting performance with PCA-based kinematic sifting

The rules that were used to evaluate walking and falling may be generalized for application to any activity class by considering the convex hull of features for each class. Although in general any set of features could be used for this purpose, generalized 2D-principle component analysis (GPCA) provides a computationally efficient approach that is easily computed for any possible motion. By applying GPCA to the

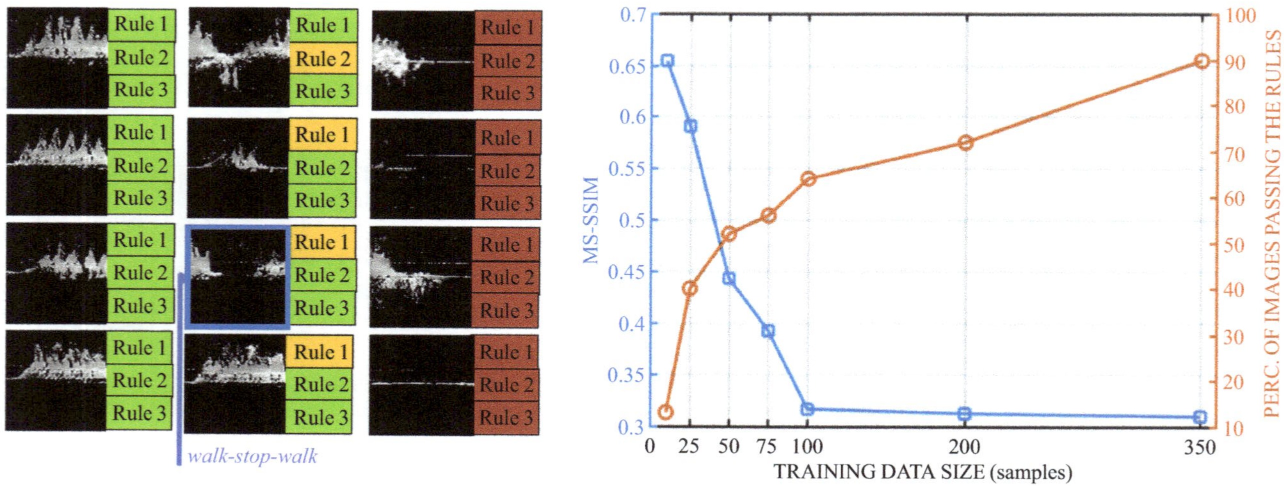

Figure 5.19   Assessment of kinematic fidelity and diversity of ACGAN-synthesized signatures for the walking and falling classes [69]

training data, the typical boundaries of features can be characterized and we can then require that any synthetic data generated fall within a certain tolerance of these boundaries to ensure fidelity with typical class characteristics.

Let us define the real training images for each class as $D_i, i = 1, 2, \dots, 8$. The objective is to find a matrix subspace set $\tilde{\mathrm{U}}_{D_i}^{(1)} \in \mathbb{R}^{I_1 \times P_1}$ and $\tilde{\mathrm{U}}_{D_i}^{(2)} \in \mathbb{R}^{I_2 \times P_2}$ that project the original tensor into a low-dimensional matrix subspace $\mathrm{Y}_m^{D_i} \in \mathbb{R}^{P_1 \times P_2}$ (with $P_1 \leq I_1$ and $P_2 \leq I_2$) defined as

$$\mathrm{Y}_m^{D_i} = \mathrm{S}_m^{D_i} \times_1 \mathrm{U}_{D_i}^{(1)^T} \times_2 \mathrm{U}_{D_i}^{(2)^T}, \tag{5.16}$$

where $\mathrm{S}_m^{D_i}$ is the real training spectrogram from class D_i. The objective function of the GPCA can be written as

$$(\tilde{\mathrm{U}}_{D_i}^{(1)}, \tilde{\mathrm{U}}_{D_i}^{(2)}) = \underset{\mathrm{U}^{(1)},\mathrm{U}^{(2)}}{\arg\max} \sum_{m=1}^{M} \|\mathrm{Y}_m^{D_i} - \overline{\mathrm{Y}_i^D}\|_F^2, \tag{5.17}$$

where $\overline{\mathrm{Y}} = (1/M)\sum_{m=1}^{M} \mathrm{Y}_m$. The core matrix for each m samples can be obtained by projecting the original images using optimized subspaces, $\tilde{\mathrm{U}}_{D_i}^{(1)}, \tilde{\mathrm{U}}_{D_i}^{(2)}$, as

$$\tilde{\mathrm{Y}}_m^{D_i} = \mathrm{S}_m^{D_i} \times_1 \tilde{\mathrm{U}}_{D_i}^{(1)^T} \times_2 \tilde{\mathrm{U}}_{D_i}^{(2)^T}. \tag{5.18}$$

Thus, the feature vector of a training sample for a specific class, m, can be constructed as $\mathrm{C}_m = \mathrm{vec}(\tilde{\mathrm{Y}}_m), \in \mathbb{R}^{1 \times D}$, where $D = P_1 \times P_2$ and vec($\cdot$) is the matrix column-wise vectorization operator. For this implementation, we assigned P_1 and P_2 a value of 2. Finally, the n-dimensional convex hull method is applied to determine the feature space boundaries of the each class. The result is illustrated for the classes of bending and falling in Figure 5.20.

To permit "reasonable" kinematic deviations from the training data, as might happen for individuals of different size, speed, and gait style, the boundaries of the

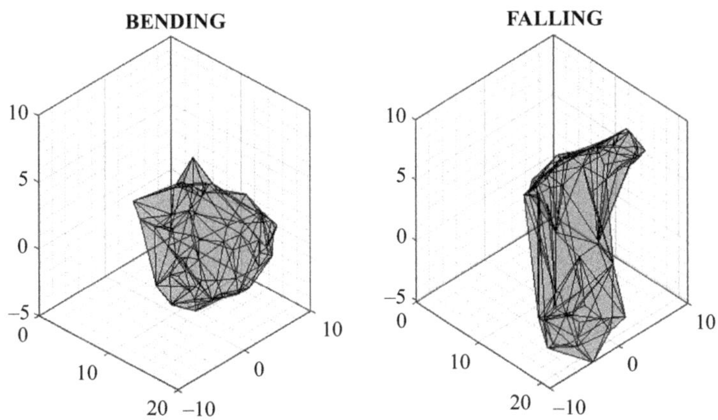

Figure 5.20 Boundaries determined by n-dimensional convex hull for bending and falling [69]

convex hull are not strictly used to evaluate the fidelity of synthetic signatures, but a tolerance is applied. Any synthetic sample falling outside the expanded convex hull boundaries are eliminated from the training dataset. For a tolerance of 0.5, this kinematic sifting resulting in the elimination of 11% of bending, 18% of falling, 8% of gesture, 33% of kneeling, 7% of reaching, 15% of sitting, 36% of standing, 22% of walking. Overall, the 40,000 sample initial synthetic database was reduced to a 31,133 sample database for training.

A comparison of the efficacy of kinematic sifting and various training approaches is shown in Figure 5.21. The 16-layer VGGnet (VGG16) performs well on this dataset providing 84.2% classification accuracy overall. This surpasses both the CVAE and ACGAN *when applied in their original form without eliminating any kinematically impossible samples*. This is a significant result because it shows even a small number of samples that deviate from the physics-based characteristics of a motion class result in degradations so significant that you actually do better pretraining on entirely unrelated optical imagery. If, on the other hand, the quality of the simulations can be improved so that the training database accurately spans the probable target feature space, a significant performance gain can be observed of roughly 9%. The significance of accurate target motion modeling is further underscored by the consideration that DivNet-15, which employed diversification of a MOCAP-based kinematic model, achieved 97% accuracy on 7 classes and 95% accuracy when generalizing to 12 classes. Both these numbers slightly surpass the 93% accuracy given by kinematically sifted ACGAN on an eight-class activity recognition dataset. In a recent study [71] comparing the training efficacy of diversified MOCAP with kinematically sifted ACGAN-synthesized signatures for the aforementioned 8-class scenario, comparable performance was attained when the synthetic datasets were used to train a 50-layer residual neural network.

5.6.6 Integrating kinematics into GAN architecture

Ultimately, the fidelity of synthetic signatures generated with adversarial learning should be improved through modifications of the network itself so that from the outset more accuracy signatures are generated. To consider how this might be accomplished, let us consider modification of the WGAN architecture for the application of gait abnormality recognition—a problem of interest for remote health monitoring. Gait abnormality recognition is a challenging problem because the motion differences that we desire to identify are on a much smaller scale, driving up the need for kinematically accurate synthetic data.

A 25-GHz software-defined CW radar, adjusted to have 2-GHz bandwidth and a 12.8-kHz sampling rate, was used to acquire the micro-Doppler signature of ten health subjects (Persons *A–J*) in an office located at the Technische Universitat Darmstadt, Germany. The radar system was positioned 1.15 m above the floor and participants asked to walk slowly back and forth between 1 and 4 m along the radar LOS. Five different classes reflecting different types of walking were acquired: normal walking, limping with one (L1) or both (L2) legs, walking with a cane in sync with one leg (CW), and out of sync with any leg (CW/oos). The same setup was used to record data of

Performance Comparison	
DNN	**Accuracy**
AlexNet	76.5%
VGG16	84.2%
CVAE	73.2%
ACGAN	82.5%
PCA-ACGAN-TOL-1.0	87.7%
PCA-ACGAN-TOL-0.5	93.2%

Confusion matrix for PCA-ACGAN-TOL-0.5

%	Bending	Falling	Gesture	Kneeling	Reaching	Sitting	Standing	Walking
Bending	100	0	0	0	0	0	0	0
Falling	0	90	3	0	2	0	0	0
Gesture	0	0	96	0	0	0	0	4
Kneeling	20	0	0	80	0	0	0	0
Reaching	0	0	0	0	84	16	0	0
Sitting	0	0	0	2	0	98	0	0
Standing	0	0	0	2	0	0	98	0
Walking	0	0	0	0	0	0	0	100

Figure 5.21 Performance comparison of various DNNs demonstrating efficacy of PCA-based kinematic sifting [69]

four additional subjects with diagnosed pathological gait disorders (Persons K–N) at Villanova University, United States. Thus, gait abnormalities of both a medical nature and nonmedical nature (L1, L2, and CW/oos) due to gait asymmetry were included in this study. A total of 271 abnormal gait measurements were collected [72].

To accurately synthesize gait signatures characteristic of medical or nonmedical abnormal gaits, capturing the motion of the legs and feet is of great importance. Any type of gait abnormality, such as due to a medical procedure (e.g., hip or knee replacement), foot injury, or neuromuscular disorder, is predominantly reflected in the kinematics of the legs, and, hence, the envelope of the micro-Doppler signature. To reinforce envelope features in the adversarial learning process, a multibranch GAN (MBGAN) architecture has been proposed [73]. The MBGAN is supplied not just the synthetic and real-time frequency data, but also the envelope of the synthetic and real signatures, as show in Figure 5.22. Before the full discriminator is trained, an envelope extractor is inserted before the second branch to determine the envelopes for both synthetic and real samples. The inputs to the full discriminator include both the spectrograms and corresponding envelopes in a dictionary format, and each input is passed separately to its corresponding branch. Prior to the softmax layer, the outputs of the two branches are concatenated, resulting in a combined loss function.

The losses are backpropagated through the two branches in parallel. Since these losses will be also used in the training of the generator, with each iteration, the generator will try to adjust the synthetic signature to better match the envelop profile. The resulting synthetic signatures generated by the MBGAN are shown in Figure 5.23. In comparison to the synthetic signatures generated by a standard WGAN, shown in Figure 5.15, the MBGAN signatures can be observed to conform to the envelop. This eliminates many of the previously observed problems relating to disconnected components of the synthetic signature, or pixels that fall beyond the kinematic extent of a real signature.

The improvement in kinematics can be visually observed by comparing the regions spanned by spectrograms for real data versus synthetic data generated by the WGAN and MBGAN, as shown in Figure 5.23. The projections of the images onto a two-dimensional space was done using the tSNE algorithm [74]. The red dots denote data points corresponding to measured signatures of gait abnormality, while the green and blue dots show the synthetic data samples generated by WGAN and the proposed MBGAN, respectively. Notice that the samples for WGAN-generated signatures for the most part do not, in fact, overlap with the space spanned by the real samples. The signatures are visually similar, but not actually representative of the real data. In contrast, the MBGAN-generated samples predominantly overlap with real data samples. While there are some MBGAN samples that deviate, the MBGAN architecture is able to replicate the bulk of the sample space spanned by the real data samples. Hence, a greater percentage of the synthetic data generated will have improved kinematic fidelity to the micro-Doppler signatures of individuals with gait abnormalities.

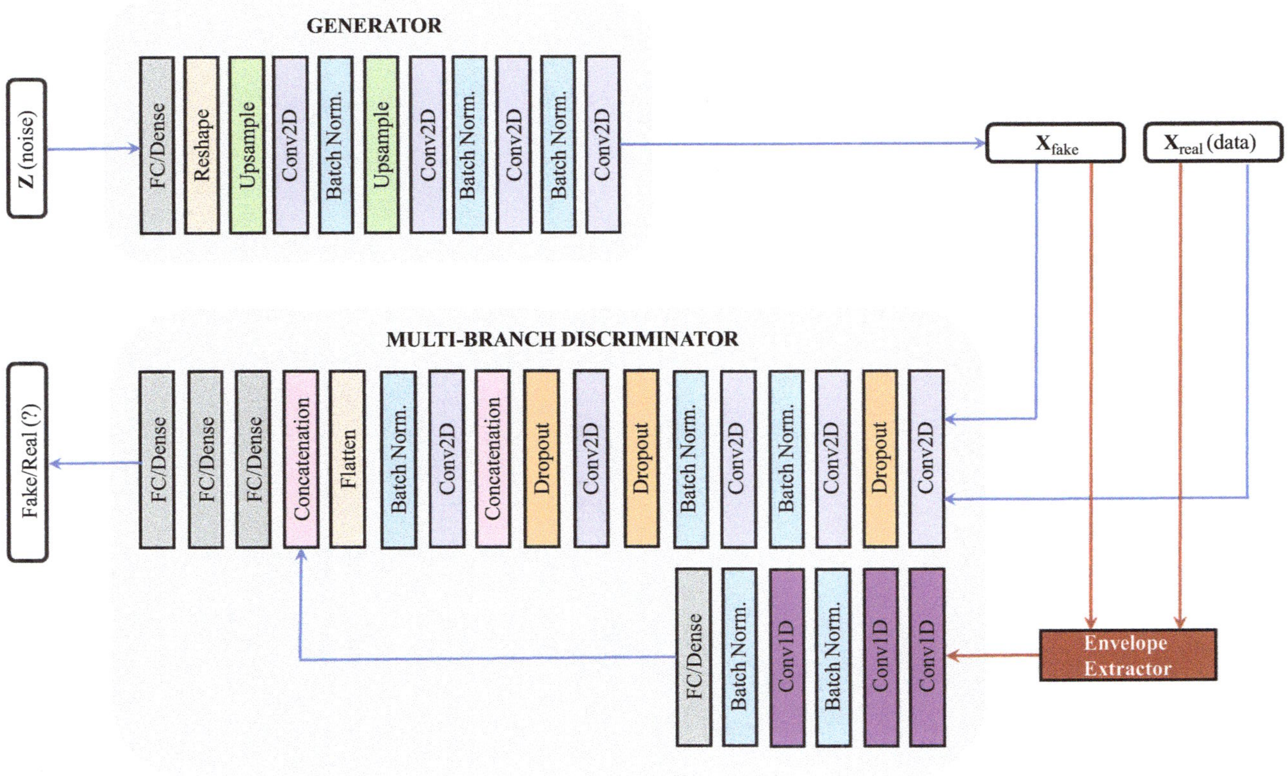

Figure 5.22 Multibranch GAN architecture with embedded kinematics [73]

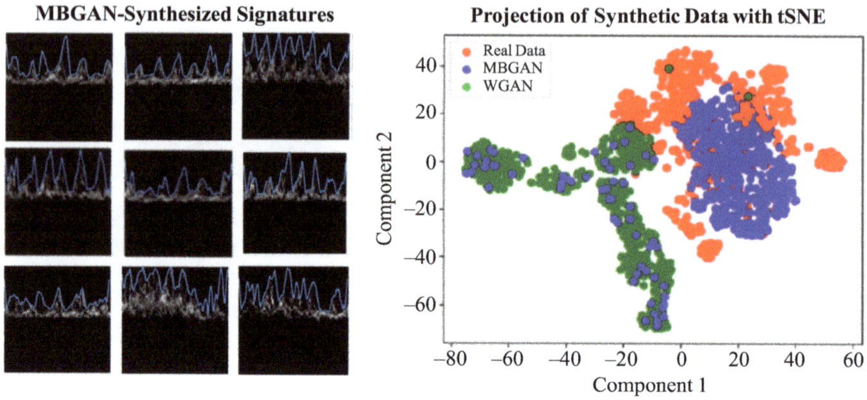

Figure 5.23 Evaluation of MBGAN signature kinematic fidelity using visual observation and tSNE [73]

5.7 Conclusion

A variety of approaches for addressing the challenges involved in training DNNs for the classification of radar micro-Doppler signatures have been presented in this chapter. The performance metrics shown throughout the chapter reveal the impact of DNN training on the accuracy and target generalization performance of the network. Although high accuracies have been attained for the classification of as much as 12 different activity classes, open areas of research remain in regards to exploitation of radar datasets of opportunity and the generation of kinematically accurate, yet diverse, synthetic data. In this regard, adversarial learning provides opportunities, but ultimately both the training strategies and network architecture for training data synthesis must be designed uniquely for radar datasets. Training approaches must consider not only accurate modeling of target and clutter signatures but also must exploit constraints imposed by the physics of electromagnetic sensing to reduce complexity and increase performance. Advances in this area have the potential to greatly expand the application of RF sensors toward human motion recognition in both civilian and military applications.

References

[1] Deng J, Dong W, Socher R, *et al.* ImageNet: A Large-Scale Hierarchical Image Database. In: *2009 IEEE Conference on Computer Vision and Pattern Recognition*, Miami, FL; 2009. p. 248–255.

[2] Abu-Mostafa YS, Magdon-Ismail M, and Lin HT. *Learning From Data.* AMLBook; 2012.

[3] Kay SM. *Fundamentals of Statistical Signal Processing, Volume I: Estimation Theory (v. 1).* Prentice-Hall, Upper Saddle River, NJ; 1993.

[4] Goodfellow I, Bengio Y, and Courville A. *Deep Learning.* MIT Press; 2016. http://www.deeplearningbook.org.

[5] Vapnik V. *The Nature of Statistical Learning Theory.* Springer, New York, NY; 2000.

[6] Shalev-Shwartz S and Ben-David S. *Understanding Machine Learning – from Theory to Algorithms.* Cambridge University Press, New York, NY; 2014.

[7] Joulin A, van der Maaten L, Jabri A, *et al.* Learning Visual Features from Large Weakly Supervised Data. CoRR. 2015;abs/1511.02251. Available from: http://arxiv.org/abs/1511.02251.

[8] Zhang C, Bengio S, Hardt M, *et al.* Understanding Deep Learning Requires Rethinking Generalization. CoRR. 2016;abs/1611.03530. Available from: http://arxiv.org/abs/1611.03530.

[9] Krueger D, Ballas N, Jastrzebski S, *et al.* Deep Nets Don't Learn via Memorization. In: *Proceedings of the International Conference on Learning Representations (ICLR)*, Toulon, France; 2017.

[10] Belkin M, Hsu D, Ma S, *et al.* Reconciling Modern Machine-Learning Practice and the Classical Bias–Variance Trade-Off. *Proceedings of the National Academy of Sciences.* 2019;116(32):15849–15854. Available from: https://www.pnas.org/content/116/32/15849.

[11] Du S, Lee J, Li H, *et al.* Gradient Descent Finds Global Minima of Deep Neural Networks. In: Chaudhuri K, Salakhutdinov R, editors. *Proceedings of the 36th International Conference on Machine Learning. vol. 97 of Proceedings of Machine Learning Research.* Long Beach, California, USA: PMLR; 2019. p. 1675–1685. Available from: http://proceedings.mlr.press/v97/du19c.html.

[12] Berrada L, Zisserman A, and Kumar MP. Training Neural Networks for and by Interpolation. CoRR. 2019;abs/1906.05661. Available from: http://arxiv.org/abs/1906.05661.

[13] He K, Zhang X, Ren S, *et al.* Deep Residual Learning for Image Recognition. CoRR. 2015;abs/1512.03385. Available from: http://arxiv.org/abs/1512.03385.

[14] Chen VC. *The Micro-Doppler Effect in Radar, 2nd Edition.* Artech House, Boston; 2019.

[15] Chen VC, Tahmoush D, and Miceli WJ. *Radar Micro-Doppler Signature Processing and Applications.* The IET, London, UK; 2014.

[16] Jokanović B and Amin M. Fall Detection Using Deep Learning in Range-Doppler Radars. *IEEE Transactions on Aerospace and Electronic Systems.* 2018;54(1):180–189.

[17] Kim Y and Toomajian B. Hand Gesture Recognition Using Micro-Doppler Signatures With Convolutional Neural Network. *IEEE Access.* 2016;4:7125–7130.

[18] Kim Y and Moon T. Human Detection and Activity Classification Based on Micro-Doppler Signatures Using Deep Convolutional Neural Networks. *IEEE Geoscience and Remote Sensing Letters.* 2016;13(1):8–12.

[19] Seyfioğlu MS, Özbayoğlu AM, and Gürbüz SZ. Deep Convolutional Autoencoder for Radar-Based Classification of Similar Aided and Unaided

Human Activities. *IEEE Transactions on Aerospace and Electronic Systems.* 2018;54(4):1709–1723.

[20] Klarenbeek G, Harmanny RIA, Cifola L, *et al.* Multi-Target Human Gait Classification Using LSTM Recurrent Neural Networks Applied to Micro-Doppler. In: *2017 European Radar Conference (EURAD)*, Nuremberg; 2017, p. 167–170, doi: 10.23919/EURAD.2017.8249173.

[21] Cao P, Xia W, Ye M, *et al.* Radar-ID: Human Identification Based on Radar Micro-Doppler Signatures Using Deep Convolutional Neural Networks. *IET Radar, Sonar Navigation.* 2018;12(7):729–734.

[22] Bengio Y and Lecun Y. In: Bottou L, Chapelle O, DeCoste D, *et al.*, editors. *Scaling Learning Algorithms Towards AI.* MIT Press, Cambridge, MA, USA; 2007.

[23] Bengio Y, Lamblin P, Popovici D, *et al.* Greedy Layer-Wise Training of Deep Networks. In: *Advances in Neural Information Processing Systems 19.* MIT Press, Cambridge, MA, USA; 2007.

[24] Erhan D, Courville A, Bengio Y, *et al.* Why Does Unsupervised Pre-training Help Deep Learning? In: Teh YW and Titterington M, editors. *Proceedings of the Thirteenth International Conference on Artificial Intelligence and Statistics. vol. 9 of Proceedings of Machine Learning Research.* Chia Laguna Resort, Sardinia, Italy: PMLR; 2010. p. 201–208. Available from: http://proceedings.mlr.press/v9/erhan10a.html.

[25] Shrestha A, Murphy C, Johnson I, *et al.* Cross-Frequency Classification of Indoor Activities with DNN Transfer Learning. In: *2019 IEEE Radar Conference (RadarConf)*, Boston, MA, USA; 2019. p. 1–6. doi: 10.1109/RADAR.2019.8835844.

[26] Yang Y, Hou C, Lang Y, *et al.* Open-Set Human Activity Recognition Based on Micro-Doppler Signatures. *Pattern Recognition.* 2018;85.

[27] Scheirer WJ, de Rezende Rocha A, Sapkota A, *et al.* Toward Open Set Recognition. *IEEE Transactions on Pattern Analysis and Machine Intelligence.* 2013;35(7):1757–1772.

[28] Cherti M, Kégl B, and Kazakçı A. Out-of-Class Novelty Generation: An Experimental Foundation. In: *2017 IEEE 29th International Conference on Tools With Artificial Intelligence (ICTAI)*, Boston, MA; 2017. p. 1312–1319. doi: 10.1109/ICTAI.2017.00197.

[29] Gurbuz SZ, Rahman MM, Macks T, *et al.* Cross-Frequency Training With Adversarial Learning for Radar Micro-Doppler Signature Classification. In: *Proceedings of the SPIE, Radar Sensor Technology XXIV*, 114080A; 11 May 2020. https://doi.org/10.1117/12.2559155. 2020.

[30] Zhu J, Park T, Isola P, *et al.* Unpaired Image-to-Image Translation Using Cycle-Consistent Adversarial Networks. CoRR. 2017;abs/1703.10593. Available from: http://arxiv.org/abs/1703.10593.

[31] Russakovsky O, Deng J, Su H, *et al.* ImageNet Large Scale Visual Recognition Challenge. CoRR. 2014;abs/1409.0575. Available from: http://arxiv.org/abs/1409.0575.

[32] Krizhevsky A, Sutskever I, and Hinton GE. Imagenet Classification With Deep Convolutional Neural Networks. In: *Advances in Neural Information*

Processing Systems, edited by Pereira F, Burges CJC, Bottou L, and Weinberger KQ. Curran Associates, Inc., Lake Tahoe, Nevada; 2012. p. 1097–1105.
[33] Simonyan K and Zisserman A. Very deep convolutional networks for large-scale image recognition; 2014. arXiv 1409.1556.
[34] Szegedy C, Liu W, Jia Y, *et al.* Going Deeper With Convolutions. CoRR. 2014;abs/1409.4842. Available from: http://arxiv.org/abs/1409.4842.
[35] Park J, Javier R, Moon T, *et al.* Micro-Doppler Based Classification of Human Aquatic Activities via Transfer Learning of Convolutional Neural Networks. *Sensors*. 2016;16:1990.
[36] Seyfioğlu MS and Gurbuz SZ. Deep Neural Network Initialization Methods for Micro-Doppler Classification With Low Training Sample Support. *IEEE Geoscience and Remote Sensing Letters*. 2017;14(12):2462–2466.
[37] Seyfioğlu MS, Erol B, Gurbuz SZ, *et al.* DNN Transfer Learning From Diversified Micro-Doppler for Motion Classification. *IEEE Transactions on Aerospace and Electronic Systems*. 2019;55(5):2164–2180.
[38] Alnujaim I, Alali H, Khan F, *et al.* Hand Gesture Recognition Using Input Impedance Variation of Two Antennas With Transfer Learning. *IEEE Sensors Journal*. 2018;18(10):4129–4135.
[39] Khanna R, Oh D, and Kim Y. Through-Wall Remote Human Voice Recognition Using Doppler Radar With Transfer Learning. *IEEE Sensors Journal*. 2019;19(12):4571–4576.
[40] Darrah CA and Luke DW. Site-specific clutter modeling using DMA digital terrain elevation data (DTED), digital feature analysis data (DFAD), and Lincoln Laboratory five frequency clutter amplitude data. In: *Proceedings of IEEE National Radar Conference*, Ann Arbor, Michigan; 13–16 May, 1996.
[41] Billingsley JB. *Low Angle Radar Land Clutter*, SciTech Publishers; 1991.
[42] Kulemin GP. *Millimeter Wave Radar Targets and Clutter*. Artech House, Norwood, MA; 2003.
[43] Dogaru T and Nguyen L. FDTD Models of Electromagnetic Scattering by the Human Body. In: *2006 IEEE Antennas and Propagation Society International Symposium*, Albuquerque, NM; 2006. p. 1995–1998.
[44] van Dorp P and Groen FCA. Human Walking Estimation With Radar. *IEE Proceedings – Radar, Sonar and Navigation*. 2003;150(5):356–365.
[45] Boulic R, Thalmann NM, and Thalmann D. A Global Human Walking Model With Real-Time Kinematic Personification. *The Visual Computer*. 1990;6(6): 344–358.
[46] Ram SS, Gurbuz SZ, and Chen VC. Modeling and Simulation of Human Motions for Micro-Doppler Signatures. In: Amin MG, editor. *Radar for Indoor Monitoring: Detection, Classification, and Assessment*. CRC Press, Boca Raton, FL; 2017.
[47] Blasch E, Majumder U, and Minardi M. Radar Signals Dismount Tracking for Urban Operations. *Proceedings of SPIE – The International Society for Optical Engineering*. 2006.
[48] Majumder U, Minardi M, Blasch E, *et al.* Radar Signals Dismount Data Production. *Proceedings of SPIE – The International Society for Optical Engineering*. 2006.

[49] Ram S and Ling H. MicroDoppler signature simulation of computer animated human and animal motions. In: *2008 IEEE Antennas and Propagation Society International Symposium*, San Diego, CA; 2008. p. 1–4.

[50] Ram SS, Christianson C, Kim Y, *et al.* Simulation and Analysis of Human Micro-Dopplers in Through-Wall Environments. *IEEE Transactions on Geoscience and Remote Sensing.* 2010;48(4):2015–2023.

[51] Carnegie Mellon University Motion Capture Library. http://mocap.cs.cmu.edu/.

[52] Karabacak C, Gurbuz SZ, Gurbuz AC, *et al.* Knowledge Exploitation for Human Micro-Doppler Classification. *IEEE Geoscience and Remote Sensing Letters.* 2015;12(10):2125–2129.

[53] Erol B, Karabacak C, Gurbuz SZ, *et al.* Simulation of Human Micro-Doppler Signatures With Kinect Sensor. In: *2014 IEEE Radar Conference*, Cincinnati, OH; 2014. p. 0863–0868.

[54] Erol B, Karabacak C, and Gurbuz SZ. A Kinect-Based Human Micro-Doppler Simulator. *IEEE Aerospace and Electronic Systems Magazine.* 2015; 30(5):6–17.

[55] Seyfioğlu MS, Erol B, Gurbuz SZ, *et al.* Diversified Radar Micro-Doppler Simulations as Training Data for Deep Residual Neural Networks. In: *2018 IEEE Radar Conference (RadarConf18)*, Oklahoma City, OK; 2018. p. 0612–0617.

[56] Ioffe S and Szegedy C. Batch Normalization: Accelerating Deep Network Training by Reducing Internal Covariate Shift. CoRR. 2015;abs/1502.03167. Available from: http://arxiv.org/abs/1502.03167.

[57] Gurbuz SZ and Amin MG. Radar-Based Human-Motion Recognition With Deep Learning: Promising Applications for Indoor Monitoring. *IEEE Signal Processing Magazine.* 2019;36(4):16–28.

[58] Creswell A, White T, Dumoulin V, *et al.* Generative Adversarial Networks: An Overview. *IEEE Signal Processing Magazine.* 2018;35(1):53–65.

[59] Lewis B, Liu J, and Wong A. Generative Adversarial Networks for SAR Image Realism. In: *Proceedings of the SPIE 10647, Algorithms for Synthetic Aperture Radar Imagery*, Orlando, FL. vol. 10647; 2018.

[60] Alnujaim I, Oh D, and Kim Y. Generative Adversarial Networks for Classification of Micro-Doppler Signatures of Human Activity. *IEEE Geoscience and Remote Sensing Letters.* 2019:1–6.

[61] Kingma DP and Welling M. Auto-Encoding Variational Bayes. CoRR. 2013; abs/1312.6114.

[62] Doersch C. Tutorial on Variational Autoencoders; 2016. arXiv 1606.05908

[63] Goodfellow IJ, Pouget-Abadie J, Mirza M, *et al.* Generative Adversarial Networks; 2014. arXiv 1406.2661

[64] Goodfellow IJ. NIPS 2016 Tutorial: Generative Adversarial Networks. CoRR. 2017;abs/1701.00160. Available from: http://arxiv.org/abs/1701.00160.

[65] Arjovsky M, Chintala S, and Bottou L. Wasserstein GAN; 2017. arXiv 1701.07875

[66] Sohn K, Lee H, and Yan X. Learning Structured Output Representation Using Deep Conditional Generative Models. In: *Proceedings of the 28th International Conference on Neural Information Processing Systems - Volume 2*, Montreal, Canada; 2015. p. 3483–3491.

[67] Erol B, Gurbuz SZ, and Amin MG. GAN-based Synthetic Radar Micro-Doppler Augmentations for Improved Human Activity Recognition. In: *2019 IEEE Radar Conference (RadarConf)*, Boston, MA, USA; 2019. p. 1–5.

[68] Odena A, Olah C, and Shlens J. Conditional Image Synthesis With Auxiliary Classifier GANs. 2016. Available from: http://arxiv.org/abs/1610.09585.

[69] Erol B, Gurbuz SZ, and Amin MG. Motion Classification Using Kinematically Sifted ACGAN-Synthesized Radar Micro-Doppler Signatures. *IEEE Transactions on Aerospace and Electronic Systems*. 2020;56(4):3197–3213.

[70] Zhou Wang, Bovik AC, Sheikh HR, *et al.* Image Quality Assessment: From Error Visibility to Structural Similarity. *IEEE Transactions on Image Processing*. 2004;13(4):600–612.

[71] Erol B, Gurbuz SZ, and Amin MG. Exploitation of Motion Capture Data for Improved Synthetic Micro-Doppler Signature Generation With Adversarial Learning. In: *Proceedings of the SPIE, Defense + Commercial Sensing Symposium*, Anaheim, CA; 2020.

[72] Seifert A, Amin MG, and Zoubir AM. New Analysis of Radar Micro-Doppler Gait Signatures for Rehabilitation and Assisted Living. In: *2017 IEEE International Conference on Acoustics, Speech and Signal Processing (ICASSP)*, New Orleans, LA, USA; 2017. p. 4004–4008.

[73] Erol B, Gurbuz SZ, and Amin MG. Synthesis of Micro-Doppler Signatures for Abnormal Gait Using Multi-Branch Discriminator With Embedded Kinematics. In: *IEEE International Radar Conference (RADAR)*, Washington, DC, USA; 2020.

[74] van der Maaten LJP and Hinton GE. Visualizing High-Dimensional Data Using t-SNE. *Journal of Machine Learning Research*. 2008;9:2579–2605.

Chapter 6

Machine learning techniques for SAR data augmentation

Benjamin Lewis[1], Theresa Scarnati[1], Michael Levy[1], John Nehrbass[2], Edmund Zelnio[1], and Elizabeth Sudkamp[1]

6.1 Introduction

Automatic target recognition (ATR) for synthetic aperture radar (SAR) is a difficult problem due to the sensitivity of recognition performance relative to operating conditions (OCs). These OCs can be divided into three categories: target conditions, sensor conditions, and environmental conditions. Examples of target conditions include target configuration, component articulations, and target pose. Sensor conditions, including sensor resolution, signal-to-noise ratio, and sample spacing all affect the image scale and composition. As a result, different radars can produce images that vary in significant ways. Environmental conditions such as background clutter, obscuration, and object adjacency can make discerning and classifying targets even more difficult. These OCs can be combined in any number of ways, which indicates that a large amount of imagery is required to represent the variability inherent in the SAR target recognition problem.

For some applications, such as traditional electro-optical (EO) camera imagery, data is plentiful and is representative of the variabilities due to the OCs or nuisance parameters. However, for target recognition applications that use SAR as a sensor, representative, measured data is not readily available and is expensive to collect. This lack of data is exacerbated when applying deep learning to the SAR target recognition problem, as deep learning learns about the variability of target, sensor, and environmental conditions by sampling data over all these conditions. Limited data does not adequately represent the OCs, which causes problems with classifier generalization.

Although deep learning algorithms have been successfully applied to images in the visible spectrum, limited attention has been given to applying deep learning to SAR. Like visible sensors, SAR is an imaging sensor, but there are some significant

[1]Air Force Research Laboratory, Sensors Directorate, Dayton Ohio
[2]Department of Electrical Engineering, Wright State University, Dayton, USA

differences that make the SAR ATR problem unique and will likely require some additional innovation for the successful application of deep learning.

Some of the differences between radar and visible wavelengths favor SAR. A major advantage of SAR is that it is an active sensor that controls its own illumination. In contrast, visible sensors are, in general, passive and must rely on a relatively uncontrolled illumination source. This illumination control allows SAR to produce images in which the scale is independent of range, whereas the scale of visible sensor images inherently depends on range.

Other differences favor EO sensors, which operate at a significantly shorter wavelength. At these shorter visible spectrum wavelengths, most surfaces are rough, whereas the majority of surfaces, including most man-made ones, are fairly smooth. Thus surfaces viewed at visible wavelengths tend to be observable from all viewing orientations, while the same surfaces exhibit specular reflection at radar wavelengths. In effect, this means that many man-made surfaces are only visible from a few viewing locations in the radar domain, generally only when the surface is perpendicular to the sensor line of sight.

The imaging geometry is also different for SAR than for EO. In the EO domain, images always project along the line of sight of the camera. When forming SAR images, the image is a product of the flight path and radar look direction, which creates additional signature variability. Fortunately for SAR, there are a few reflection mechanisms that persist over wide viewing angles and make up the predominant scattering in an SAR target signature. Despite this, the longer wavelength and specular scattering makes SAR signatures much less interpretable and stable than images in the visible spectrum.

The effect of OCs and the physics of the problem also affect the way that deep learning is applied to the SAR ATR problem. One of the common approaches to training deep learning algorithms is data augmentation, where the training imagery is augmented by translations, reflections, rotations, scaling, additive noise, etc. These techniques can help the limited SAR data problem if applied judiciously to a point, but many of the variabilities of SAR, particularly considering target and environment variability, are not amenable to image transforms to a limited amount of data. Hence, a promising form of data augmentation is to add synthetic data to the training set. The advantage of synthetic data is that a wide variety of OCs are straightforward to generate synthetically, while the same configurations are expensive and impractical to collect with measurement systems. Another key advantage of synthetic data is that the labels and accurate metadata about the target, environment, and sensing conditions are inherent in the data generation process. In contrast, the labeling and truthing process for collected data is very expensive and often very time-consuming. Human annotation of SAR imagery is significantly more difficult as the longer wavelength and specular nature of SAR makes the imagery highly nonliteral and, therefore, much more difficult to correctly label.

Unfortunately, electromagnetic data is also very difficult to model, which makes the generation of high-fidelity SAR imagery complex and computationally challenging. Substantial progress has been made on improving both the quality and speed of SAR synthetic image generation. However, even with state-of-the art synthetic SAR imagery, there remains a gap between the measured SAR imagery and synthetic SAR imagery when using these two image sources to train deep learning algorithms. This gap will be explored in Section 6.3.

This chapter investigates several strategies to close the synthetic/measured SAR data gap so that synthetic data can augment measured data to effectively allow deep learning algorithms to be used to perform SAR synthetic data augmentation in support of SAR ATR. These strategies include:

- improvements on the fidelity and speed of SAR synthetic signature prediction (Section 6.2.1.2);
- the development of a measured and synthetic dataset to be used to explore solutions to this problem, including a way to measure progress and compare results (Section 6.2);
- optimization-based preprocessing to smooth speckle and to help normalize the relative amplitudes between measured and synthetic data (Section 6.4.2);
- using joint measured and synthetic training strategies that share features at the lower layers of a deep learning architecture yet contend with the different measurement and synthetic data manifolds at the later classification layers (Section 6.4.3);
- using generative adversarial networks (GANs) to alter synthetic data to look closer to measurement data, in order to make more effective training data (Section 6.4.5); and
- using siamese and triplet networks to learn feature spaces that reduce the gap between measured and synthetic data features while still discriminating among the target classes (Section 6.4.6).

These strategies will be described and results will be presented. These approaches are not the only approaches that can be used to attack this important synthetic data augmentation challenge for SAR. Although progress has been made, many issues still remain as this key data augmentation problem remains an open problem.

6.2 Data generation and the SAMPLE dataset

A key enabler for any work in machine learning is a rich dataset. Unfortunately, extensive radar datasets are difficult to come by. One of the best known SAR datasets was collected during the late 1990s by the Moving and Stationary Target Acquisition and Recognition (MSTAR) program [1], which, despite the name, only contains imagery of stationary vehicles. This program was a collaboration among DARPA, the US Air Force Research Laboratory, and Sandia National Laboratory, as well as a number of contractors.

The publicly available portion of the MSTAR dataset has enabled a large amount of research in SAR ATR. Prior to the widespread use of machine learning algorithms, this research utilized many traditional computer vision techniques, such as engineering feature extractors, leveraging pattern recognition, and designing expert model-based and rule-based knowledge systems [2,3]. These systems perform well under certain circumstances but, as is the case when applied to traditional camera images, require expert knowledge and great effort to maintain.

The advent of convolutional neural networks (CNNs) was a disruptor in this field in the same way as the computer vision field, and much research has been devoted to achieving good SAR ATR performance using these techniques. The technology is quite mature for classifying vehicles within the MSTAR dataset, with typical classification results when training and testing on MSTAR data at about 99% [4,5].

To go further, more data is needed. As discussed in the introduction, data collection is expensive and time-consuming. The most promising and cost-effective form of SAR data augmentation is simulated data, but this is not without problems. There is an inherent divide between measured and synthetic data, mainly due to the way in which data simulation overrepresents some SAR phenomenology.

To more fully investigate these differences, the Synthetic and Measured Paired Labeled Experiment (SAMPLE) dataset was created [6]. This dataset uses a part of the MSTAR dataset and pairs synthetic data with each measured image. Using available data about the position of vehicles during the MSTAR data collects, computer models of each target were created to match this configuration. The targets included in the SAMPLE dataset are listed in Table 6.1. This section discusses methods of data generation, model truthing, and image formation techniques (IFTs) used to create the SAMPLE dataset.

6.2.1 Electromagnetic data generation

In an ideal world, all SAR images would be acquired from real-world measuring systems such as radars on board aircraft (air-to-ground), air-to-air radar, satellite systems, or from laboratory research measurement systems. Unfortunately, creating the required volume of data is generally not feasible. For example, it is not possible in many cases to adequately control radar systems to consistently collect data in the desired configuration, as the air turbulence and wind will bounce an aircraft, resulting in nonlinear measurement collections caused by speed variations.

In some cases, real-world radar measurement is not even possible. It is often desirable to have information about the electromagnetic properties of a product that is still in development, especially when those properties are a key characteristic of that product. Particularly in early stages of development, measurement-ready devices have not yet been manufactured, rendering physical assessment impossible.

In these cases, electromagnetic simulation software is an invaluable tool for the SAR ATR developer. This type of software makes use of Maxwell's equations and a computer model of the target to calculate the response of the target to electromagnetic

Table 6.1 A list of the vehicles and serial numbers included in the SAMPLE dataset

Vehicle	2S1	BMP2	BTR70	M1	M2	M35	M548	M60	T72	ZSU23
Serial	B01	9563	C71	0AP00N	MV02GX	T839	C245HAB	3336	812	D08

radiation. The consistency, speed, and low expense of simulation allow the creation of large datasets that serve to augment the physical collection of radar data and SAR imagery.

Because simulated data is generally meant to be used alongside or to inform real-world data and design, it is important that the software be able to accurately model the physics of real radar systems. As is the case in many software techniques, however, there is a trade-off between performance and fidelity.

6.2.1.1 Simulation types

Electromagnetic simulation codes fall into two different classes. Full-wave solvers, which provide very accurate electromagnetic results, are implemented by solving Maxwell's equations over the surface or volume of interest. Solutions generated by these solvers are very accurate, but the computational requirements for such fidelity are very high and scale exponentially. In contrast, approximate solvers, also known as asymptotic solvers, use methods such as ray tracing to more quickly provide information at the cost of accuracy. The computational cost of approximate methods scales linearly with the complexity of the problem, making it much more feasible for large problems.

Approximate methods become more accurate as the electrical size of the target grows, but these methods perform poorly in regions containing cavities and detailed components. Therefore, hybrid methods exist to take advantage of the strength of both approaches, balancing computational cost and accuracy. These mixed solvers implement full-wave methodology exclusively over difficult regions and use asymptotic methods elsewhere. The two independently calculated solutions are then combined.

The complexity and intricacy of the target model under simulation also influences the type of code used. This is due to the ray sampling density required to adequately simulate the response of a target to electromagnetic radiation. When using computers to model the real world, the continuous electromagnetic waveforms and the computer-based model of the target are discretely sampled. Exactly how these sampling locations are selected is dependent on the simulation code and is beyond the scope of this chapter but, in general, the Nyquist sampling rate is chosen as a fraction of a wavelength $L = c/f$ (e.g., 10 samples per wavelength), where L is the wavelength in meters, c is the speed of light in meters per second, and f is the frequency of the electromagnetic wave in hertz. The sampling rate, in conjunction with the electrical size of the target, determines the number of calculations required to simulate the response of a target to an electromagnetic wave. This computational complexity is directly proportional to the size of the target and the frequency of the radiation. For this reason, simulations of large targets (such as a school bus) simulated at high frequencies (e.g., 10 GHz) require sampling rates that are too costly to solve with full-wave solvers and instead rely on asymptotic or mixed-method software.

6.2.1.2 Simulation workflow

The output of an electromagnetic simulation of a target from one viewpoint produces the equivalent of a single radar echo from a monostatic system. Such radar pulses are a measure of the complex radar cross section (RCS) of an area of interest (AoI) over many different frequencies in both phase and magnitude. A collection of these pulses is referred to as phase history data (PHD).

Collecting multiple radar pulses from a variety of angles allows for a richer electromagnetic view of the target. The additional degrees of freedom enable the data to be assembled into higher dimensional views of the target. A few simple equations, which are summarized in Data Box 6.1, determine the resolution of this data.

A single radar pulse with a given bandwidth B (Hz) is the Fourier transform of a one-dimensional range profile. Range profiles describe the intensity of radar return as a function of distance from the radar. The down-range resolution dr (m) is inversely proportional to the bandwidth according to the equation $dr = c/2B$. If a number of discrete frequencies (N_f) are sampled over the bandwidth, the range profile provides salient data over an unambiguous range extent (m) according to $\mathrm{RE} = (N_f - 1) \times dr$.

To create two-dimensional radar images, radar pulses must be simulated over several different locations. The elevation angle is generally fixed and radar pulses are collected at several different azimuth angles over a collection aperture of θ radians. This extra degree of freedom enables the formation of 2D complex SAR images. Similarly, samples over many elevation angles introduce another degree of freedom with which to create a 3D image.

In the two-dimensional SAR case, the cross-range resolution dx (m) is dependent on the azimuth extent θ and the frequencies present in the radar pulse. If $L_{\max}$ is the wavelength of the lowest frequency of B, then the cross-range resolution is given by $dx = L_{\max}/2\theta$. The unambiguous extent that can be represented by the collection of radar pulses is dependent on the number of radar pulses in the azimuth N_s. This maximum cross-range extent is given by $\mathrm{XRE} = (N_s - 1) \times dx$; any image that is larger than this will exhibit aliasing. This implies that to form a reasonable SAR image, many closely spaced radar pulses must be simulated or collected.

Data Box 6.1

SAR resolution equations

$dr = c/(2B)$ $\quad$ $\mathrm{RE} = (N_f - 1) \times dr$ $\quad$ $L = c/f$

$dx = L_{\max}/(2\theta)$ $\quad$ $\mathrm{XRE} = (N_s - 1) \times dx$

Variable definitions

c—speed of light (m/s)
f—frequency
L—wavelength
B—bandwidth (Hz)
θ—azimuth aperture (radians)
dr—down-range resolution (m)
dx—cross-range resolution (m)
XRE—unambiguous cross-range extent (m)
RE—unambiguous range extent (m)
$L_{\max}$—wavelength of lowest frequency in B (m)
N_s—number of samples in θ
N_f—number of frequency samples in B

Although a true SAR image is a function of a collection of independent radar pulses, electromagnetic simulations may use an approximation method to create a reasonable estimate of these pulses using only the rays traced from one location at the center of the SAR aperture. This single-ray approximation method then extrapolates a sufficient number of additional pulses (the number of which is dictated by the previous equations) by bistatic scattering. This method works best when targeting well-known sensors because knowledge about the sensor can aid in the particular extrapolation method. While single-ray scattering inherently introduces inaccuracies, SAR images may be produced in minutes using this method, magnitudes faster than when computing each radar pulse from its actual geometric location.

A more accurate but costly method of forming SAR imagery is to compute a PHD dome, in which radar pulses are finely simulated over 360° in azimuth and a swath in elevation. In real-world collects, SAR phase history radar pulses are collected at defined points along the radar flight path. With synthetic data, radar pulses from these same flight path points are instead extracted from the data dome to form the phase history. This method is ideal when the target sensor is not well understood or when modeling different sensors, as the specific SAR image formation parameters may be chosen during postprocessing. In contrast, it is difficult to reform the SAR images generated by the single-ray method according to the parameters of a different sensor.

The computational burden of forming a PHD dome is not trivial, however. As an illustrative example, consider a small SAR training set consisting of 30 different targets. Suppose that the data dome for a given articulated target requires 25 samples per degree over 20° of elevation and 360° in azimuth, and that co- and cross-polarizations are required. A data dome for one target would require 18 million ($360 \times 25 \times 20 \times 25 \times 4$) radar pulse simulations. Furthermore, target articulations and changes in ground materials each require the data dome to be completely recomputed. If each of the 30 targets were placed in four articulations (which is a small number of configurations) and placed over three ground materials (e.g., tar, sand, and soil), the entire dataset would require nearly 6.5 billion ($30 \times 4 \times 3 \times 18$ million) radar pulse simulations.

The combinatoric growth of the number of pulses required to form a synthetic dataset means that such simulation quickly becomes computationally expensive. Computing a single RCS return for a given target can take several hours, especially since it is common to simulate 2^n frequencies over the simulated bandwidth (with n typically ranging from 2 to 11). In short, the development of large simulated datasets requires the use of massively parallel high-performance computers (HPCs).

Even when using HPC systems, it is desirable to find ways to accelerate simulations and reduce computational costs. One simple method is to leverage general-purpose graphics processing units to accelerate the ray tracing portion of asymptotic simulations, which constitutes approximately 80% of the computational burden. More complex techniques can reduce the number of ray traces required by making a few assumptions about the geometry of the problem.

In most cases, the variation in ray paths between closely spaced vantage points is extremely minimal. However, the relative distance along each ray path from source to target, and thus the phase of each electromagnetic waves, will change. Rather than recomputing the problem from the start, including the expensive ray tracing, it is much more efficient to instead correct the phase of each ray according to the geometry of the target. This extrapolation method is generally used for samples between computed points (e.g., between samples at 1° increments). For problems with 25 samples per degree (the resolution used to create the SAMPLE dataset), the speedup can theoretically be 625× while providing nearly the same results as computing a ray trace for every sample.

With that said, the extrapolation between points is, in general, discontinuous. For example, rays traced from a 10° azimuth angle will follow different paths than those traced from 11° azimuth. While the paths followed by rays from the 10° pulse may adequately approximate the rays seen from 10.04°, it is less of a valid approximation for the problem as seen from 10.5°. In that case, the ray geometry as seen from 11° azimuth is just as valid an approximation as that from 10°.

To avoid discontinuities as extrapolation locations change, each extrapolated point is computed as a combination of adjacent computed points with smoothing function applied. This smoothing function makes the final simulated data more cohesive. Inter-pulse extrapolation also reduces electromagnetic scintillation from the angular facets of the models, which appears as undesirable cross-range smearing in SAR images.

A useful sanity check when implementing inter-pulse extrapolation is to create and view a matrix of several adjacent pulses. Discontinuities will clearly be present at the boundaries where the smoothing function is not robust, such as in Figure 6.1(a). Likewise, good implementations will produce "fingerprint-like" patterns where there is close agreement between ray shoots and extrapolated points, as in Figure 6.1(b).

For the purposes of this chapter, the end goal of creating simulated data is to convince a detection system that there is no difference between simulated data and measured data. Once this objective has been obtained, the simulated data gains great utility for the end user. Several factors make this objective challenging. Computer-aided design (CAD) models of targets, which are at best approximate representations of real vehicles, are pristine. In contrast, real-world targets are deformed, dented,

(a) (b)

Figure 6.1 An example of the "fingerprint" produced when viewing data generated using the inter-degree extrapolation technique discussed here. The erroneous data in (a) exhibits obvious discontinuities along vertical lines, while the data in (b) is continuous across the entire extent

dirty, and subject to variable conditions. Thus a collection of several measurements of different buses of the exact same make and model will be different.

The material properties used in simulations are also often inadequate. While simulation codes are very capable at accounting for electromagnetic material properties, which can be assigned to individual parts of the computerized target model, good information about these parameters is difficult to obtain. In a similar vein, the ground plane on which targets are placed is generally represented as a trivial flat surface with statistical scattering properties. While this is obviously a poor recreation of the realities of the real world, which include rocks, trees, grass, and potholes, it is simply not feasible to model all these features.

To this end, the work of this chapter is fundamentally about exploring machine learning techniques to overcome the limitations discussed here. For best results, the simulation expert must work closely with the ATR developer throughout the process to ensure that the generated data is adequate for their purposes. Failure to do so ultimately affects the ability of such algorithms to perform adequately in deployment.

For a more complete treatment of SAR geometries and image formation, see [7].

6.2.2 Model truthing

Simulated data generation is heavily influenced by the quality of the CAD model of each target. Electromagnetic simulation works by computing the interaction of simulated radiation with the surface of each model. CAD models are generally represented as collections of triangular facets. The size of each facet impacts how much of a contribution it will make to the overall electromagnetic return of the target. For this reason, CAD models that are more finely faceted are a better representation of actual targets than coarsely defined models, and the radar return from these targets will more closely match that of targets in the real world.

The CAD model's articulation (i.e., the position of movable parts) is also a determining factor that affects the radar return from a target. While no given articulation is inherently better than any other, certain configurations may be of more interest. The radar returns from a vehicle with its door open and shut are much more distinct than returns from a vehicle with its windshield wipers in different positions. With that said, exposing the interior of a vehicle by opening a door or a hood requires much more consideration when modeling the vehicle, as the exposure of internal components (such as the engine) makes the radar return much more complex.

The SAMPLE dataset is based on measured SAR images from the MSTAR collect. In creating the simulated SAR images, it was decided that minimizing the configuration differences between the measured data and the synthetic data was of high value. By so doing, the bulk of the difference between images from the two domains is due to the inherent mismatch between simulation and measurement, rather than controllable configuration differences.

CAD models for targets were sourced from MSTAR program data products. While the MSTAR dataset contains more than the ten targets listed in Table 6.1, the provided CAD models for many of these targets were unfit for electromagnetic simulation. Obvious defects such as major missing parts, excessively coarse representation, or completely missing models disqualified the targets from inclusion in the SAMPLE dataset.

Although the MSTAR program made efforts to configure CAD models to the extent described here, these efforts were incomplete even on the viable models. Large and small differences necessitated work to bring the CAD models into the ideal configuration. This section describes the effort to do so by adjusting the CAD models to match the positions of the chosen MSTAR targets, including collecting and organizing truth data, standardizing the CAD models, and iteratively validating the CAD models with scrupulous comparison to truth data.

6.2.2.1 Truth data

The configuration of each target's CAD model is based on data from the MSTAR data collect. Information such as photographs of vehicles from the data collects, a data handbook written by the Air Force Research Laboratory, data file headers, contemporary reports, and information about baseline vehicle configurations all were collected to serve as ground truth information for each vehicle.

The MSTAR data handbook includes information about the MSTAR extended operating conditions. Of particular use were sections detailing the intentional variations in articulations and target obscurations as well as unintentional variations (such as damaged or missing parts). The handbook also contains lists of the articulation angles of major parts (such as the gun barrel or turret) for each target.

One of the most helpful sources of information was the original MSTAR CAD model reports, created by Sverdrup Technology Inc. These validation reports were created jointly with the CAD models in the late 1990s, and they served as a helpful supplement to the other information.

Perhaps the most useful information came from the contemporary MSTAR photographs (such as the example shown in Figure 6.2(b)). In general, the visual information presented in the photos allowed for rapid configuration of the CAD models. However, some images were taken under suboptimal conditions, such as poor lighting and abundant shadows, which at times made it difficult to see the details in the photos and limited their utility.

(a) (b)

Figure 6.2 An AFacet CAD model of a M60 tank (a) that has gone through the iterative validation process described here. The image in (b) is one of the collected images with which the CAD model was compared. Note the correct position of large elements, such as the turret, and small elements, such as various hatches

6.2.2.2 CAD model standardization

Before model truthing, a CAD model cleaning process was undertaken. Many vehicles exhibited a mismatch between equipment that was present on the vehicle during the collect and equipment represented on the CAD model. Accordingly, the models were adjusted to have the appropriate items. Many models also contained residual solids, or CAD objects that served as temporary stand-ins during the modeling process (such as cylinders in place of wheels) that were necessary to remove.

The CAD models were also converted to a standard file format. Initially, the computer models were stored in a format used by an outdated CAD model program. This format was difficult to work with and the software lacked documentation. To address this issue, the AFacet format [8] and associated software tools were developed in-house. In contrast to the binary format of Modelman files, the AFacet format consists of text files that describe the target and break it into separate part assemblies. Using this format, individual part assemblies can be added, removed, and articulated separately. AFacet also supports conversion to industry-standard CAD file formats, such as .stl and .obj.

To increase the utility of the models, electromagnetic material properties (such as those of glass, metal, or various man-made materials) were assigned to each surface. Many of the specific values for these properties were derived from in-house material properties libraries. Because electromagnetic waves interact with each material differently, the addition of material properties greatly increases the fidelity of simulation results.

6.2.2.3 Iterative model validation

The final step in achieving good computer models was to compare each model to the available MSTAR truth information over several iterations. This was done with a human in the loop, as automated processes were unable to effectively compare the 2D images to the 3D model at the desired fidelity. Each truthing step involved attention to ever finer details due to the wavelengths used in the electromagnetic simulation. Because simulations for the SAMPLE dataset were carried out with a radar center frequency of 9.6 GHz, the median wavelength was on the order of 1 in. (3 cm). In general, electromagnetic waves interact strongly with features that are larger than one wavelength. Thus, it was important to pay attention to the position of objects down to this size.

This stage of model truthing was the most time-consuming, as it required human judgment and repeated assessments of each model. At times, the documented truth was imprecise or self-contradictory. In these cases, a judgment call was made to determine which source of data was most reliable. Over the course of several passes, the CAD models eventually reached a point at which no major disagreements between the truth data and the modeled vehicle could be detected. Due to this validation, the fidelity of the models is very high relative to the MSTAR targets. An example of a validated model of an M60 tank can be seen in Figure 6.2. In theory, this model fidelity directly affects the quality of the electromagnetic simulation and the resulting SAR images.

At the end of this process, the CAD models were well prepared for simulation. Using simulation methods described in Section 6.2.1, a dense data dome was created for each target, consisting of 360° in azimuth and elevations from 13.5° to 18.5°, a range sufficient to cover the azimuths and elevations of the MSTAR targets of interest. Radar pulses were computed every 0.04° in both azimuth and elevation and included the 600-MHz bandwidth, centered at 9.6 GHz, that was used by the MSTAR radars.

6.2.3 Image formation

Matching the CAD models to the configuration of each target as it appeared during the MSTAR collect was a key enabler of the SAMPLE dataset. By so doing, the simulated data domes contain information about each target in the microwave domain in which radar operates. With this data, it is possible to create any number of SAR images.

As stated earlier, a simulated SAR image is formed by computing the points along a flight path at which a real radar would collect radar pulses to create the desired image. These radar pulses are then extracted from the data dome and combined into the phase history. The image is then formed from the assembled PHD.

The SAMPLE dataset includes a matching synthetic SAR image for every measured SAR image. Each of the measured SAR images in the MSTAR collect is the radar view of a target at a given elevation and center azimuth angle. The files that store these images include the complex-valued SAR image and detailed metadata about the conditions under which the data was collected. Included in this metadata are the elevation and center azimuth angle as well as the information required to compute the resolution and number of samples, as described in Section 6.2.1. This data was sufficient to reconstruct the necessary radar flight path corresponding to a given SAR image, which then was used to create the synthetic radar data phase history.

A few preprocessing steps were used to make a more appropriate synthetic SAR image. In particular, a Taylor window function with $\bar{n} = 4$ constant-level sidelobes at -35 dB was applied to the phase history in order to minimize the contribution of sidelobes to the formed SAR image. Because using such a window widens the mainlobe, the equivalent SAR aperture (and corresponding flight path) was widened by a dimensionless factor of 1.184 to compensate. The range resolution is also modified by this same factor, with a range resolution of $(c/2 \cdot 591 \cdot 10^6\text{Hz}) \cdot 1.184 \approx 0.3\,\text{m}$.

Inevitably, most of the points required to create the phase history did not lie exactly on the evenly spaced simulated points. Data at these points was well approximated using bilinear interpolation from the nearest four data dome pulses to the desired viewpoint.

Monostatic SAR images were formed from the constructed phase history using the far-field polar format algorithm [9], a widely used fast Fourier transform–based image formation method that was used during the MSTAR program to form the SAR images. Images were formed with an image size of 128 × 128 pixels with a pixel spacing of approximately 0.2 m in each direction. While most measured images were already 128 × 128 pixels in dimension, the rare larger exceptions were cropped to this smaller dimension. Other image formation parameters are listed in Table 6.2.

Table 6.2 General parameters for the MSTAR image files, which are available in the metadata of the publicly released MSTAR dataset [1]

Range resolution	0.30 m	Bandwidth	591 MHz
Range pixel spacing	0.20 m	Center frequency	9.6 GHz
Range extent	25.8 m	Image size	128 × 128
Cross-range resolution	0.30 m	Polarization	HH
Cross-range pixel spacing	0.20 m	Elevations	14°–18°
Cross-range extent	25.8 m	Taylor weighting	−35 dB

Each measured image was aligned with its corresponding synthetic image to ensure the best match of all parameters between the two images. As the position of the vehicle within the simulated image is well defined, the measured images were aligned with the synthetic images rather than vice versa. For each image, a mask around the target was formed using a nonparametric variance-based joint sparsity (VBJS) method [10]. Outlier pixels were removed with an RANSAC-based [11] algorithm. All pixels within each mask were then quantized into seven quantization levels, of which the lowest two levels were set to zero. This produced a pair of quantized images with reasonably similar shapes and magnitudes. The pixel offset between the pair of images was computed using two-dimensional cross-correlation. The measured image was circularly shifted by the pixel offset amount to align with the synthetic image.

For the purposes of the machine learning techniques described in this chapter, the complex images were converted into magnitude-only images. After converting each image to magnitude by using the absolute value function, the image was normalized to values between zero and one. An image normalization function was then applied by taking the fourth root of each pixel (e.g., $|x|^{1/4}$), which is a common SAR image normalization technique that helps to make the details of the target stand out more. Each normalized image was saved as a lossless image file for later use. One image of each measured MSTAR vehicle and the corresponding synthetic SAR image are shown in Figure 6.3.

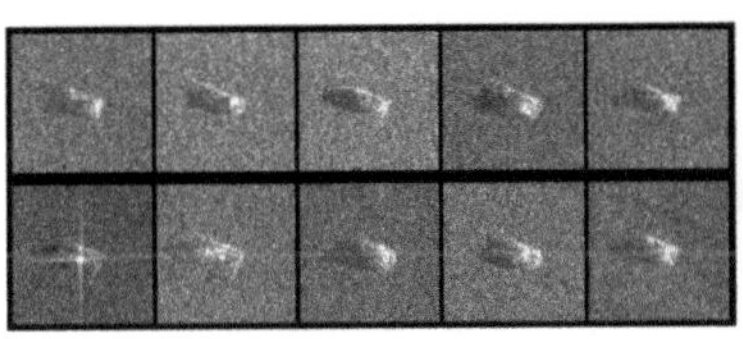

(a)

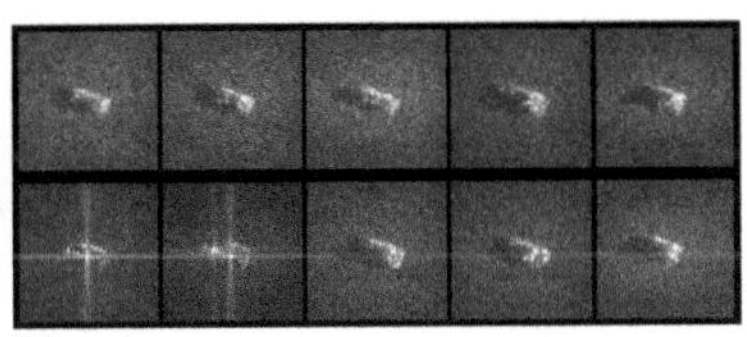

(b)

Figure 6.3 A set of measured MSTAR images (a) and corresponding synthetic SAR images (b) as seen from 17° elevation and 160° azimuth. Each small image within a larger figure represents one of the ten targets from the SAMPLE dataset, with targets presented in the same order in subfigures (a) and (b)

6.3 Deep learning evaluation and baseline

Due to the combinatorially large number of possible views of a given target, ATR in SAR imagery is a difficult task. It is simple enough to build a classifier for a subset of the space, such as the MSTAR dataset. Several papers have reported 99% classification accuracy on MSTAR classification tasks, where the training dataset consists of images collected at 15° elevation and test imagery is from 17° elevation [4,5]. However, the problem of domain adaptation quickly becomes relevant when the training and testing data are not well matched. The difference between synthetic and measured data domains is a prime example of this mismatch.

To illustrate this issue, the t-distributed stochastic neighbor embedding (t-SNE) [12] was calculated on the measured and synthetic portions of the SAMPLE dataset, which was introduced in Section 6.2. The t-SNE algorithm is designed to help visualize the clustering of data in high-dimensional datasets in a comprehensible two-dimensional plane. Points that are close together in the full-dimensional space tend to cluster together in the t-SNE projection, which makes t-SNE a convenient way to visualize the approximate separability of a dataset. The class label of individual points is not used when computing the t-SNE, but it is often beneficial to add the class labels after t-SNE is computed to aid in graph comprehension.

To visualize the measured and synthetic portions of the SAMPLE dataset (which, as the reader should recall, was designed from the beginning to be as well matched as possible), a DenseNet [13] CNN was trained to classify the measured SAMPLE images. Using the trained network, multidimensional features were extracted from the network at the layer before the fully connected classification layer. A joint dataset of features was computed for both the measured and synthetic portions of the SAMPLE dataset. All features were then used as inputs to t-SNE, which computed the two-dimensional embedding for each feature point. The embeddings were then separated according to the source of the image (measured or synthetic) and class labels were assigned. The measured t-SNEs were plotted in the graph shown in Figure 6.4(a), while the synthetic embeddings were plotted in Figure 6.4(b).

As can be seen, the points representing each class cluster fairly well for the measured images represented in Figure 6.4(a). However, this separability decays for the synthetic points in Figure 6.4(b). Because the two embeddings for the two sets of points are the output of the same network, this difference between the graphs suggests that even with great effort to match the collected imagery, the gap between the two domains is nontrivial.

To further illustrate the point, a DenseNet network implemented in PyTorch was used to classify measured SAR images. The DenseNet was modified to accommodate monochromatic images. The first MaxPool layer was also replaced with an adaptive MaxPool layer to allow for smaller input sizes. In the rest of the chapter, these same modifications were made to accommodate the SAR images from the SAMPLE dataset.

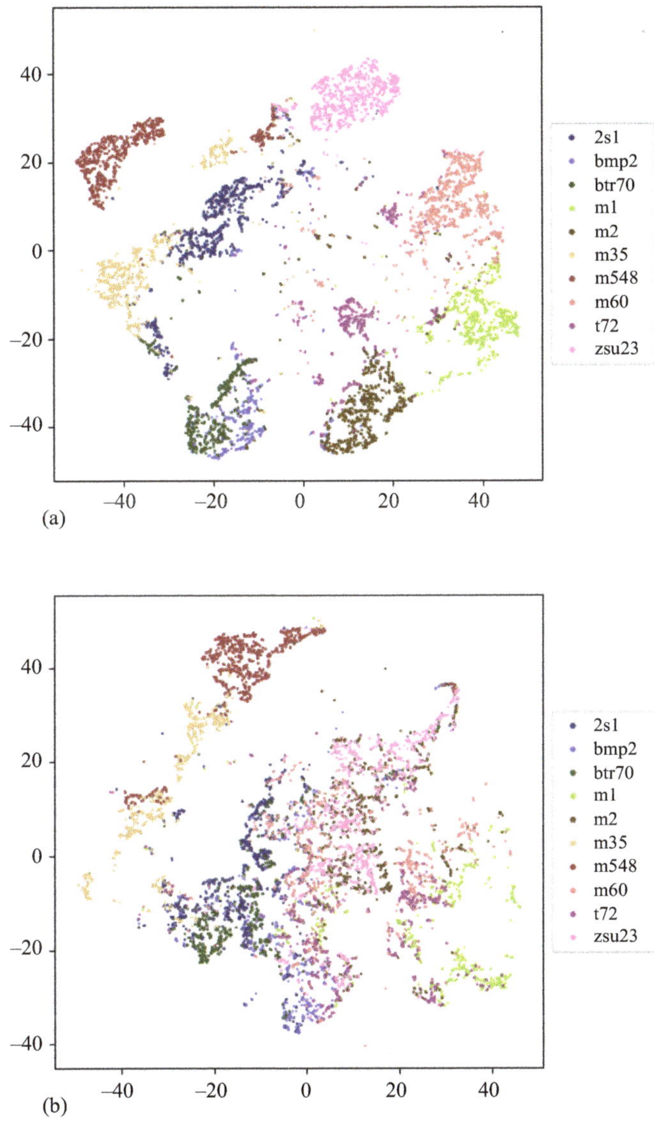

Figure 6.4 A comparison of the synthetic and measured images from the SAMPLE dataset, as represented by the t-distributed stochastic neighbor embedding (t-SNE). In (a), the t-SNEs for the measured images are plotted, while (b) shows the embeddings for the synthetic data. Figure from [16]

Two experiments were performed. In the first, the classifier was trained on measured data, while in the second, the classifier was trained on synthetic data. Measured images were then classified with these trained networks. In each experiment, the DenseNet was trained on randomly sampled batches of 64 images. The Adam optimizer [14] was used with learning rate $\alpha = 0.001$ and beta parameters $\beta_1 = 0.9$ and $\beta_2 = 0.999$. Training was carried out for 50 epochs, several epochs after the validation accuracy had reached near-perfect levels, in accordance with theoretical results in the machine learning literature [15]. Training data consisted of SAR images at 17° elevation, while test data was the measured data collected at 15° elevation.

In the first experiment, a near-perfect test classification accuracy of 97.5% was achieved, as shown in Figure 6.5(a). This accuracy is well in line with results reported in the literature [4,5]. The success of this network is due to the extreme similarities of the train and test dataset, in that both the train and the test data come from the same vehicle and the same collect. Because of this, the only major difference between the two datasets is the elevation at which it was collected. Due to the nature of the MSTAR dataset, OCs such as weather, target articulation, radar characteristics, ground type, and many more were essentially the same for the train and test sets of images.

In the second experiment, where the DenseNet was trained on synthetic data and tested on measured data, the strong performance of the first experiment was significantly degraded. The confusion matrix in Figure 6.5(b) shows that the accuracy of the network dropped to 51.5%. This illustrates the same point as the t-SNEs in Figure 6.4—namely, that the two domains are distinct and not easily bridged.

While more sophisticated and SAR-specific networks may provide better classification performance on this problem, the point is that even well-matched synthetic and measured data can have significant differences that hurt the generalization of ATR algorithms. Despite great effort devoted to matching CAD models to collected truth, there will always be a mismatch between the two domains due to judgment calls, approximations made by computational electromagnetic software, and inaccurate truthing information. These imperfections motivate the use of machine learning to try to bridge the measured/synthetic gap.

In the rest of the chapter, several techniques will be presented, each of which aims to make measured and simulated data more interchangeable. As experimental results are presented, DenseNet classification results will be presented and compared to the cross-domain performance presented in Figure 6.5(b) where appropriate. DenseNet classification can appropriately be applied to techniques in which the synthetic data is altered to look more like measured data in some way. For example, layer retraining (Section 6.4.3), autoencoder-related techniques (Section 6.4.4), and GANs (Section 6.4.5) are all well suited to direct classification methods (although the results presented with layer retraining do not use the DenseNet architecture). In contrast, siamese and triplet networks (Section 6.4.6) do not directly adjust the data but rather work to learn appropriate data representations for both domains of data. As such, DenseNet results will not be directly presented for these latter methods. Instead, ways to leverage siamese and triplet networks as classifiers will be discussed.

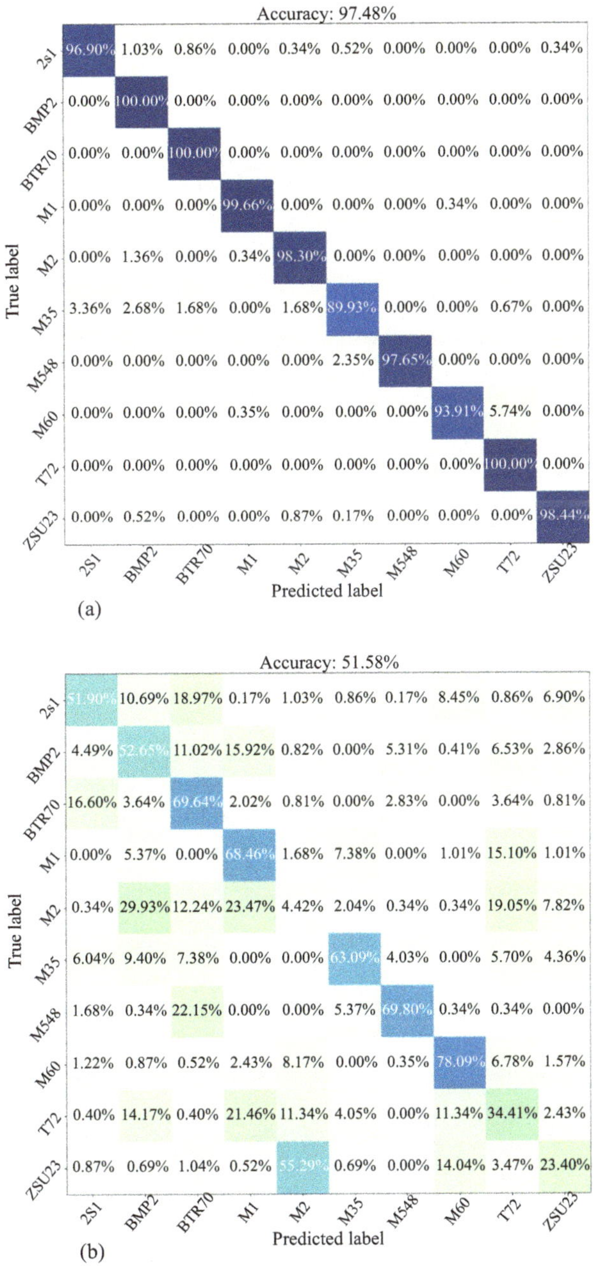

Figure 6.5 A comparison of using synthetic and measured data to train a classifier. In (a), a DenseNet was trained using measured data, while in (b) the same network was trained with synthetic data. The confusion matrix shows the probability of correct identification when classifying measured data with each network

6.4 Addressing the synthetic/measurement gap with deep neural networks

For the reasons described in the previous sections, the problem of augmenting an SAR dataset with synthetic data is nontrivial, and no claim is made that the work in this chapter fully solves the problem. In essence, bridging the difference between the two domains is a type of transfer learning problem.

Over the past few years, the authors have investigated several methods to address the measured/synthetic SAR data gap. Before discussing machine learning algorithms, a few data preprocessing routines are examined. These include a discussion of data augmentation measures that are productive and sensible for SAR imagery (Section 6.4.1) as well as a brief foray into SAR image despeckling techniques (Section 6.4.2). Removing the speckle and background from SAR images is an attractive preprocessing step that removes much of the nontarget information from SAR images; in theory, this information removal allows the network to learn target signatures without undue influence from nontarget portions of the image. On the matter of data augmentation, some typical data augmentation schemes do not make sense in the context of the physics of SAR images.

The remainder of the section discusses machine learning approaches to bridging the synthetic/measured gap. While space does not permit more than a cursory overview of each method, the authors have published conference papers on each topic, which are referenced in the appropriate section. These approaches include layerwise training of networks (Section 6.4.3), autoencoders (Section 6.4.4), GANs (Section 6.4.5), and siamese and triplet networks (Section 6.4.6). Each subsection discusses the architectural concepts of the network, including any major modifications made to support radar data, and presents results using the network to solve the synthetic/measurement gap. In the case of off-the-shelf networks, some light modifications were made to allow the network to operate on single-channel images of size 128×128 pixels.

While several approaches have been taken by the authors, much work remains to effectively use synthetic data when training machine learning classification algorithms. Future developments will be published by the authors as they become viable.

6.4.1 SAR data augmentation

In the machine learning community, a number of data augmentation techniques are generally prescribed as a way to help expand the variety of the dataset in a computationally inexpensive manner. Some of these methods are at times not appropriate for use with SAR data, because the image formation process ensures that the structure of the image data is somewhat constrained.

In particular, flipping, rotating, and scaling data augmentations should be used judiciously and only with good reason on SAR data. SAR images may be formed such that the illumination from the radar is seen from a constant direction in the image. As all images can be formed looking from the same direction, flipping or rotating images merely to create additional views of the target is not necessary. Likewise, the measured spacing between pixels is a defined parameter when forming images. Thus, scaling images up or down by large amounts is useful only in limited contexts.

Other traditional data preprocessing techniques are of more use with SAR data. The position of the target within the image and the size of the image are not essential characteristics, so cropping images around the center or translating the target within the image are valid ways to augment the data. Image cropping tends to work well when classifying images, because much of the background speckle noise is no longer available to provide nontarget information to the network. Target translation can also aid networks in learning position invariance. While there are many ways to fill in data that is removed in translation, a viable way to do so with SAR imagery is to circularly shift the image such that the edge that is removed from one side is added to the other side. This is acceptable because the background contains no significant information about the target, is often noisy enough to mask the discontinuity between the two edges, and should not produce a measurable effect on the classification of the image.

Data augmentations that modify the pixel values in images are also appropriate for use with SAR imagery. Adding random noise to images can help with classification robustness as well as aid convergence in generative techniques. Transforming the range of pixel values to lie between -1 and 1 is also good practice, as it ensures that all pixels are in the same order of magnitude and prevents extreme gradients within the network.

6.4.2 Mathematical despeckling

When viewing an SAR image, it is apparent that the brightness distribution of the image is not smooth but rather composed of complicated granular patterns of bright and dark spots. This salt- and pepper-like noise, which is called speckle, is the result of the radar illuminating a surface that contains features that are rough when compared to the length of the illuminating wavelength. For example, it is common in SAR for the resolution cell to be at least an order of magnitude larger than the transmitted wavelength. The coherent summation of returns within the imaging cell from objects much smaller than the cell creates a return pattern of varying intensity. Speckle noise varies based on the environment being imaged. From one image to the next, even if the same target is being imaged, speckle noise is independent.

Even though independent speckle noise adds a lot of information to each image, machine learning systems have a huge capacity to fit a model to data and can still overfit this noise. The nature of convolutional networks also means that these networks do not inherently possess the ability to focus on one area of an image in particular. As such, machine learning networks can easily learn to categorize images based on the background rather than the actual target, which is an obvious impediment to generalization.

One way to mitigate this problem is to eliminate the background clutter and speckle. This is an important task, as the clutter found in measured data is independent of the simulated clutter found in synthetic data.

Understanding speckle requires a detailed examination of the properties of electromagnetic waves after they have been reflected or scattered from rough objects [17]. Because of the difficulty in understanding such details of the small-scale structures of the complex wavefronts leaving the objects, it is common to instead generate a statistical speckle model. Such models predict the statistical properties of intensity over an ensemble of different rough surfaces with the same macroscopic properties, but different in microscopic detail. There are many examples of speckle models in the literature [10,18–24].

A statistical model of speckle can be derived from first principles. Speckle appears in images as a multiplicative noise term that affects the RCS of each pixel [25]. The intensity of the multiplicative speckle term is characterized on a per-pixel basis by a negative exponential distribution. If the RCSs are known, then the speckle can be fully characterized using first order statistics [25]. These statistical characteristics can then be exploited by speckle reduction techniques.

Speckle reduction methods can broadly be categorized as IFTs or postprocessing methods. Reducing speckle during image formation is typically more flexible, as the techniques are applied to unprocessed data. Even so, postprocessing is also an important signal processing technique that can help filter irrelevant information from images.

When raw data is available, image formation despeckling techniques, including regularized inversion [24], variational methods [21,23], and sparsity-based methods [10], are best able to preserve the information in the SAR image while removing speckle. These algorithms leverage the data to inform how the speckle noise should be reduced, all prior to introducing assumptions that affect the final image. Statistical speckle models and other prior knowledge about the scene may also be included when forming the image.

After the SAR image has been formed, postprocessing methods, such as filtering [19], multi-look averaging [18], and quantization [26–28], affect the individual pixels of the image in advantageous ways. These methods typically are much faster computationally but tend to reduce the amount of relevant information within an image. Often, the target in a postprocessed image loses its structure and edges become blurred, which diminishes the information.

One useful IFT is high-order regularization, which was posed in [29]. In [30], speckle reduction via high-order regularization was explicitly studied and it was shown qualitatively and quantitatively that this technique successfully reduces speckle in SAR imagery while maintaining target fidelity.

VBJS [10,32] image formation was adapted for speckle reduction of SAR imagery. This method eliminates the need to hand-tune regularization parameters by exploiting the overlapping support of multiple measurement vectors (images) in the domain where the signals are assumed to be sparse. Figure 6.6(b) displays the result of applying the VBJS technique to measured and synthetic (top and bottom images of Figure 6.6(a), respectively) SAR data.

An example of a postprocessing technique is quantization. This has been explored in many template-based classification algorithms [26–28]. There are two main steps in the quantization procedure. First, a mask is placed over the AoI. In SAR images,

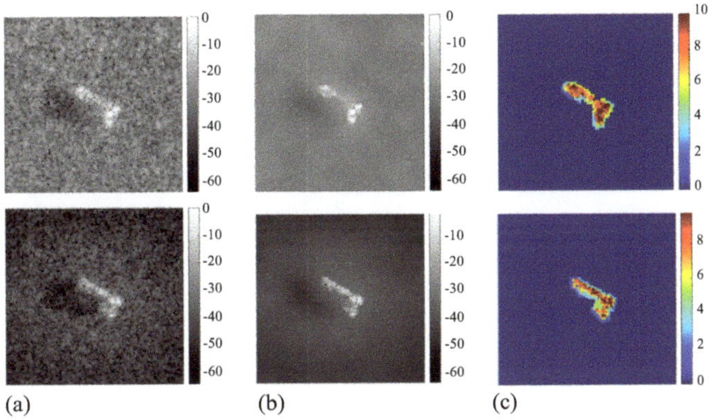

Figure 6.6 *(a) Speckled images. (b) VBSJ IFT. (c) VBJS IFT with the quantization into n = 10 bins. (top) Measured SAR data and (bottom) matching synthetic SAR data. Figure from [31]*

the AoI is typically considered the target region, but at times, the shadow region is also included in the AoI. While it is a trivial exercise in bookkeeping to generate such a mask for synthetic images, AoI masks for measured data must be hand-drawn or discovered algorithmically. Algorithmic approaches, in particular, are crucial when handling large datasets. After the AoI mask is given, the pixels within the mask area are organized into n bins. It has been shown that for SAR images $n = 4$ [28] and $n = 10$ [26,27] work well for template-based classification algorithms. However, similar research has not been done for machine learning classification algorithms.

Quantization can be applied after speckle reduction during image formation. Figure 6.6(c) displays the result of applying quantization with $n = 10$ bins after despeckling the SAR images via the VBJS. Here, the AoI masks are an intermediate output of the VBJS algorithm.

To illustrate the merit of despeckling SAR images for classification, a DenseNet [13] network was trained on four different image types (Figure 6.7). The experiment was run with images that were (i) unaltered, (ii) despeckled via VBJS, (iii) despeckled via VBJS and quantized with $n = 4$, and (iv) despeckled via VBJS and quantized with $n = 10$. The experiment was designed to show the performance of the classification algorithm as a function of the amount of measured data available during training. Various combinations of measured and synthetic data were used to train the network, which was then used to classify measured data. Each combination of data is defined by the fraction of the measured data that was included in the training set.

When reading the chart, we see that the network trained on original images performed better than those trained on despeckled images. However, this is an unfair comparison because, with the original data, the classification algorithm has the opportunity to learn not only the target signatures, but also the environment parameters that dictate the speckle noise. As all of the measured data was collected under similar conditions, there is a significant amount of shared information among these images.

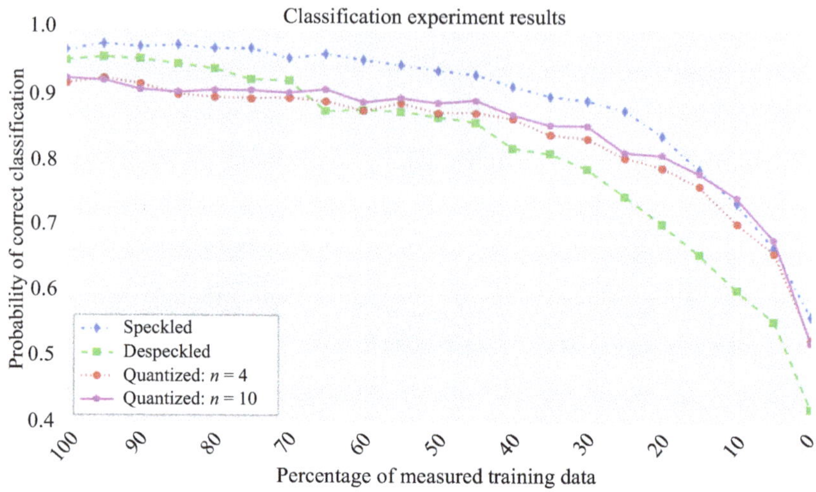

Figure 6.7 Classification performance as the percentage of measured data within the training set decreases and the percentage of synthetic data used for training increases for four different image types. Figure adapted from [31]

That shared information is attenuated by the despeckling process. Yet in the context of classification, this focused removal of information has the potential to broaden network generalization by allowing the network to solely focus on target signatures.

For each experiment, the classification performance decreased as the percentage of measured training data decreased and the percentage of synthetic training data increased. For each image type, performance dramatically dropped when half of the training data was synthetic. However, Figure 6.7 shows that despeckling with the VBJS algorithm and then quantizing the result into $n = 10$ bins yields the highest classification performance when more than 35% of the training data are synthetic. These results show that SAR image despeckling techniques are a promising preprocessing technique for deep neural networks [31].

6.4.3 Layer freezing and layerwise training

Based on the classification performance of networks that are trained on synthetic data and tested on measured data, it is clear that there is a fundamental disagreement between the two domains and how each target is represented. The scattering in SAR is predominantly local and predominantly specular, which means that flat surfaces are generally not seen in radar returns [33]. However, at the rare viewing angles where specular surfaces are visible, simulations tend to overpredict the amplitude of the return. These differences in the geometry of the model vs. the real target and the fact that the prediction codes are approximate create a gap between SAR model predictions and target measurements. Unless accommodations are taken as part of the data preparation and CNN design process, these differences between synthetic and measured data can degrade the performance of the classifier.

While CNNs have shown success on classification problems in multiple domains, limited data is still an issue. Transfer learning technique [34], in which information from a network trained with large amounts of data is adapted to a problem with limited data, is one way of dealing with this limitation. One popular technique is to use layers of a network trained on a rich dataset, such as ImageNet [35], then retrain just the fully connected classification layers on the new data while keeping the feature extraction layers frozen. Unfortunately, an analog to ImageNet is not available for SAR, so this particular method is ineffective for the problem at hand.

Despite the differences between measured and synthetic data, there are still many similarities between the two datasets, and neural networks certainly have the capacity to leverage that shared information. The experiment in this section leverages a simple CNN architecture based on LeNet [36] to perform joint domain training. The different layers of the network are presented with data from either the measured or the synthetic data in order to learn features from both domains. In general for these experiments, some or all layers of the network are trained primarily on synthetic data, while measured data is used to train the first layers. By so doing, the layers nearest the input learn information that is unique to the measured data while the later layers learn the overall structure of the targets.

Although more recent architectures such as ResNet [37] and DenseNet [13] have shown improved classification performance and faster training convergence, the separation of lower and higher layers in the network architecture is less clear. In the more advanced networks, the layers become mixed either through skip connections, such as in ResNet, or through the concatenation of early layers with later layers, as in DenseNet. LeNet's simple architecture is straightforward to analyze as the small number of layers can be more easily interpreted as corresponding to simple/local to more complex/global features. However, the LeNet derivative architecture used for this study is updated with advances since the LeNet architecture was developed, including batch norm [38], dropout [39], and layer widening [40], to provide higher performance.

During training, the network (shown in Figure 6.8) was presented with several combinations of measured and synthetic data. The training began with a simple 50/50 split of both data domains into training and testing. This split, rather than a split on

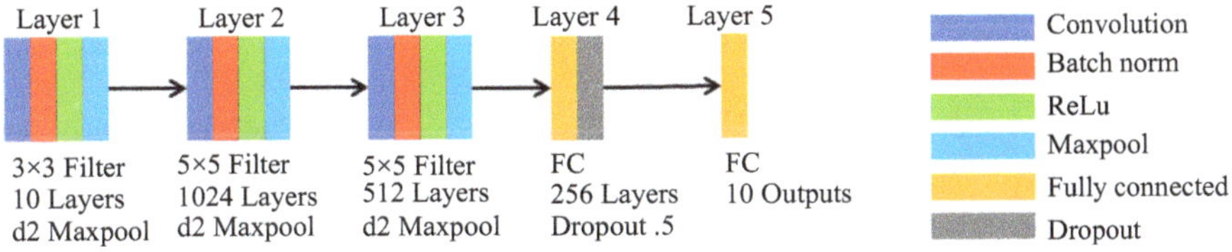

Figure 6.8 Illustration of the architecture used in this study, including the composition and parameterization of each layer

elevation, was used to allow the network to learn from the synthetic data as much as possible over a wide variety of data. Doing so is a good fit for the problem at hand.

During testing, the synthetic test data was not used. Furthermore, measured data for the T72 and ZSU23 was excluded entirely during training. By so doing, this experiment serves to test the ability of this transfer learning strategy to leverage synthetic information to correctly classify measured data.

Each experiment was defined by the set of layers that were trained only on measured data, the set trained on synthetic data, and the set of layers trained on both types of data. Each training batch was presented to the network in up to three parts (measured, combined, or synthetic data). Notationally, this procedure is defined as a combination of $M^{set\_of\_layers}$, $X^{set\_of\_layers}$, and $S^{set\_of\_layers}$, respectively, where *set_of_layers* denotes the layers of the architecture that were trained with that data. The layers not indicated by the notation were frozen during that stage of training. For example, M^{12345} denotes that all layers of the network were trained and updated by measured data. $M^{12345}S^{45}$ denotes that the CNN was trained initially with measured data on all layers then subsequently trained with synthetic data where only Layers 4 and 5 were updated and the first three layers were frozen. $S^{12345}X^{1}S^{2345}$ denotes that the network was trained with synthetic data on all layers, then the first layer was trained with mixed measured and synthetic data, and finally the last four layers were trained with synthetic data while Layer 1 was frozen.

For each training stage of the training sequence, stochastic gradient descent was used to optimize the weights, with an initial learning rate of 0.001 on a cosine weighted schedule [41]. The batch size was 21 and the number of epochs was 150. A cross-entropy loss function with soft targets [41] was used.

Thirty different training sequences were evaluated and are given in Table 6.3. Along the top of table are the headings: Layer 1; Layer 1,2; and Layer 1,2,3. These denote the layers that are frozen for synthetic data and updated for the measured or

Table 6.3 A description of the 30 different experiments performed and the classification accuracy (orange text). For the results, the first number gives the overall probability of correct identification when tested on measured data, while the second number gives the performance on the withheld T72 and ZSU23 classes.

	All layers		Layer 1		Layer 1, 2		Layer 1,2,3	
Meas—All Targets	M^{12345}	99.0, 100						
Syn	S^{12345}	69.4, 62.7						
Syn, Mea				74.7, 65.5	$S^{12345}M^{12}$	74.6, 4.35	$S^{12345}M^{123}$	80.22, 1.0
Syn, Mea, Syn			$S^{12345}M^{1}S^{2345}$	75.3, 73.4	$S^{12345}M^{12}S^{345}$	79.1, 71.7	$S^{12345}M^{123}S^{45}$	90.7, 65.9
Mea	M^{12345}	79.2, 0.0						
Mea, Syn			$M^{12345}S^{2345}$	86.1, 54.7	$M^{12345}S^{345}$	85.9, 53.7	$M^{12345}S^{45}$	90.2, 48.3
Mea, Syn, Mea			$M^{12345}S^{2345}M^{1}$	85.7, 45.7	$M^{12345}S^{345}M^{12}$	80.7, 10.3	$M^{12345}S^{45}M^{123}$	80.4, 4.6
Mix	X^{12345}	86.6, 36.4						
Mix, Syn			$X^{12345}S^{2345}$	91.2, 60.1	$X^{12345}S^{345}$	90.9, 59.4	$X^{12345}S^{45}$	91.6, 62.6
Mix, Syn, Mea			$X^{12345}S^{2345}M^{1}$	90.6, 57.6	$X^{12345}S^{345}M^{12}$	87.7, 42.5	$X^{12345}S^{45}M^{123}$	87.8, 40.3
Mix, Syn, Mix			$X^{12345}S^{2345}X^{1}$	90.8, 58.2	$X^{12345}S^{345}X^{12}$	89.3, 50.7	$X^{12345}S^{45}X^{123}$	87.6, 40.2
Syn, Mix			$S^{12345}X^{2345}$	74.8, 66.9	$S^{12345}X^{345}$	80.6, 35.1	$S^{12345}X^{45}$	88.1, 42.2
Syn, Mix, Syn			$S^{12345}X^{1}S^{2345}$	74.9, 71.4	$S^{12345}X^{12}S^{345}$	83.9, 70.4	$S^{12345}X^{123}S^{45}$	90.1, 52.2

mixed data, which is also represented in the notation in the table. The labels on the left side of the table describe the training sequences used for a particular training strategy.

For each experiment, the results are listed next to the layer code in the orange-colored pair of numbers. The first number in each cell indicates the classification accuracy across all ten measured targets, including the held-out T72 and ZSU23. The second number presents the accuracy on just the held-out classes. The results in the first row titled "Meas—All Targets" demonstrates that the CNN architecture is capable of getting near-perfect results given measured training data for all targets. The rest of the results correspond to networks that were trained on all classes of synthetic data and eight classes of measured data.

The table illustrates that the "Layer 1" column gives the highest second percentage, or transfer learning performance, for the T72 and ZSU23. It also illustrates that the "Layer 1,2,3" column gives the best total measured performance but at the expense of the transfer learning performance. In addition, the "Syn, Meas, Syn" row gives the best overall performance of all the sequences tested. The results represented by the experiments in the "Syn, Meas, Syn" row for "Layer 1" and "Layer 1,2,3" are further detailed by the confusion matrices shown in Figure 6.9.

Based on the results in Table 6.3, the first layer has the most generic and most transferable features. In contrast, when the third layer is included, the features become less transferable but more discriminative. These observations are consistent with layer complexity analysis performed using layer visualization for visible wavelength sensors [42]. Note that the M2 target was problematic. For the Layer 1 experiment in Figure 6.9(a), the M2 clearly had poor performance, and in the Layer 1,2,3 experiment in Figure 6.9(b), the M2 was a clear confuser for the ZSU23.

Although 30 different training strategies were compared and clear trends were observed, there are clearly many additional training strategies and networks that could be investigated using these training principles. The absolute numbers, though encouraging, are likely not reflective of current state of the art or future potential. Even so, care was taken in the experimentation so that relative performance numbers are informative. The lessons learned from building on the two trends of layer transferability and sequence transferability (represented by the columns and rows of Table 6.3, respectively) may provide fruitful directions for future research in transfer learning as applied to the synthetic/measured gap challenge.

6.4.4 Autoencoders

Autoencoders are one of the simplest deep learning algorithms. This type of network is designed to compress input data to a vector representation (which is generally but not always smaller than the input data) then decompress this representation back to a close reconstruction of the input data. Autoencoders are composed of a connected encoder and decoder network (see Figure 6.10), which are generally fully connected or convolutional networks. The loss metric for the network penalizes reconstruction errors between the input and the reconstructed data. Autoencoder networks are useful because they extract the most salient features of data and represent them as simple numerical vectors.

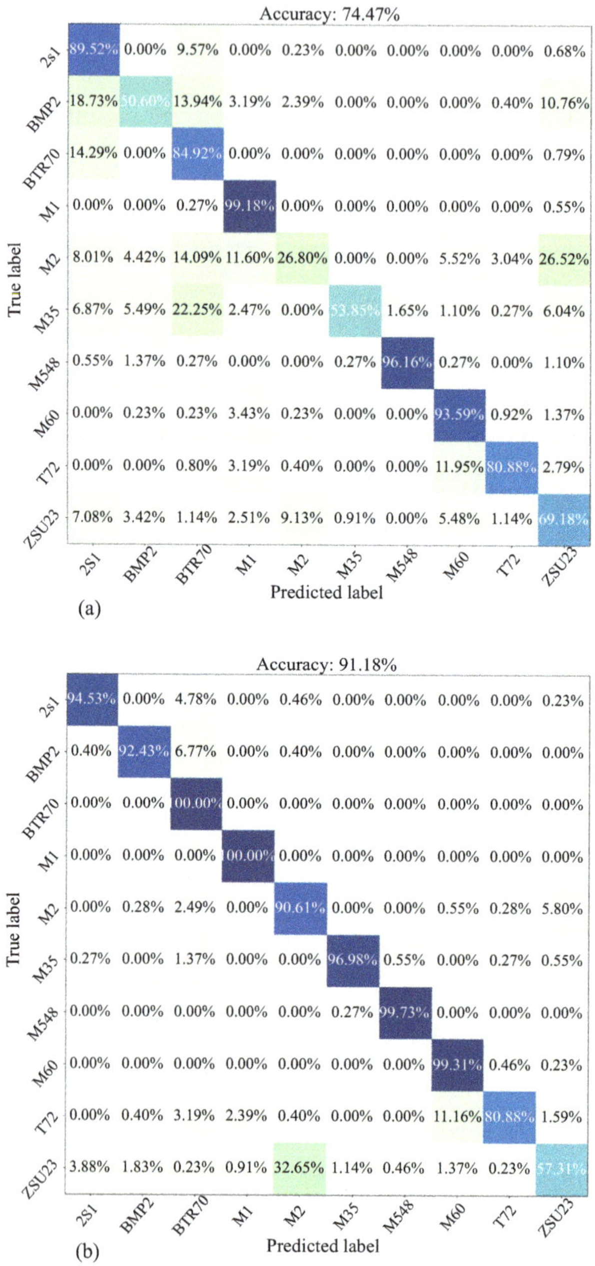

Figure 6.9 Confusion matrices corresponding to $S^{12345}M^1S^{2345}$ (a) and $M^{12345}M^{123}S^{45}$ (b). The weighted accuracy for the withheld T72 and ZSU23 classes in (a) is 73.4%, and the weighted accuracy for the same two classes in (b) is 65.9%

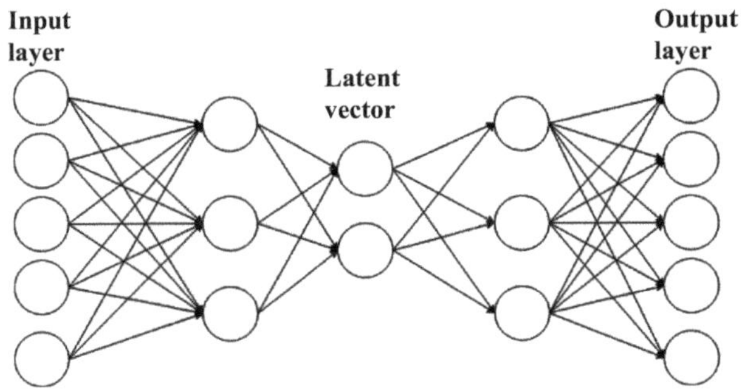

Figure 6.10 A graphical representation of an autoencoder. Autoencoders are composed of layers that compress the input to a one-dimensional latent vector then decompress the latent vector back to a reconstruction of the input. The structure of the network is generally symmetric around the latent vector

In theory, this power to extract the most salient features of an image can help represent the two domains in a single, combined way. An autoencoder trained on measured and synthetic data should learn to represent both domains, although imperfectly. While the latent representation of each domain would not necessarily have any inherent adjacency for the two domains, the decompressed output of the autoencoder would represent, in some way, a unified image space. A classifier trained on the decompressed data would then enjoy a measure of agnosticism to the source of the data.

In practice, though, this is difficult to achieve. A simple autoencoder is not sufficient to bridge the synthetic/measurement data gap. In particular, missing data (e.g., one target not being represented) in one of the two domains means that the autoencoder is unable to effectively represent that target in both domains.

To show this, a convolutional autoencoder was trained with a mix of synthetic and measured data, with the goal of learning an intermediate representation of the data. Alternating batches of the two types of data were presented to the network during training. An L2 loss was applied to the output of the network, which computed the difference between the output of the network and the input image. A second L2 loss was computed between the output and the equivalent image in the other domain, where available, to help the network learn to map to the other domain as well. By evaluating the loss against both images, the network was encouraged to learn domain independence in its representation. After training for 100 epochs, inference was performed on held-out data.

When the network was trained on all classes of data, the difference between autoencoded synthetic and measured data was minimal, as shown in Figure 6.11(a). When measured data for two targets, the T72 and BTR70, was excluded from training, artifacts from the synthetic imagery were propagated during inference (see Figure 6.11(b)).

A DenseNet classifier was then trained on synthetic autoencoded data then tested on measured autoencoded data. Although the two autoencoded manifolds should be

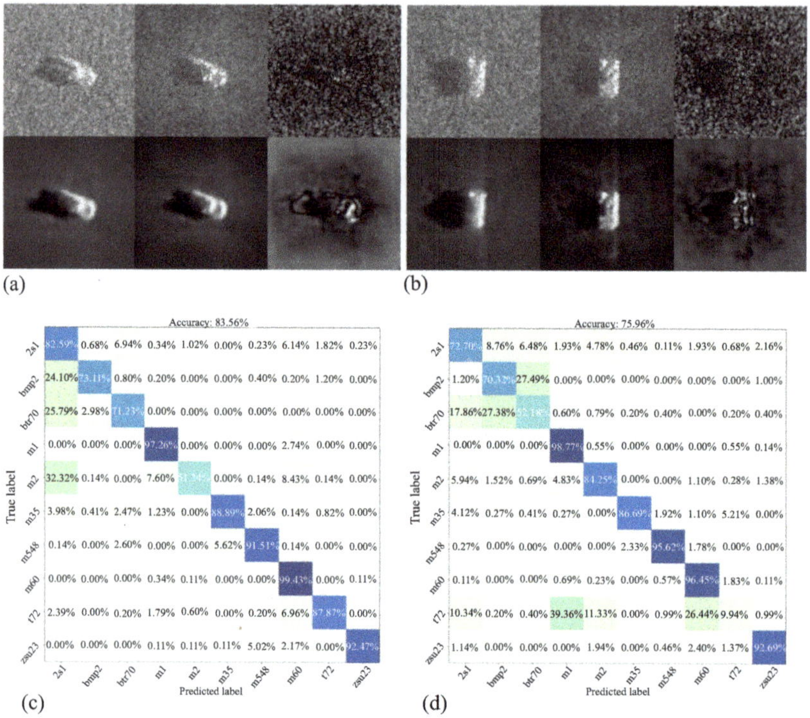

Figure 6.11 Output from an autoencoder designed to represent measured and synthetic images in an intermediate space. Blocks (a) and (b) show sample images of a T72. The top row shows the input images, while the bottom shows the autoencoded images. The left column shows real images, middle shows synthetic images, and the right shows the difference (autoscaled to emphasize differences; block (b) has greater disparity). For (a), all measured and synthetic images were present during training. In (b), the T72 and BTR70 measured data was replaced by synthetic data. Blocks (c) and (d) show confusion matrices when classifying the output of the autoencoder in experiments in (a) and (b). Note the poor performance on the withheld targets in (d)

very similar, the classification results show that this is not the case. In particular, the classification accuracy for withheld classes (Figure 6.11(d)) was unimpressive, and even as bad as chance guessing for the T72.

As a more advanced architecture is clearly needed to address missing data, a "manifold-joining autoencoder" was developed. This architecture, shown in Figure 6.12, builds on the idea of an autoencoder by adding a convolutional discriminator. The autoencoder portion of the network functions in a traditional way, i.e., by reconstructing input data. The addition of the discriminator adds an additional

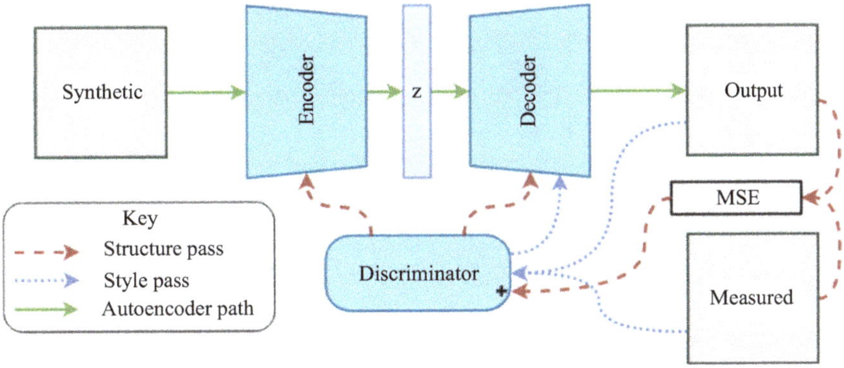

Figure 6.12 A manifold-joining autoencoder described and pictured in [16]. This architecture extends the traditional autoencoder by adding a discriminator network. The discriminator works to distinguish between an autoencoded synthetic image and measured images, which allows the network to adjust the style of the autoencoded image

requirement that the autoencoded image look like measured data. Doing so encourages the network to break the symmetry between the input and the reconstructed image. When trained, the manifold-joining autoencoder serves to convert synthetic data into the measured realm, where it can more efficiently be used to train a classifier for measured data.

The manifold-joining autoencoder was trained in three repeated steps, each designed to make a part of the architecture perform in a certain way. In the first step, the discriminator was trained, using a binary cross-entropy loss, to distinguish between unaltered measured and simulated images. By so doing, the network learned the difference between the two domains, which can provide useful gradient information at later steps.

In the second step, the autoencoder was trained on synthetic images. A mean-squared error between the input $\mathcal{S}$ and its reconstruction $\hat{\mathcal{S}}$ was computed. The output $\hat{\mathcal{S}}$ was also classified by the discriminator as measured or synthetic, with a "synthetic" designation penalized and a "measured" label rewarded. These combined losses served as a gradient for the encoder and decoder portions of the autoencoder, with the discriminator loss given a few orders of magnitude less weight. This step is represented by the red arrows in the figure.

Finally, the synthetic image was passed through the updated autoencoder. $\hat{\mathcal{S}}$ was classified by the discriminator, and a binary cross-entropy loss was computed again. In contrast to the previous steps, this loss was only used to update the weights of the decoder, serving to enforce a measured-like style to the images. This step is represented by blue arrows in the figure.

After the model was trained, it was used to convert synthetic data to the measured domain. These outputs are shown in Figure 6.13, which shows a similar experiment to that performed with the simple autoencoder. It can be seen that the manifold-joining

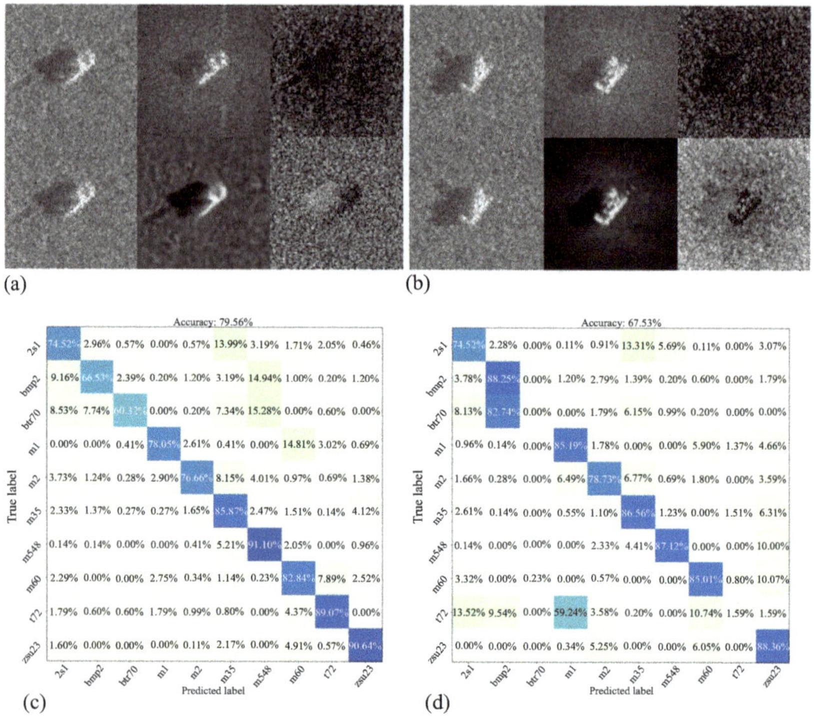

Figure 6.13 Representative output from a manifold joining autoencoder. The figures (a–d) are set up in a similar manner as in Figure 6.11. Although the encoded difference is greater than those in Figure 6.11, the stylistic attention to detail is encouraging and the output of the autoencoder looks more like a measured image

autoencoder does not blur the image nearly as much as the simple autoencoder. A DenseNet classifier was trained on the adjusted synthetic images and tested on unaltered measured images (in contrast to the simple autoencoder, where train and test images were both autoencoded). The performance of this classifier is given in the confusion matrices in Figure 6.13(c) and (d) for cases in which all ten measured classes were present (Figure 6.13(c)) and in which the measured images of T72 and BTR70 were removed from the autoencoder training (Figure 6.13(d)). In both cases, DenseNet was trained on the altered images from the autoencoder. While the classification results in the withheld data case are not impressive, there is potential for this architecture due to its ability to make synthetic images look more realistic.

In summary, autoencoders show some promise in bringing the two domains together but suffer from an inability to represent missing data. Adding a discriminator to the autoencoder encourages greater variation in the encoded output, but this is also unable to fully solve the problem. More details about the manifold-joining autoencoder are available in [16].

6.4.5 Generative adversarial networks

GANs are a relatively new method in the field of machine learning. These networks, which were introduced in 2014 by Ian Goodfellow and his collaborators [43], are designed to create new data that in some form mimics the statistical properties of a given set of training data. Given a target dataset, such as celebrity faces [44] or categories from the ImageNet [35] dataset, a GAN can be trained that generates new, unseen data that (ideally) fit comfortably and indistinguishably in the dataset. Since the introduction of GANs, several variations of the architecture [45] and many theories to help train these inherently unstable networks [46] have been developed.

In general, GANs are composed of generator and discriminator neural networks (Figure 6.14), which, for imagery data, are typically convolutional networks. Training is accomplished by repeatedly presenting the networks with data from a target dataset. The generator is tasked with learning to convert random n-dimensional vectors to data matching the dataset. The discriminator, in turn, is tasked with distinguishing between data from the dataset and the generator's output. In a descriptive analogy offered by Goodfellow *et al.*, the generator can be likened to an art forger, the goal of which is to create undetectable forgeries of the world's great artists. The discriminator plays the role of a detective, trying to discover which pieces are real and which are fakes.

The loss function for a GAN is given by

$$\begin{aligned}\min_G \max_D L(D, G) &= \mathbb{E}_{x \sim p_r(x)}[\log(D(x))] + \mathbb{E}_{z \sim p_z(z)}[\log(1 - D(G(z)))] \\ &= \mathbb{E}_{x \sim p_r(x)}[\log(D(x))] + \mathbb{E}_{x \sim p_g(x)}[\log(1 - D(x))]\end{aligned} \tag{6.1}$$

where $G(z)$ is the output of generator network, $D(x)$ is the output the discriminator network, z is a multidimensional random input to the generator, p_z is the distribution of z (usually uniform), and $p_g(x)$ and $p_r(x)$ are the probability manifold distributions for the generated data and the target dataset, respectively. Via backpropagation, these

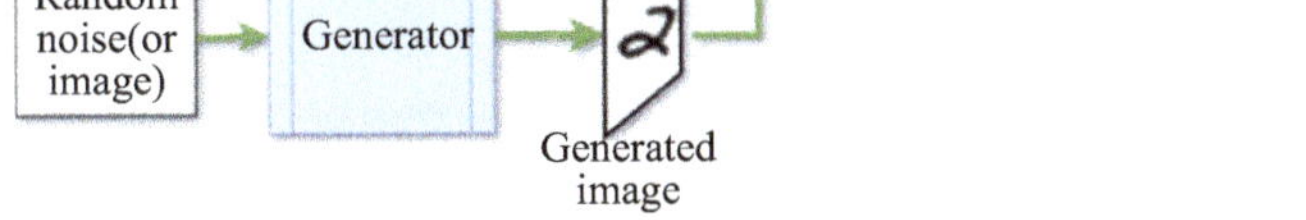

Figure 6.14 A diagram of a generic generative adversarial network. The network shown here is designed to produce new images of handwritten MNIST digits. The generator converts random noise into images that attempt to match the data from the target dataset. The discriminator distinguishes between real and generated data

objectives direct the generator to create data that fits well with the dataset, while simultaneously increasing the distinguishing power of the discriminator. It can be shown [47] that, for a GAN with sufficient capacity, this training objective minimizes the Kullback–Leibler divergence between $p_g(x)$ and $p_r(x)$. This divergence metric describes how similar two probability distributions are, with low values denoting greater similarity. In other words, training a GAN creates a generator that is able to mimic the distribution of data in the given dataset at some level.

SAR images generated entirely from scratch are likely to be physically unrealistic in comparison to data generated by computational electromagnetic software. However, there are a few varieties of GANs that are well suited to altering existing synthetic data in useful ways. In particular, the class of image-to-image translation GANs—such as Pix2Pix [48] and its extensions [49–51]—are formulated to translate images between domains, such as from grayscale images to color. This type of network can be used to translate synthetic SAR images to the measured domain and is designed to preserve the structure of the original image through the transformation. In a similar vein, methods such as SimGAN [52] aim to add realism to synthetic imagery by making small alterations to the image but do not explicitly translate images between domains. SimGAN has been used by other researchers for similar SAR image realism work [53].

As a brief example, this section explores DualGAN [49], which is based on Pix2Pix, as a way to make synthetic SAR data more realistic by translating between the two domains. Details of this experiment can be found in [54].

DualGAN [49], which is depicted in Figure 6.15, is designed to learn a mapping between two domains of imagery (e.g., real and synthetic SAR images). The architecture consists of two identical GANs, each assigned to one of the two data domains. The

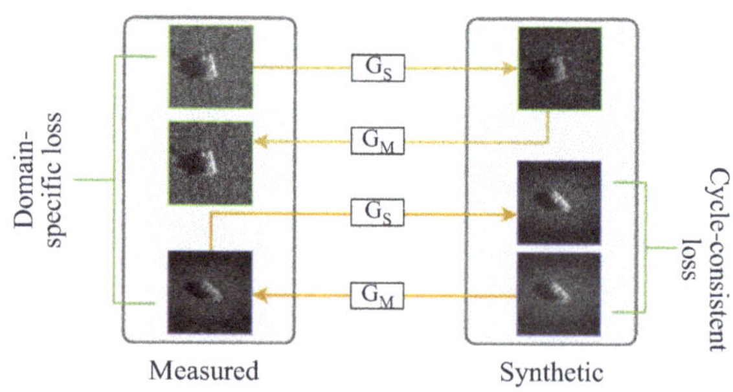

Figure 6.15 A schematic of DualGAN. The boxes denote the measured and synthetic domains. Translation operations in each direction are shown; the label on the arrow signifies a pass through a given generator. The first image in each chain is a non-altered image. The next image in the chain shows the same image translated into the other domain. Finally, the image is translated back into the original domain, completing the cycle. Loss metrics are defined between the reconstructed and original images as well as via the discriminators in the GANs

generators are composed of mirrored downsampling and upsampling convolutional layers with skip connections, in the style of U-Net [55]. In contrast to an autoencoder, the skip connections preserve information about the original structure of the image by concatenating pre-downsampled information to the upsampling layers, allowing the reproduction of important image details. The discriminators in each GAN learn to distinguish between data from each of the domains, and provide a gradient for the generators to generate data of the appropriate style. Importantly, DualGAN does not require paired data samples during training, as some image-to-image networks do. Instead, the generators learn the differences between the two domains from the dataset as a whole.

Let G_M be a generator mapping from the synthetic to the measured domain, D_M be a discriminator that determines whether an image belongs to the measured domain or not, and G_S and D_S be matching complementary networks operating on synthetic data.

DualGAN training is accomplished by converting a batch of synthetic images, S, to measured images with the measured generator, i.e., $\hat{M}_S = G_M(S)$. The discriminator D_M compares the translated images, $\hat{M}_S$, with samples of measured images, M, and computes the domain-specific loss accordingly. The translated images $\hat{M}_S$ are then translated back to the synthetic domain, $\hat{S} = G_S(\hat{M}_S)$. The difference between S and $\hat{S}$ is computed and weighted, creating the cycle-consistent loss metric. A similar path, $M \rightarrow G_S(M) \rightarrow \hat{S}_M \rightarrow G_M(\hat{S}_M) \rightarrow \hat{M}$, is followed for measured images. As both paths through the networks are repeated, the network learns a symmetric mapping between domains. Once trained, the generator G_M can translate synthetic images to measured images.

Initial experiments with the network as formulated showed that it was unable to convincingly reproduce important details (e.g., the target) in the SAR images. After further experimentation, the loss function in (6.2) was added to the generator loss. This loss metric removes the $| x |^{1/4}$ normalization applied to the images. The resulting de-normalized images have brighter spots on the area of the target and better contrast to the background. This image is then used as a mask on the L1 difference between the input image and the reconstructed image. By so doing, the network ensures that accurate reconstruction of the bright, target areas is prioritized by the network, which leads to better retention of target features.

$$\lambda_k \sum_x |(u_x)^4 \times (u_x - G_V(u_x, z)) \mid \forall x \in u, u_x \in \{0, 1\} \tag{6.2}$$

With a trained DualGAN network modified according to (6.2), the synthetic images can be transferred to the real domain using the trained generator. As an example of how well this method works, a DenseNet classifier was trained on the converted synthetic data and tested on non-altered measured images. Randomly chosen images that were translated by the trained DualGAN, as well as the confusion matrix for the DenseNet classifier, are presented in Figure 6.16.

While DualGAN is not the only GAN method by which synthetic data can be made to look more like measured data, this example is representative of this method. Further research is currently underway with other networks, including SimGAN [52,53],

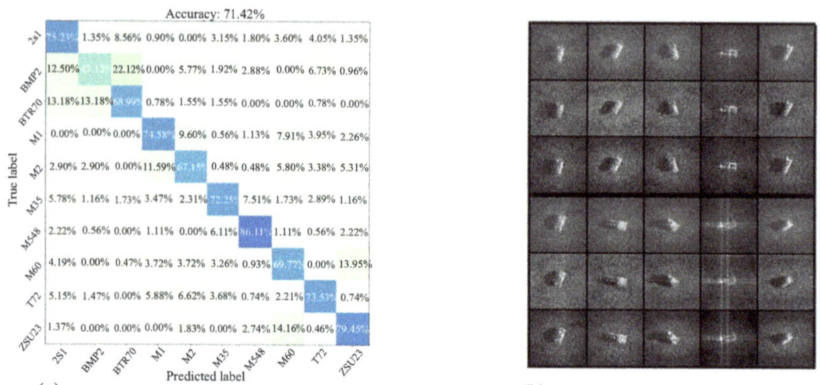

(a) (b)

Figure 6.16 (a) A confusion matrix showing the results of classifying measured images with a DenseNet trained on synthetic images that were altered by the DualGAN network as described in this section. Compare these results to those in Figure 6.5(b). All classes of measured data were used in training the DualGAN. 75% of the synthetic data were used to train the DualGAN, while the other 25% were altered using the trained DualGAN network and used to train the classifier. (b) A random sample of one image from each of ten classes. The first and fourth rows show original synthetic images, S. Rows two and five show the conversion to the measured domain, $\hat{M}_S$. The remaining rows show the conversion back to the synthetic domain, $\hat{S}$

BicycleGAN [51], and the Auxiliary Classifier GAN [56]. While space does not allow for a detailed explanation of these algorithms, the approach is similar to that presented here. For more details on the application of these networks to this problem, please see [16,54].

6.4.6 Siamese and triplet networks

While CNNs have shown great classification performance in general, their main weakness lies in their lack of ability to generalize [57].* Differences between the training and test data (such as the difference between measured and synthetic SAR data) lead to poor performance even with networks that have been carefully trained and exhibit good validation accuracy. CNNs simply are not well suited for evaluating data outside of the domain on which they were trained. The confusion matrix in Figure 6.5(b) is a prime example of this problem, showing that a DenseNet trained on synthetic data is inadequate for classifying measured data.

*Reference [58] shows an experiment in siamese neural network (SNN) generalization but does not explicitly describe the weakness of CNNs.

SNN and triplet neural networks (TNNs) are classification networks designed to learn about data from two domains. These networks are presented with sets of images from each domain and learn to identify whether these sets represent the same or different class of item. The output of both types of network indicates whether the inputs belong to the same class or not. However, this binary decision can be transformed into a more traditional classification decision. Siamese [57,59–62] and Triplet [63,64] networks consist of two and three identical neural networks, respectively, which share weights. By jointly sharing and updating the weights, the networks are able to form a shared representation of all the data. The classification layers at the end of the network enable it to identify data from matched and unmatched classes.

These networks are an attractive way of solving the measured/synthetic gap because, rather than transforming the data before training a classifier (such as the case with GANs in Section 6.4.5 and autoencoders in Section 6.4.4), they are able to learn from both data manifolds and reach a decision despite the differences in the data.

6.4.6.1 Siamese neural network

An SNN combines two neural networks (or twins) that have the same architecture. The network takes two inputs, one for each twin. The inputs represent either a true pair (e.g., the images represent the same class) or a false pair (e.g., the images represent different classes). Each twin's network has the same weights after training.

The loss function for the SNN is a similarity distance metric (such as the Euclidean distance between feature vectors), the value of which is "small" for the same type (class or model) of input and which is "large" for a different type of input. When training the network, each pair is classified by a value c that indicates whether the inputs are from the same class or not. Let $c = 1$ denote a true pair and $c = -1$ a false pair. Then for a given threshold τ for the margin of separation, the hinge-contrastive loss function (a common siamese network loss function) [65] is calculated as

$$l = \left(\frac{1+c}{2}\right)x + \left(\frac{1-c}{2}\right)\max\{0, \tau - x\}, \tag{6.3}$$

where x is a distance metric between the embedded features produced by each twin network. It is evident that $l = x$ for a true pair and $l = \max\{0, \tau - x\}$ for a false pair. The parameter τ ensures that the network learns to separate true and false pairs by at least a given margin.

The number of pairs presented to the network is a function of the number of data points presented to each twin network, as well as the distribution of those data points into classes. Let there be N classes. Let m_i, $i = 1, \ldots, N$ be the number of images presented to twin 1, and let n_i, $i = 1, \ldots, N$ be the number of images presented to twin 2. Then the total set of images for twin 1 is $\mathcal{X}$, the size of which is $|\mathcal{X}| = \sum_{i=1}^{N} m_i$, and the set presented to twin 2 is $\mathcal{Y}$, the size of which is $|\mathcal{Y}| = \sum_{i=1}^{N} n_i$. Assume that any true-pair combination of $x \in \mathcal{X}$ and $y \in \mathcal{Y}$ is not a duplication. Then the number of true pairs is

$$M_{\text{true}} = \sum_{i=1}^{N} m_i n_i, \tag{6.4}$$

and the number of false pairs is

$$M_{\text{false}} = \sum_{i=1}^{N} m_i \left(\sum_{j=1, j \neq i}^{N} n_j \right). \tag{6.5}$$

If the number of images for each target i in $\mathcal{X}$ and $\mathcal{Y}$ is balanced, then $m_i = n_i$ for all $i = 1, \ldots, N$. Then (6.4) and (6.5) become

$$M_{\text{true}} = \sum_{i=1}^{N} m_i^2 \tag{6.6}$$

and

$$M_{\text{false}} = \sum_{i=1}^{N} m_i \left(\sum_{j=1, j \neq i}^{N} m_j \right), \tag{6.7}$$

respectively. It is clear that the number of inputs $M_{\text{true}} + M_{\text{false}}$ for training an SNN grows approximately quadratically (depending on the distribution of images within classes) with $|\mathcal{X}| + |\mathcal{Y}|$.

Training the network involves generating pairs of SAR images for all targets; each pair is labeled as true (same class) or false (different classes). Ideally, all pairs would be composed of measured SAR data, as it is more self-consistent. Because this is generally not possible, one can augment the input data with synthetically generated SAR data. The full dataset need not be presented to both paired networks, which would be an exhaustive pairing of every image in the dataset with every other image. Doing so results in $\binom{n}{2}$ pairs, which can grow into the millions of pairs for more than a few thousand images and, in turn, increase the network training time dramatically. Instead, sets of data for each twin network should be chosen to present the data to the networks in some balanced fashion. For example, some pairs can be formed with one measured and one synthetic image of the same target, with a given ratio of measured and synthetic images, or to represent each class equally among all the pairs. After the pairs are constructed, the network is trained by presenting each pair to the twin networks and computing the forward pass through the convolutional layers and the similarity metric. Because the twins share network weights, the weight update is a function of the output of both networks.

Once the SNN is trained, there are two options for testing. The first, straightforward option involves matching an unknown image to one representative image from each trained class. Each of the N known images is paired with the unknown image, and each pair is presented to the network. The N scores from the SNN are compared with each other to determine which, if any, of the known targets is a good match for the unknown image. Zero, one, or many targets may be a good match for the unknown image, and further logic (e.g., best score) could be used to make a class determination. Because the choice of representative image could affect the scoring, this mechanism could be repeated with a set of k representative images, producing $k \times N$ pairs and evaluations, and incorporate more complex decision logic. An example of such a classification system is shown in Figure 6.17.

The other option utilizes one of the twins (as each twin network has the same weights) and adds one or more fully connected layers to the end. The preexisting weights are then frozen and the newly added fully connected layers are trained on the existing data to create a traditional classifier.

As an example of how a siamese network may be used, this architecture was trained on pairs of data from the SAMPLE dataset [58]. A modified AlexNet [66] CNN was used for each twin's architecture, with each convolution's weights initialized with a zero-mean Gaussian distribution with a small variance (0.01 or 0.05 in different experiments). This experiment consisted of data from five measured targets for training and testing and had a classification accuracy of 80.96% (see Figure 6.18). Experiments using synthetic data have not achieved classification accuracies that are better than chance. Experiments using all ten targets have had a similar lack of success.

6.4.6.2 Triplet neural network

A TNN combines three neural networks (or triples) that have the same architecture. Triplet networks follow a similar training and data paradigm as siamese networks—triples of data are constructed for training, and each network takes one of the images in the triple during training. As in siamese networks, the three networks all share

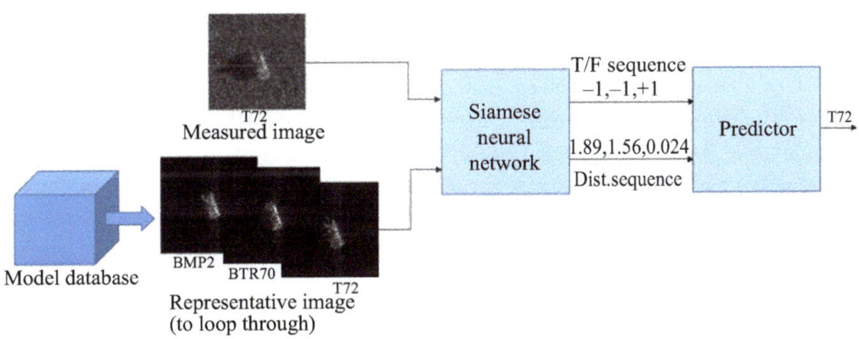

Figure 6.17 An example of a siamese-based SAR ATR in which an example image of each vehicle type is compared to an unknown image. The distance sequence scores below a threshold of 1 *indicate a match (i.e., "+1" in the true/false (T/F) sequence). The best match results from taking the minimum score from all the matches. Finally, the predictor identifies the class corresponding to the best match, in this case the T72. This figure is from Reference [58]*

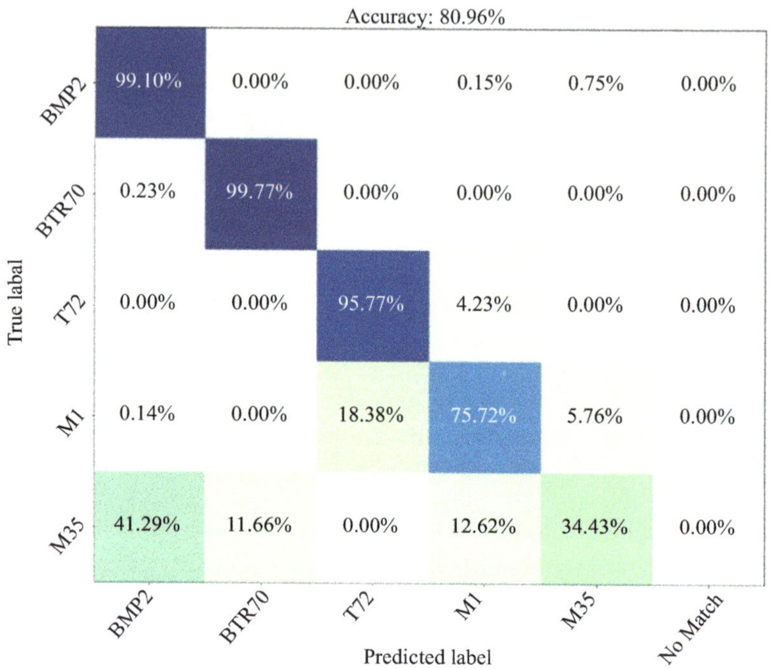

Figure 6.18 Siamese neural network classification results for a five-class problem. The network was trained on the first three targets and tested on the five targets shown in the confusion matrix. Note that this type of classifier also features a "No Match" label, which is assigned to images that do not sufficiently match any of the positive examples

weights. The three images within the triple are designated the "anchor," "positive," and "negative" [67]. The triple combines the positive and negative pairs seen in siamese networks into one group, where the anchor–positive pair is a true pair and the anchor–negative pair is a false pair. As is the case with siamese networks, the three networks share the same weights. Both the anchor–positive pair and the anchor–negative pair are combined by a distance metric, and these two intermediate values are then combined to create an overarching loss function. Just like for the SNN, the distance should be small for the anchor-positive pair and large for the anchor-negative pair. As described in [68], an advantage of a TNN over an SNN is not having to know from which class two images are; rather, the network just needs to have information on whether the classes are the same or not.

As opposed to an SNN loss function, a TNN loss function generally takes into account the Euclidean distances between the output of the anchor network and the positive and negative networks. That is, as in [64], suppose Net(X) represents the output of a triplet network given one image in the triple (X_a, X_p, X_n). The Euclidean distances for the anchor–positive network pair and the anchor–negative network pair are, respectively,

$$d_{ap} = \|\text{Net}(X_a) - \text{Net}(X_p)\|_2 \qquad \text{and} \tag{6.8a}$$

$$d_{an} = \|\text{Net}(X_a) - \text{Net}(X_n)\|_2. \tag{6.8b}$$

These Euclidean distances are then used to define a triplet loss function. One common iteration of this loss is

$$l = d_{ap}^2 - d_{an}^2 + \alpha \tag{6.9}$$

for some margin α [67]. This loss function encourages a large distance d_{an} between the X_a and X_n images, which are dissimilar, and a small distance d_{ap} between the X_a and X_p images from the same class. The α parameter drives a distance of $\sqrt{\alpha}$ between the two metrics.

A second common triplet loss function is

$$l = \|d_p, d_n - 1\|_2^2, \tag{6.10}$$

where

$$d_p = \frac{e^{d_{ap}}}{e^{d_{ap}} + e^{d_{an}}} \qquad \text{and} \tag{6.11a}$$

$$d_n = \frac{e^{d_{an}}}{e^{d_{ap}} + e^{d_{an}}} \tag{6.11b}$$

as provided in [64]. The loss function in (6.10) has a similar aim as (6.9), pushing d_n toward a value of 1 and d_p closer to a value of 0. Like the first loss function, this encourages the metric between the anchor and positive images to be small and the metric between the anchor and negative images to be large. In contrast to SNN loss functions, the loss functions (6.9) and (6.10) do not require explicit positive and negative labels; rather, these labels are inherent to the network.

Additional loss functions for TNNs are defined and investigated in [68], which describes two categories of loss functions as "difference losses" and "ratio losses," such that the positive and negative distances between the anchor–positive and anchor–negative networks, respectively, are compared by either differences or ratios, respectively.

At the time of writing, the authors of this chapter have yet to perform significant, reportable experiments with a TNN, as they are still performing experiments with siamese networks. Nevertheless, a discussion of the theory of these networks is advantageous in this context for completeness. The number of possible triples (X_a, X_p, X_n) is also a limiting factor to experimentation. For N classes each with n images, the number of possible experimental triples is $N \times (n!/((n-2)!)) \times (N-1) \times n$. For this reason, it is even more computationally intensive to experiment with this type of network than a siamese network.

As with an SNN, there are two options for TNN testing. At the time of this writing, the writers have not fully investigated discrimination with triplet networks, due to complexities in choosing anchor–positive and anchor–negative images and the size of the problem for nontrivial numbers of classes. However, classification may be performed much the same as with a siamese network. For classification, a few fully connected layers can be appended to one of the trained triplet networks, and the fully connected layers can be trained to perform classification. In order to leverage the inherent functionality of the triplet network, however, further exploration is required. Assuming that the unknown image is the anchor image, the prime difficulty lies in identifying a viable positive match to an unknown image. As anchor–negative–negative triples are far more probable than the ideal anchor–positive–negative triples, additional logic must be developed to effectively utilize the functionality of TNNs for classification.

6.5 Conclusion

The work of ATR for SAR is still a work in progress. The difficulties posed by this particular imaging method, including poor resolution, limited data, and the challenge of accurately creating simulated data mean that this problem may defy a complete solution for a time to come. In particular, the gap between synthetic and measured SAR data poses many challenges for deep learning classification systems, particularly due to the small size of available datasets.

Although the approaches in this chapter do not fully solve these challenges, they nevertheless represent an encouraging direction in which deep neural networks can help solve the measured/synthetic gap. This work with mathematical despeckling methods, autoencoders, layer retraining, GANs, and siamese networks represents a wide variety of possible approaches, and many more deep networks are available in the literature. Despite limited success to this point, these other approaches provide fertile ground on which to expand this work.

Acknowledgments

The authors acknowledge the hard work of many interns on these various projects, including (alphabetically) Julia Arnold, Michael Cline, James Cunningham, Omar DeGuchy, Devon Dollahon, Robert Friedlander, John Kaminski, Jennifer Liu, Brendan O'Connor, Joseph Sebastian, and Amy Wong.

This publication is Distribution A: Approved for public release. Distribution unlimited. PA Approval #88ABW-2019-2698.

References

[1] Sandia National Laboratory. MSTAR Overview; 1995. [Online; accessed 19-May-2017]. https://www.sdms.afrl.af.mil/index.php?collection=mstar.

[2] Han P, Wu R, Wang Y, *et al.* An Efficient SAR ATR Approach. In: *IEEE International Conference on Acoustics, Speech, and Signal Processing.* vol. 2; 2003.

[3] El-Darymli K, Gill EW, Mcguire P, *et al.* Automatic Target Recognition in Synthetic Aperture Radar Imagery: A State-of-the-Art Review. *IEEE Access.* 2016.

[4] Chen S and Wang H. SAR Target Recognition Based on Deep Learning. In *2014 International Conference on Data Science and Advanced Analytics (DSAA)*; 2014. p. 541–547.

[5] Chen S, Wang H, Xu F, *et al.* Target Classification using the Deep Convolutional Networks for SAR Images. *IEEE Transactions on Geoscience and Remote Sensing.* 2016.

[6] Lewis B, Scarnati T, Sudkamp E, *et al.* A SAR Database for ATR Development: Synthetic and Measured Paired Labeled Experiment (SAMPLE). In: *Proc. SPIE Algorithms for Synthetic Aperture Radar XXVI.* Baltimore; 2019.

[7] Oliver C and Quegan S. *Understanding Synthetic Aperture Radar Images.* SciTech Publishing, Raleigh, NC; 2004.

[8] Rosencrantz S, Nehrbass J, Zelnio E, *et al.* "AFacet": A Geometry Based Format and Visualizer to Support SAR and Multisensor Signature Generation. In: *Proc. SPIE Algorithms for Synthetic Aperture Radar XXV.* Orlando; 2018.

[9] Jakowatz CV, Wahl DE, Eichel PH, *et al.* *Spotlight-Mode Synthetic Aperture Radar: A Signal Processing Approach.* Springer Science & Business Media, Berlin, Germany; 2012.

[10] Scarnati T and Gelb A. Variance Based Joint Sparsity Reconstruction of Synthetic Aperture Radar Data for Speckle Reduction. In: *Proc. SPIE Algorithms for Synthetic Aperture Radar Imagery XXV.* Orlando; 2018.

[11] Fischler MA and Bolles RC. Random Sample Consensus: A Paradigm for Model Fitting With Applications to Image Analysis and Automated Cartography. *Communications of the Association for Computing Machinery.* 1981.

[12] Van Der Maaten L and Hinton G. Visualizing Data Using t-SNE. *Journal of Machine Learning Research.* 2008.

[13] Huang G, Liu Z, Van Der Maaten L, *et al.* Densely Connected Convolutional Networks. In: *Proc. IEEE Conference on Computer Vision and Pattern Recognition.* Honolulu; 2017.

[14] Kingma DP and Ba J. Adam: A Method for Stochastic Optimization. arXiv preprint arXiv:1412.6980. 2014.

[15] Shwartz-Ziv R, and Tishby N. Opening the Black Box of Deep Neural Networks via Information. arXiv preprint arXiv:1703.00810. 2017.

[16] Lewis B, DeGuchy O, Sebastian J, *et al.* Realistic SAR Data Augmentation using Machine Learning Techniques. In: *Proc. SPIE Algorithms for Synthetic Aperture Radar XXVI.* Baltimore; 2019.

[17] Pinel N and Boulier C. *Electromagnetic Wave Scattering from Random Rough Surfaces: Asymptotic Models.* John Wiley & Sons, Indianapolis, IN; 2013.

[18] Argenti F, Lapini A, Bianchi T, *et al.* A Tutorial on Speckle Reduction in Synthetic Aperture Radar Images. *IEEE Geoscience and Remote Sensing Magazine.* 2013.

[19] Mansourpour M, Rajabi M, and Blais R. Effects and Performance of Speckle Noise Reduction Filters on Active Radar and SAR Images. In: *Proc. International Society for Photogrammetry and Remote Sensing*. vol. 36; 2006. p. W41.

[20] Chen DQ and Cheng LZ. Spatially Adapted Total Variation Model to Remove Multiplicative Noise. *IEEE Transactions on Image Processing*. 2012.

[21] Aubert G and Aujol JF. A Variational Approach to Removing Multiplicative Noise. *SIAM Journal on Applied Mathematics*. 2008.

[22] Lu J, Shen L, Xu C, *et al.* Multiplicative Noise Removal in Imaging: An Exp-model and Its Fixed-Point Proximity Algorithm. *Applied and Computational Harmonic Analysis*. 2016.

[23] Aubert G and Aujol JF. A Nonconvex Model to Remove Multiplicative Noise. In: *International Conference on Scale Space and Variational Methods in Computer Vision*. Springer, Berlin, Germany; 2007. p. 68–79.

[24] Rudin L, Lions PL, and Osher S. Multiplicative Denoising and Deblurring: Theory and Algorithms. In: *Geometric Level Set Methods in Imaging, Vision, and Graphics*. Springer, New York, NY; 2003. p. 103–119.

[25] Goodman JW. *Statistical Optics*. John Wiley & Sons, Indianapolis, IN; 2015.

[26] Irving WW and Ettinger GJ. Classification of Targets in Synthetic Aperture Radar Imagery via Quantized Grayscale Matching. Algorithms for Synthetic Aperture Radar Imagery VI. In: *AeroSense '99*; 1999. p. 320–331.

[27] Paulson C, Wilson J, and Lewis T. Synthetic Aperture Radar Quantized Grayscale Reference Automatic Target Recognition Algorithm. In: *Proc. SPIE Algorithms for Synthetic Aperture Radar Imagery XXV.* Orlando; 2018.

[28] Horvath M and Rigling B. Performance Prediction of Quantized SAR ATR Algorithms. *IEEE Transactions on Aerospace and Electronic Systems*. 2016.

[29] Sanders T, Gelb A, and Platte RB. Composite SAR Imaging Using Sequential Joint Sparsity. *Journal of Computational Physics*. 2017.

[30] Scarnati T, Zelnio E, and Paulson C. Exploiting the Sparsity of Edge Information in Synthetic Aperture Radar Imagery for Speckle Reduction. In: *Proc. SPIE Algorithms for Synthetic Aperture Radar Imagery XXIV.* Anaheim; 2017.

[31] Scarnati T and Lewis B. A Deep Learning Approach to the Synthetic and Measured Paired Labeled Experiment (SAMPLE) Challenge Problem. In: *SPIE Algorithms for Synthetic Aperture Radar XXVI*. Baltimore; 2019.

[32] Gelb A and Scarnati T. Reducing Effects of Bad Data using Variance Based Joint Sparsity Recovery. *Journal of Scientific Computing*. 2019.

[33] Knott EF, Schaeffer JF, and Tulley MT. *Radar Cross Section*. SciTech Publishing, Raleigh, NC; 2004.

[34] Tan C, Sun F, Kong T, *et al.* A Survey on Deep Transfer Learning. In: *International Conference on Artificial Neural Networks*. Springer, Cham, Switzerland; 2018. p. 270–279.

[35] Russakovsky O, Deng J, Su H, *et al.* ImageNet Large Scale Visual Recognition Challenge. *International Journal of Computer Vision*. 2015.

[36] LeCun Y, Bottou L, Bengio Y, and Haffner P. LeNet-5, Convolutional Neural Networks; 2015. http://yann.lecun.com/exdb/lenet.
[37] He K, Zhang X, Ren S, *et al.* Deep Residual Learning for Image Recognition. In: *Proc. IEEE Conference on Computer Vision and Pattern Recognition*; 2016. p. 770–778.
[38] Ioffe S and Szegedy C. Batch Normalization: Accelerating Deep Network Training by Reducing Internal Covariate Shift. arXiv preprint arXiv:1502. 03167. 2015.
[39] Srivastava N, Hinton G, Krizhevsky A, *et al.* Dropout: A Simple Way to Prevent Neural Networks from Overfitting. *The Journal of Machine Learning Research.* 2014.
[40] Zagoruyko S and Komodakis N. Wide Residual Networks. arXiv preprint arXiv:1605.07146. 2016.
[41] Xie J, He T, Zhang Z, *et al.* Bag of Tricks for Image Classification with Convolutional Neural Networks. arXiv preprint arXiv:1812.01187. 2018.
[42] Zeiler MD and Fergus R. Visualizing and Understanding Convolutional Networks. In: *European Conference on Computer Vision.* Springer, Cham, Switzerland; 2014. p. 818–833.
[43] Goodfellow I, Pouget-Abadie J, Mirza M, *et al.* Generative Adversarial Nets. In: *Advances in Neural Information Processing Systems*; 2014. p. 2672–2680.
[44] Liu Z, Luo P, Wang X, *et al.* Deep Learning Face Attributes in the Wild. In: *Proc. of the IEEE International Conference on Computer Vision*; 2015.
[45] Hindupur A. The GAN Zoo; 2018. Online; accessed 03-Oct-2018. https://deephunt.in/the-gan-zoo-79597dc8c347.
[46] Creswell A, White T, Dumoulin V, *et al.* Generative Adversarial Networks: An Overview. *IEEE Signal Processing Magazine.* 2018.
[47] Weng L. From GAN to WGAN; 2017. Online; accessed 23-Aug-2018. https://lilianweng.github.io/lil-log/2017/08/20/from-GAN-to-WGAN.html.
[48] Isola P, Zhu JY, Zhou T, *et al.* Image-to-Image Translation with Conditional Adversarial Networks. arXiv preprint arXiv:1611.07004. 2016.
[49] Yi Z, Zhang H, Tan P, *et al.* DualGAN: Unsupervised Dual Learning for Image-to-Image Translation. arXiv preprint arXiv:1704.02510. 2017.
[50] Zhu JY, Park T, Isola P, *et al.* Unpaired Image-to-Image Translation Using Cycle-Consistent Adversarial Networks. arXiv preprint arXiv:1703.10593. 2017.
[51] Zhu JY, Zhang R, Pathak D, *et al.* Toward Multimodal Image-to-Image Translation. In: *Advances in Neural Information Processing Systems*; 2017. p. 465–476.
[52] Shrivastava A, Pfister T, Tuzel O, *et al.* Learning from Simulated and Unsupervised Images Through Adversarial Training. In: *Proc. IEEE Conference on Computer Vision and Pattern Recognition.* 2017.
[53] Cha M, Majumdar A, Kung HT, *et al.* Improving SAR Automatic Target Recognition Using Simulated Images Under Deep Residual Refinements. In: *IEEE International Conference on Acoustics, Speech, and Signal Processing.* Calgary, Alberta, Canada; 2018.

[54] Lewis B, Liu J, and Wong A. Generative Adversarial Networks for SAR Image Realism. In: *Proc. SPIE Algorithms for Synthetic Aperture Radar XXV.* Orlando; 2018.
[55] Ronneberger O, Fischer P, and Brox T. U-Net: Convolutional Networks for Biomedical Image Segmentation. arXiv preprint arXiv:1505.04597. 2015.
[56] Odena A, Olah C, and Shlens J. Conditional Image Synthesis With Auxiliary Classifier GANs. arXiv preprint arXiv:1610.09585. 2016.
[57] Koch G. *Siamese Neural Networks for One-Shot Image Recognition.* 2015.
[58] Friedlander RD, Levy M, Sudkamp E, *et al.* Deep Learning Model-based Algorithm for SAR ATR. In: *SPIE Algorithms for Synthetic Aperture Radar Imagery XXV*; 2018.
[59] Bertinetto L, Valmadre J, Henriques JF, *et al.* Fully-Convolutional Siamese Networks for Object Tracking. *European Conference on Computer Vision*, Springer. 2016. p. 850–865.
[60] Khalil-Hani M and Sung LS. A Convolutional Neural Network Approach for Face Verification. In: *International Conference on High Performance Computing & Simulation*; 2014.
[61] Han X, Lueng T, Jia Y, *et al.* MatchNet: Unifying Feature and Metric Learning for Patch-based Matching. In: *IEEE Conference on Computer Vision and Pattern Recognition*; 2015. p. 3279–3286.
[62] Shaham U and Lederman RR. *Common Variable Learning and Invariant Representation Learning Using Siamese Neural Networks.* 2016.
[63] Tzeng E, Devin C, Hoffman J, *et al.* Adapting Deep Visuomotor Representations With Weak Pairwise Constraints. In: *Workshop on the Algorithmic Foundations of Robotics*; 2016.
[64] Hoffer E and Ailon N. Deep Metric Learning using Triplet Network. In: *International Conference on Learning Representations*; 2015.
[65] Leal-Taixé L, Canton-Ferrer C, and Schindler K. Learning by Tracking: Siamese CNN for Robust Target Association. In: *IEEE International Conference on Computer Vision and Pattern Recognition Workshops (CVPRW).* Deep Vision: Deep Learning for Computer Vision; 2016.
[66] Krizhevsky A, Sutskever I, and Hinton GE. ImageNet Classification With Deep Convolutional Neural Networks. In: Pereira F, Burges CJC, Bottou L, *et al.*, editors. *Advances in Neural Information Processing Systems.* Curran Associates, Inc, Red Hook, NY; 2012. p. 1097–1105.
[67] Schroff F, Kalenichenko D, and Philbin J. FaceNet: A Unified Embedding for Face Recognition and Clustering. In: *2015 IEEE Conference on Computer Vision and Pattern Recognition (CVPR)*; 2015. p. 815–823.
[68] Liu Y and Huang C. Scene Classification via Triplet Networks. *IEEE Journal of Selected Topics in Applied Earth Observations and Remote Sensing.* 2018.

Part III

Applications

Chapter 7
Classifying micro-Doppler signatures using deep convolutional neural networks

Youngwook Kim[1]

7.1 Micro-Doppler and its representation

7.1.1 Micro-Doppler theory

An electromagnetic signal transmitted by a radar is reflected from a target then returns to the radar with the information of the target characteristics. Doppler information is commonly used to detect moving objects while suppressing clutter. In particular, the micro-Doppler signatures from nonrigid body motions contain diverse information regarding target movement [1–3]. Accordingly, the use of micro-Doppler signatures has a variety of defense, security, surveillance, and biomedicine applications, including airborne target classification, human detection, human activity classification, and drone detection.

Micro-Doppler signatures arising from micromotion dynamics such as mechanical vibrations or rotations can be mathematically described. The Doppler received from a target is modeled as a function of time [2]:

$$s(t) = A \cdot \exp(j(2\pi f_o t + \phi(t))) \tag{7.1}$$

where A is the magnitude, f_o the carrier frequency of the transmitted signal, and $\phi(t)$ the time-varying phase. For a vibrating scatterer set to a radian frequency oscillation of ω_v, the time-varying phase is

$$\phi(t) = \beta \sin(\omega_\gamma t) \tag{7.2}$$

where β is the vibration's magnitude. With vibration, the original signal $s(t)$ becomes

$$s(t) = A \cdot \exp(j(2\pi f_o t + \beta \sin(\omega_\gamma t))) \tag{7.3}$$

Fourier transform allows $s(t)$ to be written in the frequency domain as

$$s(t) = A \sum_{n=-\infty}^{\infty} J_n(\beta) \exp[j(2\pi f_o + n \cdot w_v)t] \tag{7.4}$$

where $J_n(\beta)$ is the Bessel function. Vibration-induced micro-Doppler is a sinusoidal function of time at vibrating frequency ω_v. Because detecting vibration in the

[1]Electrical and Computer Engineering, California State University, Fresno, USA

frequency domain is often difficult when the vibrating modulation is small, it is necessary to distinguish radar return produced by the target body from that produced by its vibrating structure.

Another common micromotion is rotation. For a rotating scatterer, the phase is defined as

$$\phi = \beta \sin(\Omega t + \theta_0) \tag{7.5}$$

where Ω is the rotation frequency and θ_0 is the initial phase. Therefore, the analytical received signal becomes

$$s(t) = A \exp[j(2\pi f_o t + \beta \sin(\Omega t + \theta_0))] \tag{7.6}$$

By applying the Fourier transform, the following equation is obtained:

$$s(t) = A \sum_{k=0}^{N-1} J_k(\beta) \exp[j(2\pi f_o t + \beta \sin(\Omega t + k2\pi/N))] \tag{7.7}$$

Notably, the rotating frequency modulates the main frequency in an oscillatory manner. Clearly, micro-Doppler signatures are evident in radar returns and are likely amenable to further exploitation for target detection, classification, and recognition.

7.1.2 Representing micro-Doppler signatures

Because micro-Doppler changes with time, joint time–frequency analysis has been used to observe its time-varying signatures, Doppler signals. The short-time Fourier transform (STFT) is the most commonly used time–frequency representation [4] It focuses on a signal's frequency components at a particular point in time. This is achieved by applying a narrow sliding time window $h(t)$ centered at the time of interest prior to FT. Representing the time-domain signal as $x(t)$ and the time window as $h(t)$, the STFT can be calculated by the following equation:

$$STFT(t,f) = \frac{1}{2\pi} \int x(\tau) \cdot h(\tau - t) \cdot e^{-j \cdot 2\pi \cdot f \cdot \tau} \cdot d\tau \tag{7.8}$$

Time and frequency resolution in the STFT depends on the size of the window function Increased time window size is used to improve frequency resolution while yielding poor time resolution, and vice versa.

High-order transforms such as the Wigner–Ville Distribution (WVD)—one of the most popular high-resolution time–frequency transform methods—have been suggested as a way of addressing this issue [5]. For a given signal $x(t)$, WVD is defined as

$$W(t,f) = \int x^*\left(t - \frac{\tau}{2}\right) \cdot x\left(t + \frac{\tau}{2}\right) \cdot e^{-j \cdot 2\pi \cdot f \cdot \tau} d\tau \tag{7.9}$$

where * denotes the complex conjugate. Instead of requiring the use of time windows, computation of the distribution for a given time depends on past and future samples. WVD does not focus on a certain time period—unlike STFT with its time-domain window. However, WVD's nonlocal nature also suffers from cross-term interferences that are artifacts representing false power concentrations in the time–frequency spectrum.

To reduce the cross-term interference, the reduced interference distribution, a smoothed WVD, has been suggested. Employing a filter can mitigate the effect of the cross-term interference. For example, the application of a time-domain window during WVD calculations can help one to mitigate cross-term inference in what is known as pseudo-WVD (PWVD). Smoothed PWVD (SPWVD) improves on PWVD by using a time-smoothing window $h(t)$ as well as a frequency-smoothing window $g(t)$ [6]:

$$W_P(t,f) = \int g(u) \cdot h(\tau) \cdot x^*\left(t - \frac{\tau}{2}\right) \cdot x\left(t + \frac{\tau}{2}\right) e^{-j \cdot 2\pi \cdot f \cdot \tau} d\tau \cdot du \qquad (7.10)$$

Unlike WVD, SPWVD mitigates the artifacts created by cross-term interference. Furthermore, its two windows allow fine-tuning of the resolutions of time and frequency. It thus incorporates the clarity that sets WVD above STFT while avoiding distortion from suppressed cross-term interference.

Unlike for WVD and SPWVD, the parametric frequency estimation technique can be used to construct a spectrogram. A model of a given signal is assumed and then the model parameters are estimated to maximize fit performance for data samples. A signal received from Doppler radar through I/Q demodulation follows the exponential model:

$$x(t) = \sum_{i=1}^{D} A_i \cdot e^{-j \cdot 2\pi f_i \cdot t} + n(t) \qquad (7.11)$$

where A_i is the amplitude of the signal, D the number of frequency components, and $n(t)$ the noise. The estimation of exponential parameters proceeds the same as for finding the signal's frequencies. The exponential parameter estimation has been the subject of much investigation, sharing as it does the physics of direction-of-arrival estimation [7]. Accurate high-resolution estimation of the parameters is generally achieved using super-resolution algorithms like multiple signal classification (MUSIC). MUSIC's pseudo-spectrum shows a high peak when testing frequency matches the actual signal frequency. The use of a sliding window allows the production of a spectrogram, which called short-time MUSIC (ST-MUSIC) [8]. Figure 7.1 depicts the micro-Doppler signatures of a measured quadcopter as represented by various joint time–frequency methods, with the micro-Doppler signatures in the spectrogram offering clues for target classification.

7.1.3 Conventional approaches to classify micro-Doppler signatures

In the past, there have been several approaches to classifying micro-Doppler signatures. In the early study stage, spectral analysis of micro-Doppler signatures has sought to distinguish targets without using temporal signatures [9]—a method able to identify targets having a particularly notable Doppler shift. Next, time-varying micro-Doppler signatures were extracted from a spectrogram to permit recognition of different human activities in [10]. Empirical mode decomposition has also allowed recognition of target types [11]. Principal component analysis and linear discriminant analysis have also been employed to extract feature vectors [12]. With a view

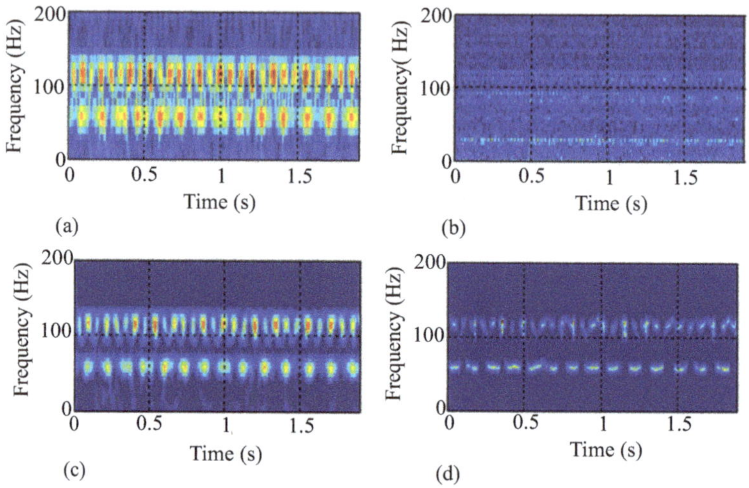

Figure 7.1 Micro-Doppler signatures of a quadcopter: (a) STFT, (b) WVD, (c) SPWVD, and (d) ST-MUSIC

to significantly reducing computational costs, linear predictive code has also been suggested for real-time processing [13], along with template-based micro-Doppler signature classification [14], micro-Doppler feature reduction using PCA [15], use of bicepstrum [16], and dictionary learning [17]. Although these approaches produced highly accurate results for the problems at hand, most did not go beyond the conventional supervised learning paradigm so that identification of discriminative features needed for classification algorithms first required preprocessing of raw micro-Doppler signals. The scalability of the proposed algorithms to other research problems is also restricted by reliance on the domain knowledge of micro-Doppler signals [18]. More importantly, the micro-Doppler classification accuracy has been improved by advanced deep learning algorithms.

7.2 Micro-Doppler classification using a deep convolutional neural network

Nonrigid body motion has become central to the classification of targets based on their micro-Doppler signatures using a powerful image classification algorithm such as a deep convolutional neural network (DCNN). In recent years, deep learning algorithms based on hierarchical neural networks have provided a revolutionary alternative by significantly outperforming once state-of-the-art schemes reliant primarily on domain knowledge-based features. The success of these deep learning algorithms comes from their ability to derive both features and classification boundaries directly from raw input data, identifying the data most salient to a given classification problem while also capturing features likely to be otherwise missed. As a result, micro-Doppler signatures have provided a means for classifying human activities and senior fall detection; recognizing human gaits, hand gestures, and voices; and detecting drones. Here, the classification examples of using micro-Doppler signature are given.

7.2.1 Detecting and classifying human activities

The prospect of classifying human activities through radar measurements has long fascinated researchers, especially in view of the high priority already placed on detecting, tracking, and recognizing humans for security, surveillance, and bioinformatics applications. Different human activities are associated with different micro-Doppler signatures, which can be recognized using DCNNs [19]. The first trial of applying DCNN to micro-Doppler signatures was testing the classification performance using the same dataset as in [10]. Twelve human beings moving directly toward the radar operating at 2.4 GHz were measured indoors under line-of-sight conditions for 3 s as they performed seven activities. Each activity was measured four times for each subject, with three spectrograms extracted from each measurement, for a total of 1,008 data samples. The size of spectrograms was 300×140. Figure 7.2 shows the diagram of measurements; Figure 7.3 shows the sample spectrograms measured.

Cross-validation (CV) was used to empirically optimize the hyper-parameters, such as the number of convolution filters in each layer, the size of convolution filter, and the number of hidden nodes in the fully connected layer. Four-fold CV was used to evaluate the DCNN's classification ability, with each fold's training and test set containing 756 (75%) and 252 (25%) samples, respectively. After a heuristic search, the best model was identified, having three convolution layers, each with twenty 5×5 filters, and two fully connected layers, the first of which featured 500 hidden nodes. The use of the rectified linear unit (ReLU) for the activation function was followed by 2×2 max-pooling. Regarding the implementation of DCNN, the open-source toolkit Caffe [20], which uses NVIDIA's GPU and CUDA library cuDNN to speed computation [21], was used. For the training, the stochastic gradient descent (SGD) was employed with a learning rate of 0.001 and a batch size of 84. The momentum method was also used, at a weight of 0.9, and dropout applied for the final fully connected layer, at probability 0.5. An NVIDIA GeForce GTX Titan Black GPU (with 6 GB of VRAM) and a 2.5 GHz Intel Xeon E5-2609 v2 CPU were used

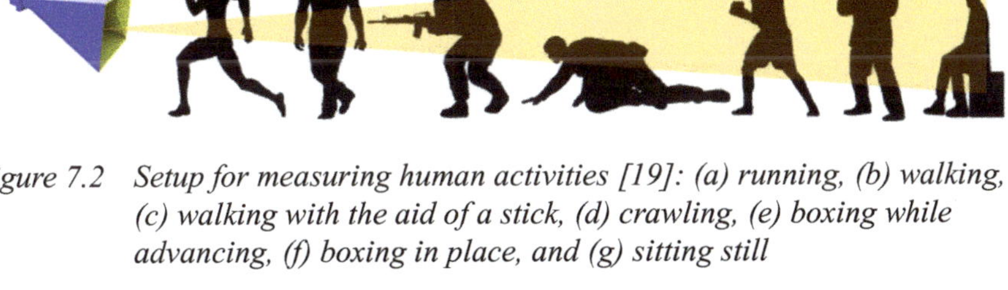

Figure 7.2 Setup for measuring human activities [19]: (a) running, (b) walking, (c) walking with the aid of a stick, (d) crawling, (e) boxing while advancing, (f) boxing in place, and (g) sitting still

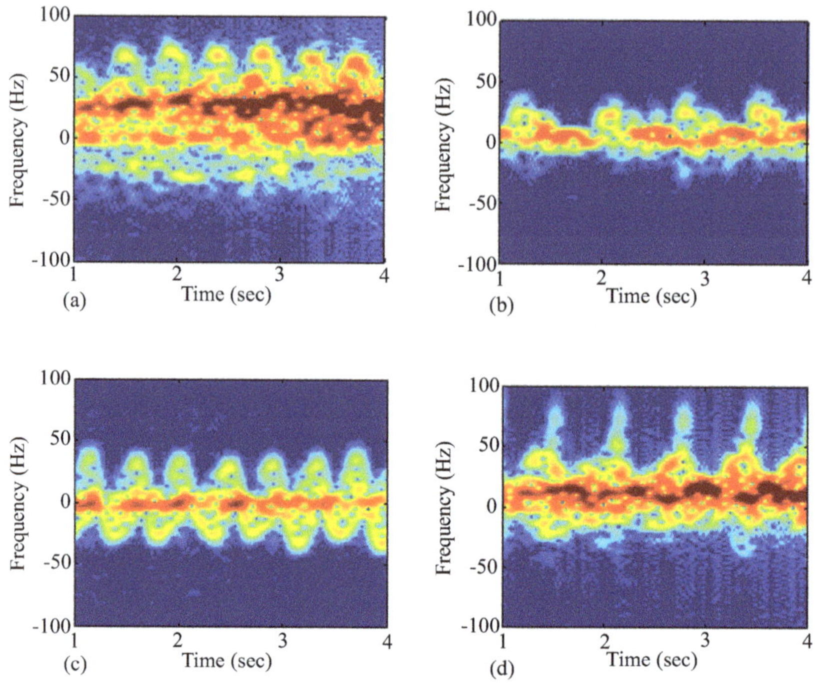

Figure 7.3 Sample spectrograms of human activities: (a) running, (b) crawling, (c) boxing in place, and (d) boxing while advancing [19]

in the experiments. Each fold with 400 epochs took about 1,420 s for training. The average accuracy of classification was 90.9%, a result that—though comparable to that of [10]—showed DCNN's potential for micro-Doppler-based target classification, considering that neither preprocessing for feature extraction nor extensive hyper-parameter tuning, such as random searches, were used. Figure 7.4 depicts the test-error curves of Fold 1, indicating that about 200 epochs, at minimum, are needed if test errors are to converge while avoiding significant overfitting.

A similar approach has been applied to the senior fall detection problem [22,23]. A majority of elderly live by themselves at homes, and one out of three elderly experience falls every year, which can result in injuries, reduced quality of life, and even death. After a fall, most seniors cannot get up by themselves. It was reported that there is a high correlation between the fall and the death within 6 months after it happened.

In [22], an approach utilizing deep learning was suggested for fall detection using radar. This approach is based on the use of a deep neural network that consists of stacked auto-encoders and a softmax layer output. The spectrogram power distributions are regarded as images. They are used as input to the network. The role of the auto-encoders is to provide a sparse representation of the spectrogram

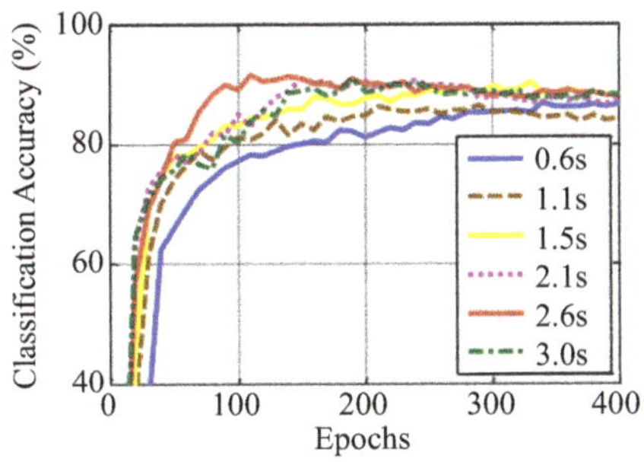

Figure 7.4 Test error curves for Fold 1 with varying time windows in spectrogram. The x-axis index corresponds to training epochs, and the y-axis corresponds to the accuracy [19]

through learning the most significant features. These features are inputted to the softmax layer that determines the class of the data. With experimental data, it has been demonstrated that the classification accuracies of the deep learning approach are higher over previous methods where the handcrafted features are used. In [23], two domains, the time–frequency domain and the range domain, were utilized for analyzing radar return signals. Two different methods for integrating the information were suggested, and the deep learning scheme that includes two stacked sparse auto-encoders and a logistic regression classifier was employed. Experimental results confirm that deep learning improves fall detection over traditional approaches and that incorporating the range along with the Doppler signatures yields higher success classification rates than the case when using a single domain.

7.2.2 Using simulated data to classify human activities

This study investigated DCNN's effectiveness at using an electromagnetic model to classify simulated micro-Doppler signatures [24]. DCNN training requires data, but the measurement of radar data is laborious and expensive. Instead, simulation data can be used to train a DCNN. To this end, the MOtion CAPture (MOCAP) database, a public dataset supplied by Carnegie Mellon University (CMU) [25], was used to generate micro-Doppler signatures. The motions featured in this free library of human MOCAP data, built by the CMU Graphics Lab, were captured using a Vicon motion capture system consisting of 12 infrared MX-40 cameras whose images were integrated to obtain bone trajectories. The dataset comprises 2,605 trials in 6 primary categories and 23 subcategories. For this study, seven common activities were chosen for classification: walking, jumping forward, running, boxing, standing, crawling, and creeping. With a view to obtaining generalized sampled data, scenarios were selected from different subjects or different motions.

Because the MOCAP data convey the positions of human bones at given times, the ellipsoid-based human backscattering model of human motion was employed to simulate EM response, with 31 joints used to model the human body. Every two adjacent joints define a body segment, and all segments are visible from any aspect angle, not taking into account the shadowing effect between different body parts. Approximation of the radar cross section of the resulting ellipsoid is then possible using the classical ellipsoid RCS formula in [26]. Part (a) of Figure 7.5 shows the position of bones, whereas part (b) shows the ellipsoid model-based human target. After modeling of the whole human body as an extended target comprising multiple ellipsoids, each segment's reflected amplitude is given by the approximated ellipsoid RCS. By coherently summing echoes from various ellipsoids, human range profiles can be generated. The simulated UWB signal's operational frequency band is 3–5 GHz, with a pulse repetition frequency of 600 Hz. The calculated radial velocity of body parts is presented in part (c) and the generated example spectrogram is shown in part (d) of Figure 7.5.

To generate sufficient number of spectrograms for each activity, scenarios exceeding 4 s were selected for simulation. Ultimately, 500 images were generated for each activity, which were then randomly assigned to a training set of 400 and a testing set of 100 images. The structures of the DCNN were established empirically, with the network architecture comprising a combination of three convolution layers and one fully connected layer. Before being imported into the network, each spectrogram

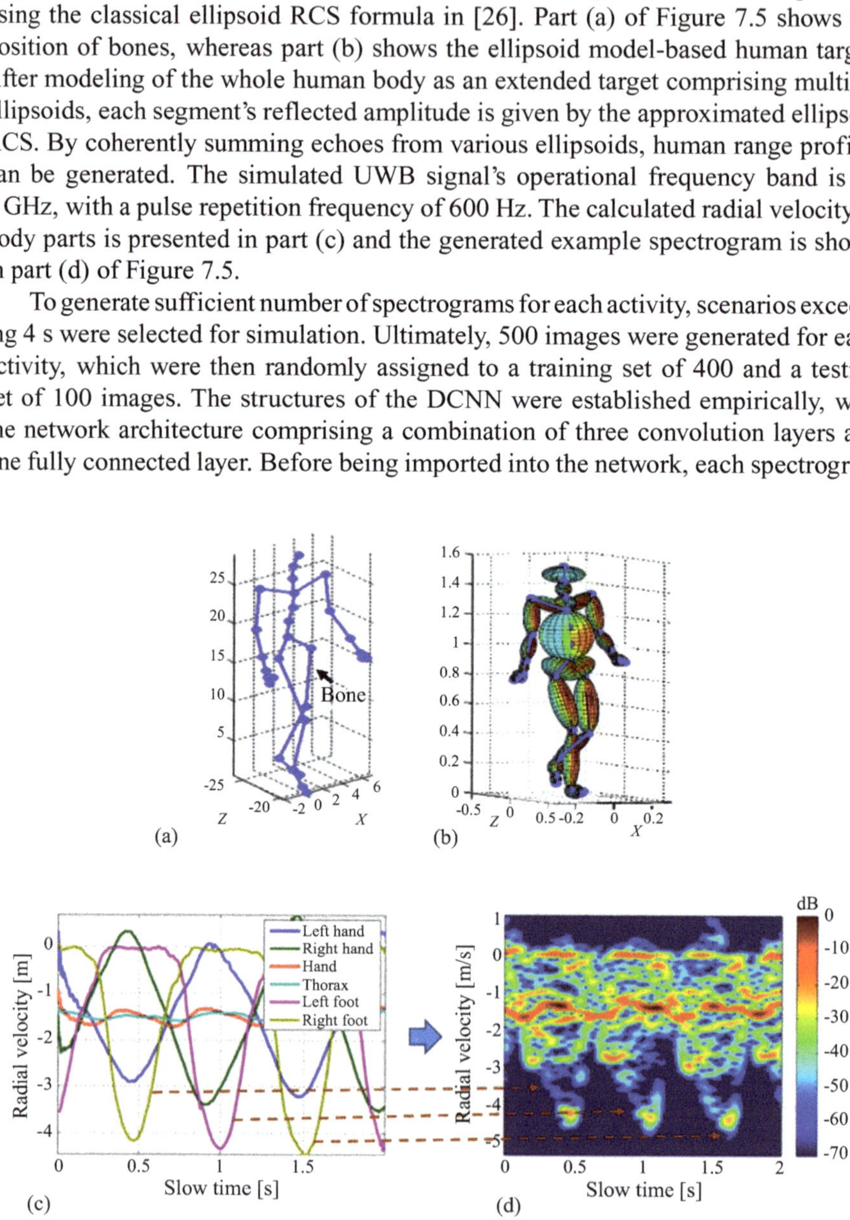

Figure 7.5 Model of human motion: (a) human body joints, (b) ellipsoid-based, (c) radial velocity of body parts, and (d) walking micro-Doppler [24]

was resized to 100×100. The convolution kernel's size was 9×9, with a stride of 1 pixel. Twenty feature maps, produced from the convolution layer, were processed sequentially using a ReLU and a 2×2 max-pooling layer, after which feature maps generated from the last convolution layer were fed into a fully connected layer followed by a softmax classifier. CNN hyper-parameters were optimized empirically to produce maximum test accuracy. Batches of 256 training images were used to train the network, with the basic learning rate set to 0.001. The open-source Caffe toolkit was used to train the network, a process accelerated by the use of the NVIDIA CUDA library and a Titan X GPU.

To preserve test accuracy amid fluctuations, the dataset was divided into five folds. The approach produced the test accuracies in the range of 98.00%–98.71%. The DCNN network trained on the generated dataset achieved an average test accuracy of 98.34% when classifying seven activities, among which boxing and creeping were the most difficult ones to classify. Excluding these two activities, the model classified the other five activities with 100% accuracy. Those misclassifications were derived from the similarities among the spectrograms of boxing, jumping, and creeping. Notably, the use of RGB versus grayscale spectrograms did not significantly affect classification accuracy. The accuracy depending on the size of sliding window is shown in Figure 7.6.

7.2.3 Classifying human gaits in multiple targets using DCNN

Notwithstanding successful classification of a single human gait, multi-target classification remains an open problem [27]. Because most current classification methods assume that a solitary moving target is within the radar beam, the multi-human problem is an interesting topic for deep learning, with classification involving a vast increase in the degrees of freedom dictating received time frequency response—e.g., number and height of persons and their walking pace and phase—and thus in complexity.

To investigate this problem, simulated micro-Doppler signatures were employed. Micro-Doppler simulation was based on the human gait model, with data generated using the model found in Boulic Thalmann [28]. A dataset was built comprising 2,000

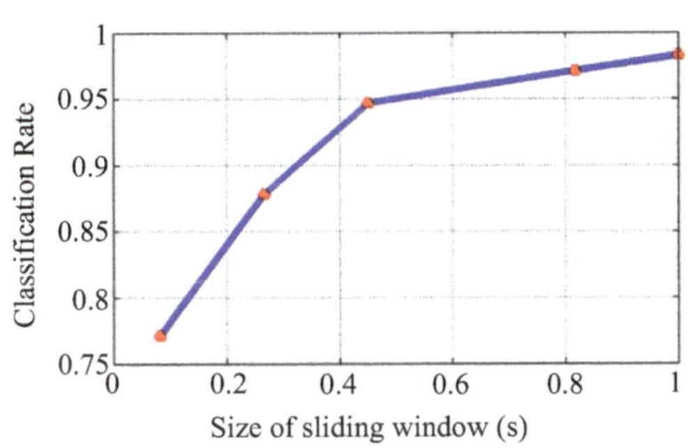

Figure 7.6 Classification accuracy of differently sized sliding windows [24]

unique time-signal examples of solitary walking individuals of heights ranging from 1.65 to 1.95 m, with subjects walking both toward and way from a simulated CW radar stationed about 1 m high. Datasets were produced at radar frequencies of 2.4, 3, 5, 7, and 10 GHz and the resulting radar signals were sampled with a Doppler sampling rate of 2 kHz. A six-class dataset was formed, with Class C0 representing the absence of human gait, as simulated by white Gaussian complex noise, and Classes C1–C5 representing one to five people walking, respectively. Samples for each class were obtained through the summation of an appropriate number of single-gait samples, randomly selected from each particular radar frequency's original datasets. Finally, varying degrees of white Gaussian complex noise were added to the composite signal to obtain datasets having signal-to-noise ratio (SNR) levels of 0, 5, 15, and 25 dB.

The resulting DCNN comprised fourteen layers, including eight convolutional layers with filter size 5×5, three 2×2 max-pooling layers that are connected to the first three convolutional layers each, two fully connected layers, and a final softmax layer for classification. All except the pooling and softmax layers used the ReLU as activation function. Dropout of probability 0.5 was applied to the two fully connected layers, and the network was trained using a fixed learning rate of 10e-4, 0.9 Nesterov momentum, and batch size 10. An NVIDIA GeForce GTX 980 Ti GPU with 6 GB of VRAM was used to train the network for 500 epochs, with each experiment taking about 12 h.

Table 7.1 gives the results of DCNN classification of model data for RF = 10 GHz and SNR = 25 dB, indicating that DCNN achieved the accuracy of 96.1%. Classes C4 and C5 are relatively hard to classify, but predicted number of gaits and true number scarcely differ. The simulation data were also used to test the effects of carrier frequency and SNR, with the use of higher frequency with high SNR found to yield more accurate classification. To verify DCNN effectiveness, the experiment was repeated using a three-nearest-neighbor classifier and a support vector machine (SVM) having a third-degree polynomial kernel, an approach that produced less accurate results.

For the verification of practical scenarios, measurements were conducted with an X-band CW radar used to measure 29 subjects (25 males and 4 females, aged 18–47) walking alone or side by side in groups of two or three on a radial trajectory away from and toward the radar. In addition, a non-gait class represented by various

Table 7.1 Classification accuracy depending on SNR and operating frequency [27]

Frequency (GHz)	**SNR (dB)**			
	25	**15**	**5**	**0**
10	96.1%	96.0%	91.0%	86.1%
7	92.9%	91.8%	88.8%	84.5%
5	91.8%	91.7%	86.4%	82.7%
3	86.9%	87.1%	86.0%	83.5%
2.4	84.9%	85.7%	81.3%	78.2%

other targets (cars, cyclists, rustling trees, etc.) was measured, so that four classes were created, corresponding to classes C0–C3 of the model data. The baseband radar signal, after sampling at 8 kHz and decimation to 2 kHz, was high-pass filtered to suppress static clutter, after which 128×192 spectrograms were created for the model data as described. For the multiple gait classes, training and test sets did not overlap in their unique combinations of subjects.

Similar results were obtained for the measurement data. For the feature extraction, PCA was applied to the spectrogram, and the resulting PCA coefficients served as a feature vector. The three-nearest-neighbor yields very high accuracy for the non-gait and single-gait classes but exhibiting poor and mediocre performance on the two-gait and three-gait classes, respectively, achieving 68.3% accuracy overall. The SVM was associated with good performance for the non-gait class and with poor to fair performance otherwise, for an overall accuracy of 60.3%, with confusion patterns much the same as for the DCNN. The accuracy of the DCNN exceeded 86%.

7.2.4 Recognizing hand gestures

The ability to recognize human hand gestures has a variety of essential applications in the fields of electronic device control, biomechanical research, computer gaming, and defense. The ability to control devices without physical contact is very convenient for users and can also increase safety, as when allowing the use of hand gestures for input in automobiles rather than requiring physical contact with buttons. Such an input modality would also be suitable for small electronic devices rather than requiring the use of uncomfortably small buttons—an approach that can do much to enhance reliability and flexibility.

Pulsed radar or frequency-modulated continuous-wave (FMCW) radar was developed to track finger motion. FMCW radar achieved an accuracy of approximately 89% for ten gestures [29], but the use of Doppler sensors offers a simpler, more cost-effective, and easier way of capturing radial velocity response. Therefore, the feasibility of recognizing hand gestures using only micro-Doppler signatures, i.e., no range information was utilized [30]. Because micro-Doppler signatures are depicted as overlapping signatures in the joint time–frequency domain when several scatterers exist—fingers, for example—careful investigation of spectrogram is needed to allow distinguishing of the signatures associated with gestures. Because spectrograms of such gestures differ only subtly, a powerful classifier is needed, inviting DCNN for application to spectrogram-based hand gesture classification.

A Doppler radar operating at 5.8 GHz, producing quadrature output at an average output power of 4.5 dBm, was used to capture hand gestures. The radar was fixed to a table, and ten hand motions were executed within the main lobe of its antenna at an average distance of approximately 10 cm: (a) swiping from left to right, (b) swiping from right to left, (c) swiping from up to down, (d) swiping from down to up, (e) rotating clockwise, (f) rotating counterclockwise, (g) pushing, (h) holding, (i) double pushing, and (j) double holding. Figure 7.7 depicts each gesture, with starting posture above ending posture. Because double pushing and double holding were repeats of the corresponding single motions, their pictures were omitted.

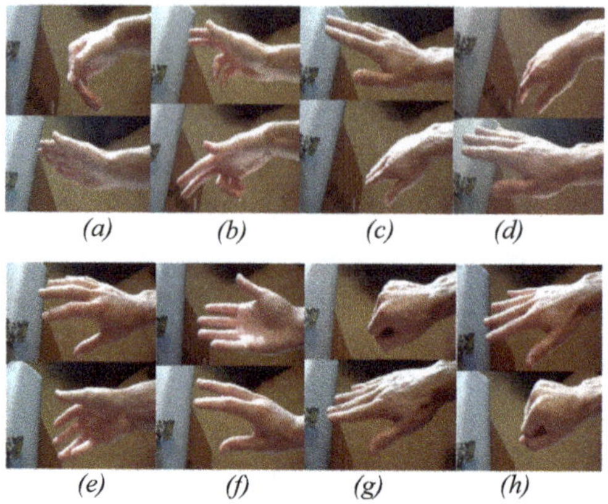

Figure 7.7 Eight of the ten hand gestures measured using Doppler radar: (a) swiping from left to right, (b) swiping from right to left, (c) swiping from up to down, (d) swiping from down to up, (e) rotating clockwise, (f) rotating counterclockwise, (g) pushing, and (h) holding [30]

With short-time fast Fourier transform (FFT), FFT size was set to 256 ms and the time step of nonoverlapping samples to 1 ms. Figure 7.8 depicts spectrograms of the ten gestures, with the micro-Doppler signatures of a single hand motion occurring within 200–300 ms and exhibiting marginally different features in the joint time–frequency domain. Each gesture was measured 50 times, for 500 total pieces of data. Differentiating among the subtle differences observed in the figure requires an effective image recognition technique rather than mere classification methods based on handcraft features.

A 2-s time window was used to crop the spectrograms for generation of the training dataset, after which each spectrogram was resized to 60×60 and values normalized from zero to one. Valid accuracy was obtained through 5-fold validation, with measured data divided into five different training datasets and test datasets. Three layers were used for the DCNN structure, having five, four, and two convolutional filters; for each the convolutional filter measured 5×5; in all pooling layers, the reduction ratio was 2:1. The numbers of layers and filters were empirically optimized for maximum accuracy. The number of epochs used was 90 in the training process because the error saturates after it.

Average classification accuracy for the ten activities considered was 85.6% after 5-fold validation. The confusion matrix shows gestures (a) and (b) as being highly inaccurate—an expected result, considering the similarities between activities (a), (b), (c), and (d). When DCNN performance was tested once more for eight activities, excluding activities (a) and (b), accuracy increased to 91.4%. When omitting (a), (b), and (c), accuracy reached 93.1%.

DCNN's generalization capabilities are particularly important in this context. Previous measurements assumed a DCNN trained on data from a particular user and were used to recognize that user's hand gestures only, in a controlled environment.

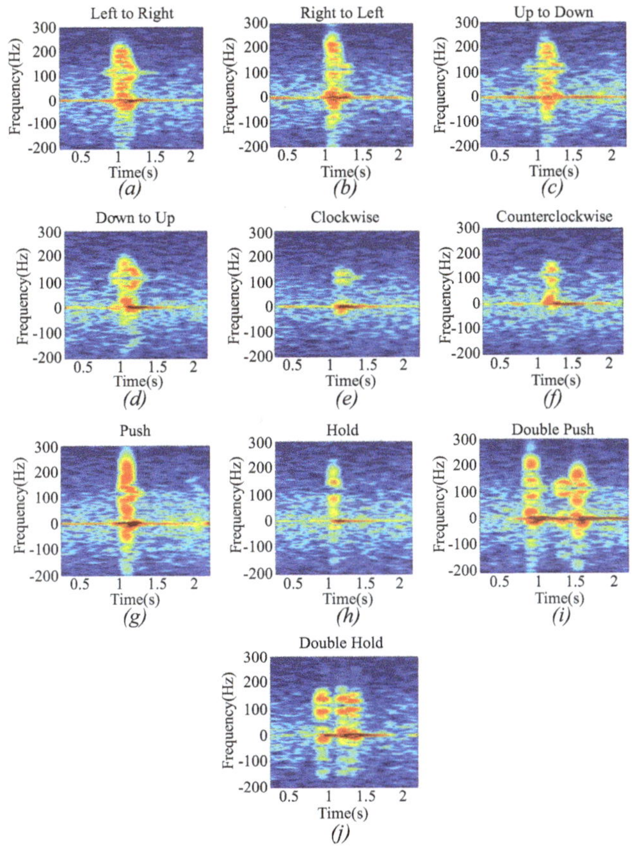

Figure 7.8 Spectrograms of ten hand gestures: (a) swiping from left to right, (b) swiping from right to left, (c) swiping from up to down, (d) swiping from down to up, (e) rotating clockwise, (f) rotating counterclockwise, (g) pushing, (h) holding, (i) double pushing, and (j) double holding [30]

By contrast, the way in which a trained DCNN would effectively recognize hand gestures in uncontrolled environments is an interesting question. Four scenarios were considered: measurements with (a) different incident angles, (b) different aspect angles, (c) different distances, and (4) a second participant.

In the first scenario, as Figure 7.9(a) shows, participants' hand gestures were measured at azimuth angles of incidence to the radar ranging from −45 degrees to +45 degrees. Doppler SNR was found to be decreased because the hand was not aligned with the main lobe of the radar antenna, which has a limited beam width of 60 degrees, so that received power decreases with an angle of incidence.

The second scenario, as depicted in Figure 7.9(b), measured hand gestures for nonzero aspect angles, with the position of the hand offset in position to produce the

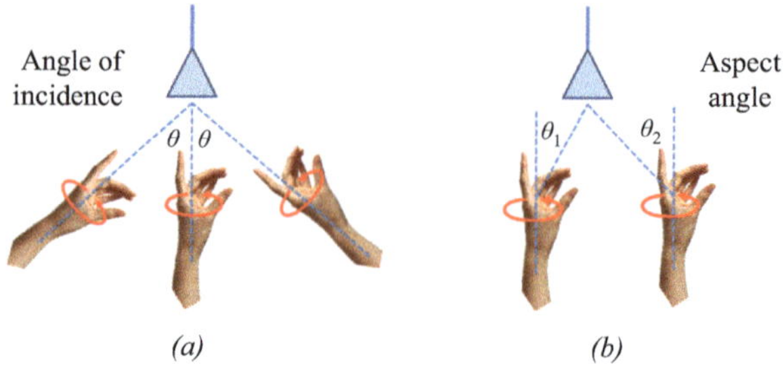

Figure 7.9 (a) With zero aspect angle and (b) with nonzero aspect angle [30]

desired aspect angle. Participant data were measured with θ_1 at 15 degrees and θ_2 at 45 degrees and their accuracy calculated: 81.12% and 57.5%, respectively, when using the previously trained DCNN.

The third scenario measured gestures at different distances with a hand aspect angle of zero. At distances of 15, 30, 45, and 60 cm, accuracies were 85.8%, 90.5%, 71.5%, and 66.7%, respectively. In general, the accuracy of classification decreased with distance owing to low SNR and variation in micro-Doppler signatures.

The fourth scenario measured seven gestures made by the second participant under the same conditions as before. Using the previously trained DCNN produced a classification accuracy of 71.5%, a reduction in accuracy that reflected differences in gesture between users.

7.2.5 Recognizing human voices

Speech is the most important mode of daily human communication. Detecting and recording the human voice are fundamental operations of audio communication, so that many different acoustic sensors have been developed through extensive research. Because the human vocal cord's vibrations modulate a radar wave when illuminated, voice information is contained in the returned radar signal, with the Doppler frequency linearly proportional to the frequency of acoustic vibration. Although detecting a single human voice in noisy surroundings—such as at a stadium, on a construction site, in a helicopter, or on a battlefield—can be a difficult endeavor, the use of a high-gain antenna can indeed allow radar to extract a solitary voice. In addition, the use of high power can allow voice detection over distances too far for an acoustic wave [31,32]. The feasibility of performing speech recognition tasks by means of Doppler radar in conjunction with DCNNs was investigated in [33].

A participant uttered the letters of the alphabet, from *A* to *Z*, 50 times, with pronunciations separated by 2 s intervals. During this process, vibrations from the neck and the mouth were captured. Three measurement scenarios were used: (a) no barrier, at a distance of 12.5 cm; (b) through a wooden panel 0.6 cm thick, at a distance of 12.5 cm, with the radar near the barrier; and (c) through the wooden panel at a distance of 35 cm, with the radar 22.5 cm from the panel. In the second scenario, unlike the third, the distance from radar to barrier was minimized to reduce reflection from the barrier, as shown in Figure 7.10.

The measured signal was converted to a spectrogram to represent the time-varying signature of the micro-Doppler signatures using the STFT. Figure 7.11 shows spectrograms for the letters A, C, D, G, J, K, M, V, and Y, which include voiced as well as unvoiced sounds. The horizontal lines are 60-Hz AC noise and its harmonics from the power source.

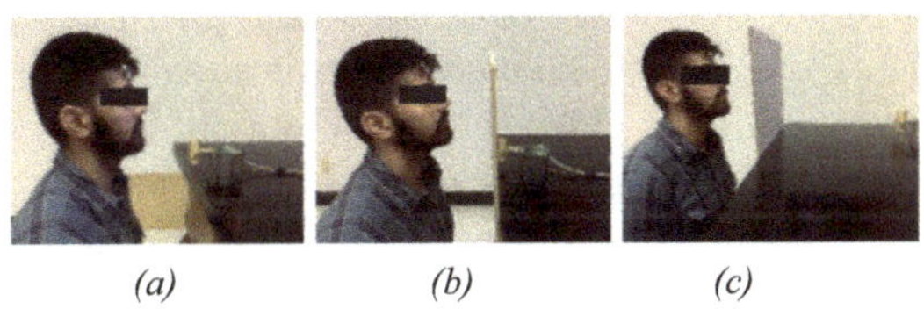

Figure 7.10 Measurement setup for the voice measurement using Doppler radar: (a) without barrier, (b) with barrier and nearby radar, and (c) with barrier and distant radar [33]

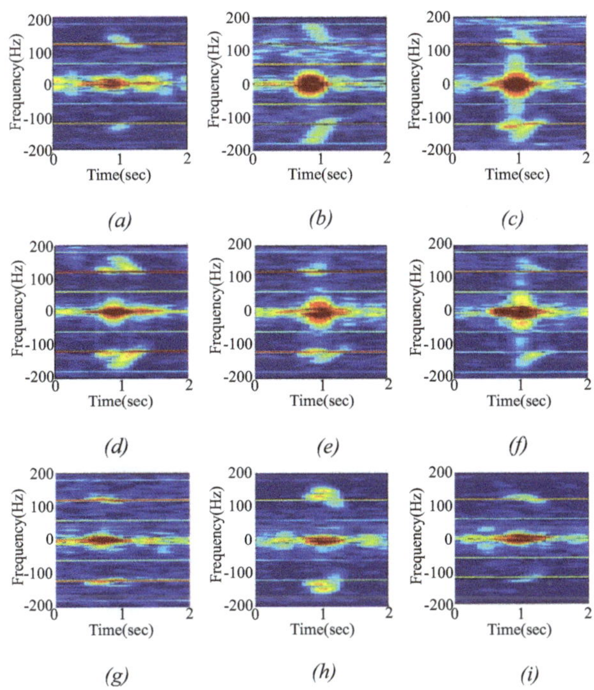

Figure 7.11 Measured letters in free space: (a) A, (b) C, (c) D, (d) G, (e) J, (f) K, (g) M, (h) V, and (i) Y [33]

The DCNN designed for this study has four convolutional layers, called DCNN-4 as shown in Figure 7.12. It comprises multiple layers, including image input, convolution, pooling, dropout, and fully connected layers, with hyper-parameters, including the number of filters, filter size, and dropout rate, heuristically determined. To allow the injection of spectrograms into DCNN-4, spectrogram size was adjusted from 426×245 to 50×50 to match the input layer, with a view to decreasing the computational complexity of neural network convolution. The first convolution layer had 64 convolution filters with a size of 2×2. The maximum pooling layer was then used, also with a size of 2×2. The second convolution layer included 128 filters of size 3×3, the third 256 filters of size 3×3, and the fourth 64 filters of size 2×2. Maximum pooling layers connected to each convolution layer were used to reduce the complexity of the computation, and the following dropout layer was used to mitigate the overfitting issue by reducing network size. Dropout rate was set to 0.5% and a fully connected layer constructed using 26 outputs, a total class being 26. Figure 7.12 depicts the DCNN-4's structure.

Each letter was measured 50 times, for a total of 1,300 images (26 letters×50 iterations). The measured data were divided into groups for training (60%), validation (10%), and testing (30%) and the initial learning rate set to 0.001. An NVIDIA GPU and MATLAB® were used to increase the speed of training. Classification was made using the SGD with momentum algorithm. The batch size was set to 16. The classification accuracy of 94% was achieved using the dataset from the first scenario but 95% when the DCNN was newly trained using the second dataset. Training the DCNN using the dataset for the third scenario yielded an accuracy of 91%. It should be noted that the data are measured from a single human subject, and the training set and the test set are completely separated in all scenarios.

7.2.6 Classifying drones using DCNN with merged Doppler images

Being much less expensive—and much easier to operate—than traditional aircraft, micro-drones are increasing in popularity for a variety of purposes, including surveillance, logistics, and aerial recordings. However, drones can also be misused in unsafe or even criminal ways—an issue made all the more serious because drones have a low radar cross section [34]. Although the micro-Doppler signatures generated by their

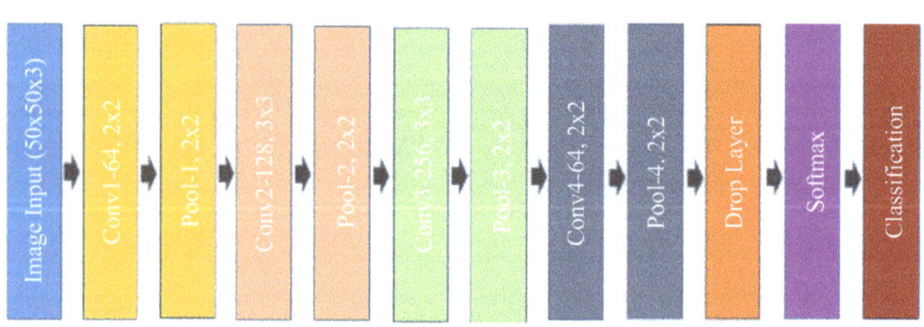

Figure 7.12 The network structures of the DCNN-4 [33]

blades provide important clues—yet they become invisible when the aspect angles of the radar and the rotating blades align. The micro-Doppler signatures of a drone pose a possibility of classifying the type of drone.

Periodic signatures in spectrogram are an important consideration when analyzing drone micro-Doppler signatures as the target's micromotions can be seen repeated spectrographically. Frequency signature analysis is possible using a cadence-velocity diagram (CVD), produced by Fourier transform of a spectrogram, iteratively for each row [35]. Because the micro-Doppler signatures depict the Doppler signal in the time domain, and the CVD in the frequency domain, a drone classification method was proposed on the basis of a Doppler image concatenating micro-Dopplers and CVD to reveal both time- and frequency-related Doppler features [36].

Two different drones were considered in [36]: a quadcopter produced by the DJI Corporation and a hexacopter produced by HobbyLord Corporation. The FMCW radar used operates in the band 14.03–14.180 GHz, at 1 W (in an anechoic chamber) or 20 W (for outdoor experimentation) transmitting power. Its dynamic range is 65 dB, from -120 to -55 dBm. Sweep time is 250 μs, and measured data output is 2,000 chirps. The measurement of one set of data takes 500 ms, at 100 M samples per second. For spectrogram generation, the STFT was applied using a Hamming window and an overlapping length of 31 samples, 7.7 ms.

The application of DCNN to the measured data allowed the classification of drone type in various situations using Szegedy's GoogLeNet DCNN model, designed to significantly boost classification rate while staying within computational budget constant by using 22 layers of inception modules. Consistent with GoogLeNet requirements, images output from measured radar signal were saved with dimensions of 256×256 pixels. The open-source CNN training system NVIDIA DIGITS, based on the Caffe toolkit and cuDNN library, was employed for the operation of GoogLeNet. Using a heuristic approach, it was found to be best with a beginning learning rate of 0.001, when using the SGD algorithm and a batch size of 4. An NVIDIA GeForce GTX 970 with 4 GB of VRAM was used for this analysis. The measurement setup is shown in Figure 7.13.

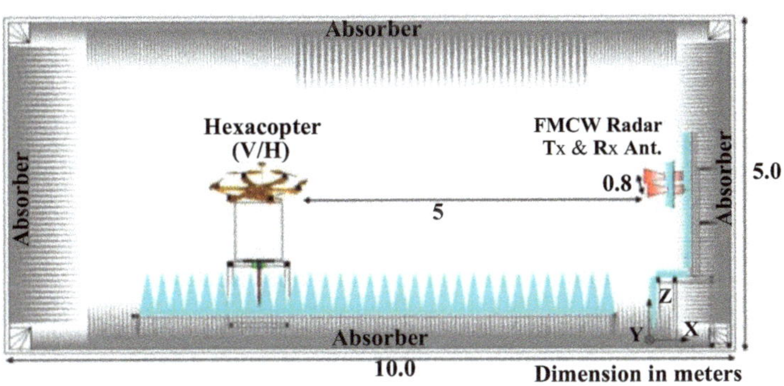

Figure 7.13 Hexacopter anechoic chamber measurement setup, indicating directions and the number of operating motors [36]

The FMCW radar's transmitting and receiving antennas were separated by 0.8 m, with the target drone 5 m from the antenna; sufficient data were obtained by changing distances between the radar and the drone within 1 m. To account for the environment, the radar's output power was set to 1 W. Classification was then made by micro-Doppler signature and CVD in various situations, with returned signal recorded for ten different situations involving the hexacopter, with the number of operating motors (1–4, 6) adjusted to test classification rate for each direction. Altogether, 50,000 images were generated from the anechoic chamber and the 4-fold validation was used. For evaluation purposes, the images in the dataset were assigned either to a training set or to a testing set, at a 4:1 ratio. With the application of a DCNN classification model based on micro-Doppler signatures, classification accuracy averaged about 99.38%, a figure that dropped to 98.11% when using CVD images. To overcome the disadvantages of both the spectrogram and CVD approaches, a new image was created that would concatenate the images produced by both, as shown in Figure 7.14. Using this approach, average classification accuracy increased to 99.59%.

The outdoor experiment measured two types of drones, quadcopter and hexacopter, hovering at a constant height. The transmitting and receiving antennas were installed normally at ground level, 20 m apart, with the FMCW radar system transmitting at 20 W. Classification was performed using various drone types at different measurement heights, so that measurement data fell into four categories. A total of 10,000 images were generated from the outdoor measurement and 4-fold validation was used. The images in the dataset were assigned either to a training set or to a testing set, at a 4:1 ratio. Three image types were classified with 100% accuracy owing to obvious differences between drones and returned power. CNN classification results revealed the proposed approach's suitability for use in actual drone-detecting radars, as each drone type produced different micro-Doppler signatures and CVDs. As the number of rotors increases, more various frequency components could be observed in CVDs.

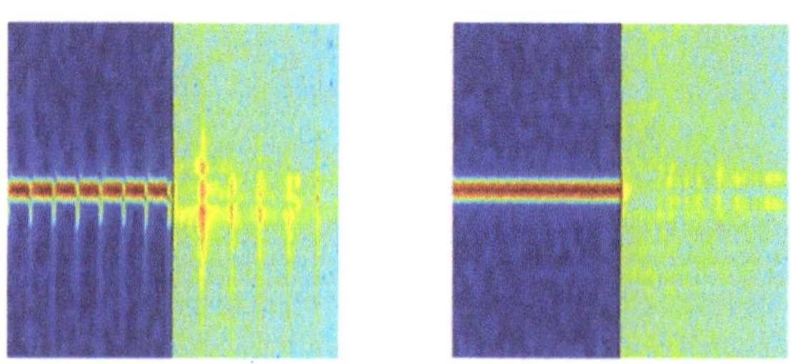

(a) (b)

Figure 7.14 Merged image, based on the measurement of the hexacopter in anechoic chamber with one motor operating, at (a) vertical and (b) horizontal fixation [36]

7.3 Classification of micro-Doppler signatures using transfer learning

Although a DCNN trained from scratch using the collected spectrogram was in these instances able to learn useful features of spectrogram and classify with a high degree of accuracy, limited training data ranging from a few hundred to a few thousand pieces of data can hardly be expected to reveal the full potential of a DCNN. Deep neural networks generally require large datasets for full training of a model's many parameters. Indeed, for optical applications millions of images are used for training of deep neural networks containing millions of parameters. The DCNN used for micro-Doppler classification thus had little chance of being fully trained, leaving room for further improvement with availability of still more data.

For overcoming issues arising from deficiencies of data, in particular for human motion analysis problem, three methods have been suggested: generation of radar image through simulation, transfer learning, and data augmentation using the generative adversarial networks (GANs). Electromagnetics simulators or analytical models can construct computer-simulated radar images. Micro-Doppler signatures were simulated using a human gait model [37], human animation files [38], or Kinect-based motion capture data [39,40]. Variances in human size, posture, and activity, as well as of radar frequency, aspect angle, and location, can allow the production of diverse micro-Doppler signatures after the development of a simulator. However, such an approach requires the use of analytical target modeling skills derived from electromagnetic theory. Therefore, simulator implementation should correspond to the specific problem to be solved.

Transfer learning [41], by contrast, involves transference of knowledge gained or models constructed during a certain task to a similar task (i.e., a target task) that lacks sufficient training data, with a view to improving accuracy on the target task. Of the various transfer learning techniques, DCNN transfer learning is particularly suitable for radar applications. Such an approach involves replacing an already trained DCNN's feature extraction layers to accommodate the new task, then updating DCNN parameters using the training data from the target task. Only the input layer and the last layer of the network need retraining for small datasets, large convolutional layers being fixed, so relatively little computational complexity is involved. The use of transfer learning to classify micro-Doppler signatures has seen success in many applications [42,43].

Two large neural networks have achieved notable popularity: AlexNet and VGG-16. The AlexNet conventional neural network, which demonstrated the highest classification accuracy seen at the 2012 ImageNet Large Scale Visual Recognition Challenge (ILSVRC), catalyzed recent research into DCNNs [44]. In seeking to classify ImageNet data containing 1.5 million images from 1,000 categories for the ILSVRC competition, AlexNet employed eight main layers and about 60 million parameters; five of the layers were convolutions, followed by three fully connected layers, with a ReLU activation function layer added after every convolutional and fully connected layers to improve performance. A max-pooling layer was also used to downsample the output and reduce the size of the network. A dropout layer followed the first two fully connected layers to reduce the number of neurons and reduce overfitting. A softmax layer for the classification of the input data followed the last fully connected layer. Figure 7.15(a) shows AlexNet's structure.

VGG-16, one of the best known trained DCNNs, outperformed all competitors at ILSVRC 2014. It comprises 16 main layers, 13 of which are convolutions that are followed by 3 fully connected layers as shown in Figure 7.15(b) [45]. VGG-16 has 138 million parameters, and all of its convolutions layers have the same filter size. It uses ReLU layers, max-pooling layers, fully connected layers, and dropout layers. The examples of using AlexNet and VGG-16 are described in the following sections.

7.3.1 Classifying human aquatic activities using transfer learning

Many researchers have investigated the use of micro-Doppler signatures for ground-level detection of humans and activities, recognition of gaits, and motion analysis, but detection of humans and classification of activities on water has not been extensively studied yet, despite their potential utility for rescue operations and border patrols.

The feasibility of classifying human aquatic activity based on measurements was investigated in [46]. The measurement of a human subject in a swimming pool allowed to focus on detecting human motion only—unlike in a sea or lake, where waves produce hard-to-distinguish interactions with the Doppler signal. Doppler radar operating at 7.25 GHz was used with a power output of 15 dBm. Vertical polarized Vivaldi antennas were employed on the assumption that human arm motion effectively interacts with the vertically polarized EM waves. The measurement pictures and their corresponding spectrograms are shown in Figure 7.16.

In constructing the training and test datasets for our DCNN-based approach, a single human subject was measured five times for each activity. For each measurement, five spectrograms are generated with 2-s intervals (100 pixels) with random

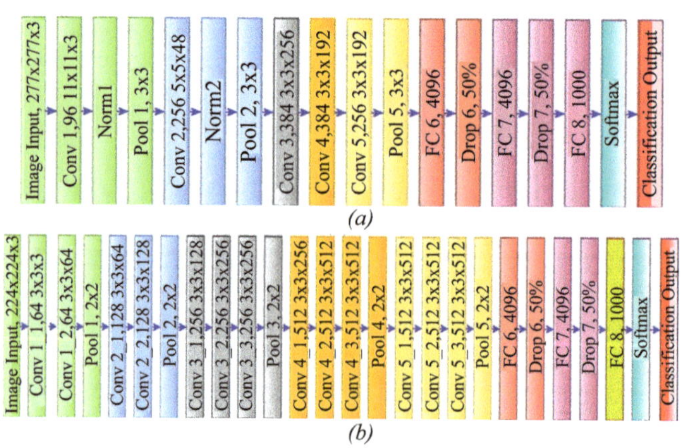

Figure 7.15 Structure of (a) AlexNet and (b) VGG-16

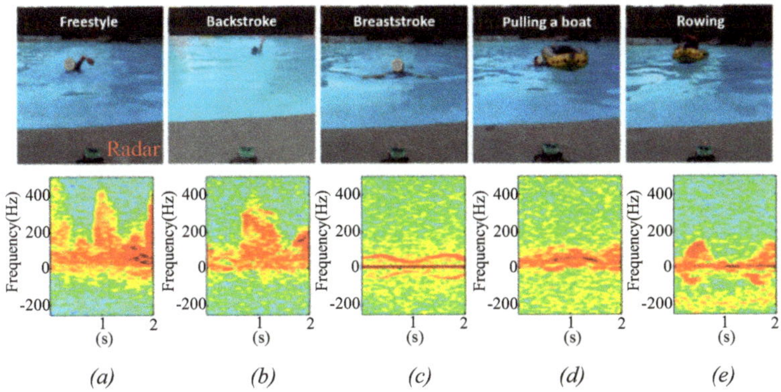

Figure 7.16 Pictures and spectrograms of (a) freestyle, (b) backstroke, (c) breaststroke, (d) towing a boat, and (e) rowing a boat [46]

starting points. In the cropped spectrogram, Doppler frequency was between -500 and 500 Hz (256 pixels). A total of 625 data samples (i.e., spectrograms), representing five actions, with 25 samples for each action for every five subjects, were constructed. Each spectrogram measured 252 (frequency) $\times$ 100 (time).

Acutely aware that the paucity of our water-based training data (625 samples) might prevent realization of the DCNN's full potential, AlexNet and VGG-16 were used to experiment with *transfer learning* DCNNs. The data from the spectrogram's single channel was copied to each of AlexNet's three channels because AlexNet is designed for RBG color images. To fine-tune network parameters, AlexNet's final softmax layer was replaced with a new perceptrons having five classes then updated all the network parameters with the spectrogram data. The same procedures were used for transfer learning with VGG-16, which are called *DCNN-TL-AlexNet* and *DCNN-TL-VGG16*.

The 5-fold CV was used. For the DCNN models, with their model configurations (e.g., the number of layers, number of convolution filters), only one hyper-parameter was determined via CV, the early stopping parameter, thus determining which SGD iteration produced the best average test score. Caffe was used to implement the DCNN and an Intel Xeon E5-2600-v3 CPU and NVIDIA GTX Titan X GPU for our experiments.

We verified the effectiveness of the transfer learning by classifying the resulting spectrograms using a classifier similar to those developed in [17], taking the conventional SVM method as a baseline. Two DCNNs were also designed to allow comparison with the performance achieved by transfer learning.

Table 7.2 summarizes the CV results, showing that the baseline SVM, using handcrafted features that have been used for [10], achieved an accuracy of 45.1%. Although this result clearly exceeded what could be expected as a result of random guessing among five activities (i.e., 20%), handcrafted features are clearly not sufficiently generalized to water-based activities. By contrast, DCNN-Scratch-I and DCNN-Scratch-II achieved accuracies of 64.0% and 66.7%, respectively—significantly exceeding the results obtained using the baseline SVM (for ~40% improvements). These results demonstrate DCNN's robustness when used for micro-Doppler classification: instead of requiring the design of a separate set of features for each different classification task, DCNNs can learn features directly from a new task's raw data and still achieve high accuracy. What is more, the transfer-learning models, DCNN-TL-AlexNet and DCNN-TL-VGG16, were 74.6% and 80.3% accurate, respectively—significantly more accurate than DCNN models that learned from scratch. What is more, DCNN-TL-VGG16 was 78% more accurate than the baseline SVM. Clearly, then, DCNN-learned features can be of great use in RGB image classification, even when transferred to a micro-Doppler human activity classification problem having few spectrographic data for use in training. Because the DCNN learns features hierarchically, low-level features learned for RGB image classification, such as edge or texture detection, can be used to detect useful micro-Doppler signatures for purposes of classification.

Figure 7.17 visualizes sample input spectrogram data passed through the convolution layer of DCNN-TL-VGG16, the most accurate model. Part (a) of Figure 7.17 shows raw spectrographic input for a "freestyle" motion, whereas part (b) depicts visualizations of the feature maps (i.e., the results of each filter's convolution) after

Table 7.2 Five-fold CV results for the compared schemes [46]

Accuracy (%)	**Fold 1**	**Fold 2**	**Fold 3**	**Fold 4**	**Fold 5**	**Ave**
SVM	27.2	36.0	56.8	70.4	35.2	45.1
DCNN-ScratchI	57.6	53.6	98.4	58.4	52	64.0
DCNN-ScratchII	68.0	51.2	98.4	63.2	52.8	66.7
DCNN-TL-Alex	60.0	79.2	93.6	69.6	70.4	74.6
DCNN-TL-VGG16	70.4	72.0	99.2	82.4	77.6	80.3

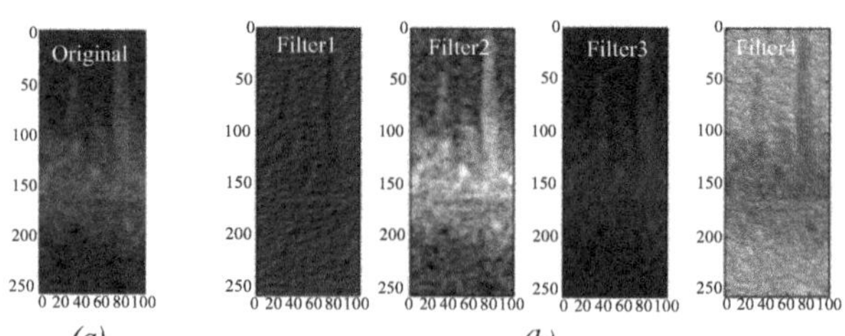

Figure 7.17 Visualization of feature maps: (a) raw spectrogram for "freestyle" and (b) visualization of some DCNN-TL-VGG16 feature maps after the first convolution layers [46]

the first convolution layer. The experiments found that DCNN-TL-VGG16's convolution filters are able to capture various aspects of the input spectrogram, including textures and edges, and thus allow a high level of accuracy in classification using those captured features.

7.3.2 Recognizing human voices using transfer learning

The problem of human voice recognition, already addressed in the previous subsection, was revisited to assess the effectiveness of transfer learning. The VGG-16 was employed for the same dataset used in [30]. Training time for transfer learning was reduced because most of the DCNN was already trained. For transfer learning, the number of nodes for the fully connected layer was set equal to the total number of classes—26 in this study. As VGG-16 requires the input image with the size of 224×224, 50×50 images were linearly interpolated to the 224×224 images so that they would contain the same information. As Table 7.3 shows, transfer learning using VGG-16 produced an accuracy of 97% for the first scenario and accuracies of 96% and 94% for the second and third, respectively.

To allow the emulation of a practical scenario, the classification accuracy of through-wall data was classified by the DCNN trained using free space data, so that the training dataset did not include through-wall data. Because barrier data are not available in most real cases—barrier information being generally unknown—the use of free space data was reasonable. This approach achieved accuracies of 94% and 88% for the second and third scenarios as shown in Table 7.4, respectively, using VGG-16. Although classification accuracy was reduced as expected, these results demonstrated the feasibility of recognizing speech in through-wall environments.

Table 7.3 Classification accuracy of the 26 letters [33]

Accuracy	Free space (%)	Barrier near the radar (%)	Barrier far from the radar (%)
Designed DCNN-4	94	95	91
VGG-16	97	96	94

Table 7.4 Classification accuracy of 26 letters measured when using the barrier, with VGG-16 trained by free space data [33]

	Barrier near the radar (%)	Barrier far from the radar (%)
Accuracy	94	88

7.3.3 Classifying moving targets using transfer learning with DCNN and micro-Doppler spectrogram

Transfer learning can also be used to classify ground-moving targets by their micro-Doppler signatures [47]. It was validated using the RadEch database published by the Military Academy, Belgrade, Serbia, which contains radar echoes of various targets, collected in controlled test environments on the premises of the Military Academy. The pretrained AlexNet model was used, which extracts features from a DCNN layer to train a softmax classifier.

In particular, this study utilized a data augmentation technique to construct a large data set. Many simple techniques are available for data augmentation, such as rotations at different angles, vertical and horizontal flips, scaling, addition of noise, and color manipulation, and these can reduce overfitting. Vertical flipping was employed to the spectrograms for data augmentation and performed a circular shift of spectrograms' original echo signals by 2, 3, 4, 5, 6, 10, 20, 30, 40, 50, 100, 150, 200, 250, and 300 samples. The original and augmented spectrograms constructed the training data set. A 70% of the dataset was used for training and the remaining 30% for testing.

Table 7.5 depicts softmax classification accuracies using the output from each layer of AlexNet, which—as expected—improved after data augmentation: the best result achieved without augmentation was 94.8%, whereas conv2, conv3, and conv5 achieved a maximum accuracy of 99.9% using augmented data. The result indicated that the use of deeper layers and data augmentation is effective for classification. However, the spectrogram augmentation should be carefully addressed because the physics of the Doppler signal needs to be incorporated in augmenting spectrogram. For example, rotation or horizontal flip, which violates underlying physics, would not be appropriate to augment micro-Doppler data.

7.3.4 Comparing transfer learning to data augmentation using generative adversarial networks

Although transfer learning produced successful results in many applications, that the feature extraction layers borrowed were optimized for optical images begs the question of whether transfer learning is ideal for use with radar images. Another way of solving the data deficiency problem involves augmentation of data using GANs. GANs were proposed by Goodfellow [48] and it is regarded as one of the most innovative models of the past decade: a neural network designed to generate similar data distributions when fed training data. If the statistical properties of GAN-generated images are

Table 7.5 Classification accuracy achieved using each state of AlexNet, with and without data augmentation [47]

Layer	Conv1	Conv2	Conv3	Conv4	Conv5	Fc6	Fc7	Fc8
Accu. w/o augm. (%)	81.5	88.9	94.8	94.1	91.9	89.6	89.6	88.1
Accu. with augmen. (%)	99.2	99.9	99.9	99.7	99.9	99.6	91.8	91.0

similar to the properties of the training data, then such images should be able to help train DCNNs. Accordingly, GANs have been used to create simulation-based spectrograms of human activity [49], open-set data [50], and measurement-based spectrograms [51].

A GAN's two neural networks, the generator and the discriminator as depicted in Figure 7.18, operate according to a game-theoretic approach in which they "play" against each other, with the generator producing a fake dataset from random noise and the discriminator using traditional supervised learning to evaluate the output data by assessing the probability that the data belong to the real dataset. The discriminator is first trained to distinguish real data from synthetic data using samples from the real dataset, which—along with its fake counterpart—will also be used later. If generated data appear highly authentic, then the discriminator's accuracy diminishes. Accordingly, the generator, having been trained to decrease the discriminator's accuracy, becomes able to produce, at little computational cost, large quantities of fake data that closely resemble real data.

7.3.5 GANs for augmentation of simulated human data

The first micro-Doppler data augmentation using GANs was addressed by classifying human micro-Doppler signatures generated by simulation. DCNNs require a sufficient quantity of training samples for accurate classification. Accordingly, it was suggested to use a deep convolutional GAN (DCGAN) to increase the number of sample data and to employ DCNN for the classification [49]. In particular, the pre-trained VGG-16 was used to implement the transfer learning. Three types of human activity using the Boulic kinematic model that can describe the human walking model were simulated. When using the DCGAN, the input image batch size was set to 64; a learning rate of 0.0005; and 260, 200, and 115 epochs for three types of human activities. Input and output images were sized 600×600 and 512×512, respectively. Figure 7.19 presents the example of generated spectrograms.

For Experiment I, the training sample comprised 60% of the total sample, with the remaining 40% serving as the test sample; both samples comprised real data only. The achieved accuracy of 99.82% confirmed VGG-16's effectiveness. In Experiment II,

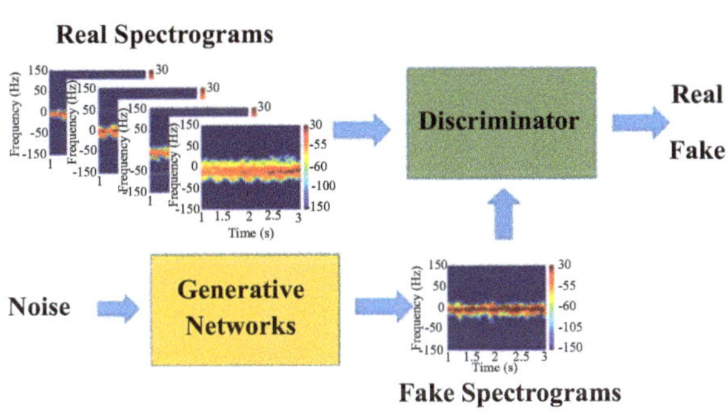

Figure 7.18 Structure of GANs [51]

the training set comprised two parts, with the real samples the same as in Experiment I but the fake samples generated by DCGAN. Each class contained 1,500 samples. To increase training speed, the same number of samples was used for each class. Using the same testing set as for Experiment I produced an accuracy of 99.91%—indicating that DCGAN can enrich training sets and boost network accuracy.

7.3.6 Use of GANs for open-ended applications

Use of radar data to recognize human activities ultimately seeks to achieve automatic recognition based on radar spectrograms. Most classifiers currently available are for "closed-set" recognition, in which training set and testing set share categories; open-set recognition is a relatively novel topic. The real world, however, comprises countless elements, making enumeration and labeling of all targets impractical. Much the same is true for the recognition of human activities, with the caveat that a lack of large-scale micro-Doppler datasets makes the use of DCNN inapplicable to micro-Doppler data. Because manual collection of negative samples is both time-consuming and labor-intensive, the problem of recognizing open-set radar data persists.

To address this open-set radar classification, micro-Doppler signatures from measurements and simulations were prepared and used to train the GANs, after which human activity data were collected using the PulsON 440 (P440) ultra-wideband (UWB) radar module, which operates between 3.1 and 4.8 GHz using two directional antennas [50]. The experiments took place indoors, with the radar 1.2 m high, the better to align with the human center of gravity. Four subjects performed seven activities facing the radar in its line of sight: (a) boxing in place, (b) crawling, (c) creeping, (d) jumping forward, (e) running, (f) standing (with slight movement), and (g) walking. More than 900 images obtained from each activity were resized to 120×120, with 400 images then chosen as the training set and 100 as the testing set for each class.

Micro-Doppler spectrograms were also simulated from MOCAP data, using an ellipsoid-based human backscattering model in which 31 joints describe the human body, with every two adjacent joints defining a body segment. The simulated UWB module's frequency band was set to 3–5 GHz, with a pulse repetition rate (PRF) of 600 Hz. Each activity had 400 spectrograms in its training set and 100 in its testing set.

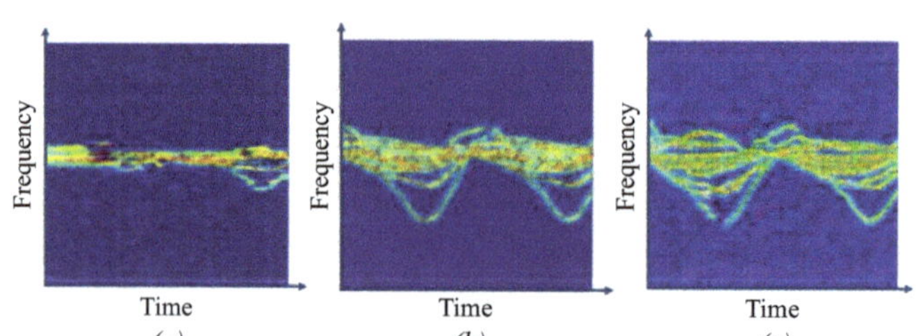

Figure 7.19 Generated samples of different human activities based on DCGAN: (a) slow walking, (b) fast walking, and (c) extremely fast walking [49]

The novel OpenGAN model, designed for open-set detection and recognition of human activities, features a generator and discriminator composed of two primary parts: (a) dense blocks for feature extraction and for the combination of multilevel features and (b) transition layers connecting two blocks. Network parameters were empirically determined. The model, inspired as it is by generative adversarial networks, uses synthesized samples to automatically construct a negative set from the generator, with no manual collection required. The discriminator is then modified through reasonable modulations to adapt it to open-set tasks, using both the measurement radar dataset and the MOCAP dataset to verify the model's effectiveness. In extensive experimentation, OpenGAN outperformed several comparison algorithms when dealing with open-set human activities, revealing the robustness of this method compared to existing approaches. Exploratory experiments conducted on the network structure also found that a network having one sub-layer in each dense block produced optimal performance under most circumstances.

7.3.7 Human activity classification with the aid of GANs

The use of GAN is an interesting topic, but it raises a pressing question: is transfer learning or a GAN-based approach more effective when radar databases are few? Because transfer learning and GANs are applicable to any radar image classification problem without the need for physical knowledge involved in radar return simulation, a comparison is provided between GANs' performance and that of transfer learning for the classification of micro-Doppler signatures related to human activities, using the dataset featured in [19] which is in order. The data set consists of 7 (activities)$\times$12 (subjects)$\times$4 (iterations)$\times$3 (extractions)=1,008 spectrograms. Extracted spectrograms measured 300$\times$140 but were reduced to 64$\times$64 to increase the computational speed while preserving classification accuracy.

Part (a) of Table 7.6 depicts the architecture of the generator neural network, whereas part (b) depicts that of the discriminator. In contrast to the original resolution of 300$\times$140, GAN output size was set to 64$\times$64 to minimize computational complexity while preserving classification accuracy [51]. When training the GANs, Xavier initialization was used to initialize neural networks' weight and biases, with the Adam optimizer used to minimize the generator and discriminator loss function. A sigmoid function was used both as the activation function and to calculate generator and discriminator cross-entropy. A mini-batch size of 64 and a learning rate of 0.005 were selected, and an Intel i7 processor and an NVIDIA GeForce GTX 770 GPU were used to train the GANs, with Python and TensorFlow used for coding. All GANs' specifications were identical except for epoch selection, the number of iterations required for GAN saturation being different. Visual inspection was also used to evaluate GAN-produced data.

One GAN was trained for each activity using data from all human participants, which totaled 144 images (12 humans$\times$4 iterations$\times$3 extractions). After training, the GANs produced 1,440 pieces of data for each activity. Figure 7.20 shows visualizations of fake data, with produced micro-Doppler signatures notably similar to the original data.

Table 7.6 Architecture of (a) the generator neural network and (b) the discriminator neural network [51]

Generator neural network	Discriminator neural network
Noise, size=100	Image, size=64×64×1
Fully connect	Conv2d, stride=2, filters=64, size=4×4
Leaky rectification unit	Batch normalization
ConvTranspose2d, stride=2, filters=256, size=4×4	Leaky rectification unit
Batch normalization	Dropout (0.4)
Leaky rectification unit	Conv2d, stride=2, filters=128, size=4×4
Dropout (0.4)	Batch normalization
ConvTranspose2d, stride=2, filters=64, size=4×4	Leaky rectification unit
Active function	Dropout (0.4)
	Fully connect
	Leaky rectification unit
	Fully connect
	Active function
(a)	(b)

The original images and the images produced by the GAN were then used to train DCNN-6 with its six main layers: three convolution layers and three fully connected layers. The DCNN features 16 filters of size 8×8 for the first convolution layer, 32 of size 4×4 for the second, and 64 of size 2×2 for the third. DCNN structure and hyperparameters were heuristically optimized. Proof sets comprising 10% of the original data were employed after training, with a different set of randomly selected images used in each fold. Four-fold validation was used to evaluate the suggested strategy fairly. For training, the dataset was divided into three sets: 80% for training, 10% for validation, and 10% for testing, with the validation set used to provide early stop capabilities so as to avoid overfitting. A batch size of 64 and initial learning rate of 0.0005 were used in conjunction with an SGD and momentum optimizer.

Altogether, (1,440+144)×7=11,088 samples were used for the DCNN training, validation, and test process, producing a dataset 10 times bigger than the original. DCNN-6 achieved average accuracies of 100% over the training set, 99.55% over the validation set, and 99.71% over the testing set, as well as of 97.60% over the proof set (the original data not used for training). The resulting accuracy exceeded the results of other methods used to date [17,25].

To compare the performance of transfer learning with that of the suggested approach, AlexNet and VGG-16 were used. The dimensions of the input layer and the last fully connected layer were changed to accommodate the new data set. An early stop algorithm was implemented to prevent overfitting. AlexNet achieved an accuracy of 94.34% and VGG-16 an accuracy of 95.56%—the latter is slightly higher perhaps because of the high complexity of the network. Both results were exceeded,

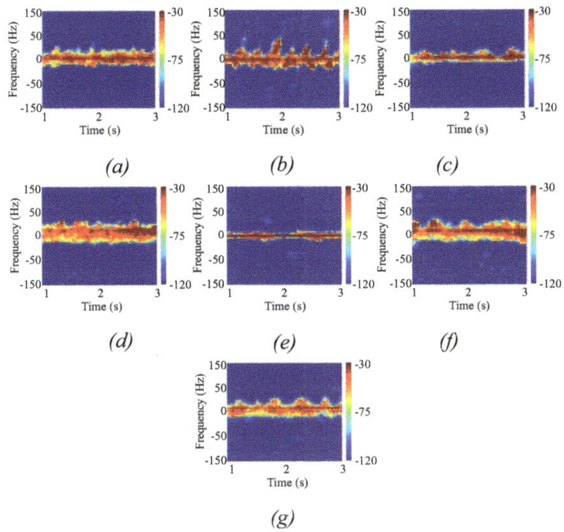

Figure 7.20 Example images produced by GANs: (a) boxing while advancing, (b) boxing in place, (c) crawling, (d) running, (e) sitting still, (f) walking, and (g) walking with the aid of a stick [51]

however, by the accuracy of using the GANs images. For this application, the use of GANs was more effective than the use of transfer learning.

7.4 Conclusion

This chapter discussed applications of DCNN for classification of micro-Doppler signatures. Through joint time–frequency analysis, temporal micro-Doppler signatures can be represented as an image so that the powerful image classification tool DCNN can extract meaningful features through the convolutional layers, allowing construction of the decision boundary through the fully connected layers. DCNN's outperformance of conventional classification methods has been verified for scenarios ranging from human activity classification, human gait analysis, and hand gesture recognition to drone detection and voice recognition. Accordingly, the use of DCNN for target classification based on micro-Doppler signatures is well suited to achieving high accuracy.

However, a few considerations remain. Because classification accuracy is a function of spectrogram quality, investigation of which joint time–frequency analysis contributes most to DCNN's efficacy is needed. The use of high-resolution images should yield high classification accuracy but has not yet been confirmed to do so. In addition, DCNN's real-time processing capabilities might give rise to problems related to increased computational complexity. In particular, DCNNs using transfer learning based on AlexNet and VGG-16 consist of large networks containing millions of parameters. Embedded systems' memory and computational power would not be sufficient to conduct the needed calculations in real time. Accordingly, the DCNN should be designed to process data within a reasonable time considering available memory and computational power.

References

[1] Chen V.C. 'Analysis of radar micro-Doppler with time-frequency transform'. *Proceedings of the Tenth IEEE Workshop on Statistical Signal and Array Processing* (Cat. No. 00TH8496), IEEE, Pocono Manor, PA, USA; Aug 2000.

[2] Chen V.C., Li F., and Ho S.-S., Wechsler H. 'Analysis of micro-Doppler signatures'. *IEE Proc.—Radar Sonar and Navig.* 2003, vol. 150, pp. 271–6.

[3] Chen V.C., Li F., and Ho S.-S., Wechsler H. 'Micro-Doppler effect in radar: phenomenon, model, and simulation study'. *IEEE Trans. Aerosp. Electron. Syst.* 2006, vol. 42, pp. 2–21.

[4] Chen V.C. and Ling H., *Time-frequency transforms for radar imaging and signal analysis*. Artech House, Boston; 2002.

[5] Tan C.M. and Lim S.Y. 'Application of Wigner–Ville distribution in electromigration noise analysis'. *IEEE Trans. Device Mater. Reliab.* 2002, vol. 2(2), pp. 30–35. doi:10.1109/TDMR.2002.802114.

[6] Tan R., Lim H.S., Smits A.B., Harmanny R.I.A., and Cifola L. 'Improved micro-Doppler features extraction using smoothed pseudo Wigner–Ville distribution'. *IEEE Region 10 Conference*, Singapore, Singapore, Nov 2016, pp. 730–3. doi:10.1109/TENCON.2016.7848099.

[7] Tuncer E. and Friedlander B. *Classical and modern direction of arrival estimation*. Elsevier, Amsterdam; 2019.

[8] Godana B., Leus G., and Barroso A. 'Estimating indoor walking velocity using a software radio-based radar'. *First International Conference on Sensor Device Technologies and Applications*, Venice, Italy, Jul 2010, pp. 44–51. doi:10.1109/SENSORDEVICES.2010.16.

[9] Stove A.G. and Sykes S.R. 'Doppler-based automatic target classifier for a battlefield surveillance radar'. *Int. Radar Conf.*, Edinburgh, Scotland, Oct 15–17, 2002, pp. 419–23.

[10] Kim Y. and Ling H. 'Human activity classification based on micro-Doppler signatures using a support vector machine'. *IEEE Trans. Geosci. Remote Sens.* 2009, vol. 47(5), pp. 1328–37.

[11] Fairchild D. and Narayanan R. 'Classification of human motions using empirical mode decomposition of human micro-Doppler signatures'. *IET Radar Sonar Navig.* 2014, vol. 8, pp. 425–43.

[12] Li J., Phung S., Tivive F., and Bouzerdoum A. 'Automatic classification of human motions using Doppler radar'. *Int. Joint Conf. Neural Networks* (IJCNN), Brisbane, Australia, Jun 10–15, 2012, pp. 1–6.

[13] Rios J. and Kim Y. 'Application of linear predictive coding for human activity classification based on micro-Doppler signatures'. *IEEE Geosci. Remote Sens. Lett.* 2014, vol. 11, pp. 1831–4.

[14] Smith G., Woodbridge K., and Baker C. 'Template based micro-Doppler signature classification'. *Presented at European Radar Conference*, IEEE, Manchester, UK; Sep 2006.

[15] Zabalza J., Clemente C., Caterina G., Ren J., Soraghan J., and Marshall S. 'Robust PCA micro-Doppler classification using SVM on embedded systems'. *IEEE Trans. Aerosp. Electron. Syst.* 2014, vol. 50, pp. 2304–10.

[16] Molchanov P.O., Astola J.T., Egiazarian K., and Totsky A. 'Classification of ground moving targets using bicepstrum-based features extracted from micro-Doppler radar signatures'. *EURASIP J. Adv. Signal Proc.* 2013. doi:10.1186/1687-6180-2013-61.

[17] Vishwakarma S. and Ram S. 'Dictionary learning for classification of indoor micro-Doppler signatures across multiple carriers'. *Presented at IEEE Radar Conference*, IEEE, Seattle, WA, USA; May 2017.

[18] Amin, M. *Radar for indoor monitoring: detection, classification and assessment*. CRC Press, Oxford; 2017.

[19] Kim Y. and Moon T. 'Human detection and activity classification based on micro-Dopplers using deep convolutional neural networks'. *IEEE Geosci. Remote Sens. Lett.* 2016, vol. 13, pp. 2–8.

[20] Jia Y., Shelhamer E., Donahue J. 'Caffe: convolutional architecture for fast feature embedding'. *Proceedings of the 22nd ACM International Conference on Multimedia, Association for Computing Machinery*, pp. 675–678, 2014.

[21] Chetler S., Woolley C., Vandermersch P. *et al.* 'cuDNN: efficient primitives for deep learning'. *arXiv:1410.0759*, Cornell University. 2014.

[22] Jokanovic B., Amin M., and Ahmad F. 'Radar fall motion detection using deep learning'. *Proceedings of the IEEE Radar Conference (RadarConf)*, Philadelphia, May 2–6, 2016, pp. 1–6.

[23] Jokanovic B. and Amin M. 'Fall detection using deep learning in range-Doppler radars'. *IEEE Trans. Aerosp. Electron. Syst.*, 2018, vol. 54, pp. 180–9.

[24] Lang Y., Hou C., Yang Y., and Huang D. 'Convolutional neural network for human micro-Doppler classification'. *Presented at European Microwave Conference*, German; 2017.

[25] re3data.org: CMU Graphics Lab Motion Capture Database; editing status 2016-12-16; re3data.org – Registry of Research Data Repositories. Available from http://doi.org/10.17616/R3N35M [Accessed 10 September 2020].

[26] Ram S.S., Christianson C., Kim Y., and Ling H. 'Simulation and analysis of human micro-Dopplers in through-wall environments'. *IEEE Trans. Geosci. Remote Sens.* 2010, vol. 48(4), pp. 2015–23.

[27] Trommel R.P., Harmanny R.I.A., Cifola L., and Driessen J.N. 'Multi-target human gait classification using deep convolutional neural networks on micro-Doppler spectrograms'. *Presented at European Radar Conference*; Oct 2016.

[28] Boulic R., Thalmann N., and Thalmann D. 'A global human walking model with real-time kinematic personification'. *Vis. Comput.* 1990, vol. 6(6), pp. 344–58. http://dx.doi. org/10.1007/BF01901021.

[29] Molchanov P., Gupta S., Kim K., and Pulli K. 'Multi-sensor system for driver's hand-gesture recognition'. *Proc. IEEE Int. Conf. Workshops Autom. Face Gesture Recognit.*, May 2015, pp. 1–8.

[30] Kim Y. and Toomajian B. 'Hand gesture recognition using micro-Doppler signatures with convolutional neural network'. *IEEE Access.* 2016, vol. 4, pp. 7125–30.

[31] Jiao M., Lu G., Jing X., Li S., Li Y., and Wang J. 'A novel radar sensor for the non-contact detection of speech signals'. *Sensors* 2010, vol. 10, pp. 4622–33.

[32] Li S., Tian Y., Lu G., *et al.* 'A 94-GHz millimeter-wave sensor for speech signal acquisition'. *Sensors* 2013, vol. 13, pp. 14248–60.

[33] Khanna R., Oh D., and Kim Y. 'Through-wall remote human voice recognition using Doppler radar with transfer learning'. *IEEE Sensors* 2019, vol. 19, pp.4571–6.

[34] Ritchie M., Fioranelli F., Griffiths H., and Torvik B. 'Micro-drone RCS analysis'. *Proc. IEEE Radar Conf.*, Oct 2015, pp. 452–6.

[35] Björklund S., Johansson T., and Petersson H. 'Evaluation of a microDoppler classification method on mm-wave data'. *Proc. IEEE Radar Conf.*, Atlanta, GA, USA, May 2012, pp. 0934–9.

[36] Kim B., Kang H., and Park S. 'Drone classification using convolutional neural networks with merged Doppler images'. *IEEE Geosci. Remote Sens. Lett.* 2017, vol. 14, pp. 38–42.

[37] van Dorp P. and Groen F.C.A. 'Human walking estimation with radar'. *IET Radar Sonar Navig.* 2003, vol. 150, pp. 356–65.

[38] Ram S.S., Christianson C., Kim Y., and Ling H. 'Simulation and analysis of human micro-Dopplers in through-wall environments'. *IEEE Trans. Geosci. Remote Sens.* 2010, vol. 48(4), pp. 2015–23.

[39] Erol B., Karabacak C., Gürbüz S., and Gürbüz A. 'Simulation of human micro-Doppler signatures with Kinect sensor'. *IEEE Radar Conference*, IEEE, Cincinnati, OH, USA, May 2014.

[40] Seyfioğlu M., Erol B., Gurbuz S., and Amin M. 'DNN transfer learning from diversified micro-Doppler for motion classification'. *IEEE, IEEE Transactions on Aerospace and Electronic Systems*, vol. 55, pp. 2164–2180; Oct 2019.

[41] Pan S. and Yang Q. 'A survey on transfer learning'. *IEEE Trans. Knowl. Data Eng.* 2010, vol. 22, pp. 1345–59.

[42] Du H., He Y., and Jin T., 'Transfer learning for human activities classification using micro-Doppler spectrograms'. *IEEE International Conference on Computational Electromagnetics*, Chengdu, China; Oct. 2018.

[43] Du H., Jin T., Song Y., Dai Y., and Li M., 'Efficient human activity classification via sparsity-driven transfer learning'. *IET Radar, Sonar & Navigation*, vol. 13, pp. 1741–1746; Oct. 2019.

[44] Krizhevsky A., Sutskever I., and Hinton G.E. 'ImageNet classification with deep convolutional neural networks'. *Proc. Adv. Neural Inf. Process. Syst.* 2012, pp. 1097–105.

[45] Simonyan K. and Zisserman A. 'Very deep convolutional networks for large-scale image recognition'. *arXiv:1409.1556* [cs.CV], Sep 2014.

[46] Park J., Rios J., Moon T., and Kim Y. 'Micro-Doppler based classification of human activities on water via transfer learning of convolutional neural networks'. *Sensors* 2016, vol. 16, pp. 1990.

[47] Mufti E. and Werghi B. 'Transfer learning with convolutional neural networks for moving target classification with micro-Doppler radar spectrograms'. *Presented at International Conference on Artificial Intelligence and Big Data*, IEEE, Chengdu, China; May 2018.

[48] Goodfellow I., Pouget-Abadie J., Mirza M. *et al.* 'Generative adversarial nets'. *Proc. Adv. Neural Inf. Process. Syst.* 2014, pp. 2672–80.

[49] Shi X., Li Y., Zhou F., and Liu L. 'Human activity recognition based on deep learning method'. *Int. Conf. Radar* 2018.

[50] Yang Y., Hou C., Lang Y., Guan D., Huang D., and Xu J. 'Open-set human activity recognition based on micro-Doppler signatures'. *Pattern Recognit.* 2019, vol. 85, pp. 60–9.

[51] Alnujaim I., Oh D., and Kim Y., 'Generative adversarial networks for classification of micro-Doppler signatures of human activity'. *IEEE Geosci. Remote Sens. Lett.* 2019, pp. 1–6.

Chapter 8

Deep neural network design for SAR/ISAR-based automatic target recognition

Simon Wagner[1] and Stefan Brüggenwirth[1]

The automatic recognition of targets with radar is an ongoing research field for many years already. Since 2014, a new methodology based on deep neural networks is becoming more established within this field. This chapter gives a short overview with some examples of this short history of target recognition using deep learning (DL) and some comparative results with a basic implementation of a convolutional neural network (CNN). This network is applied to the commonly used Moving and Stationary Target Acquisition and Recognition (MSTAR) dataset and to an inverse synthetic aperture radar (ISAR) dataset measured with the Tracking and Imaging Radar (TIRA) of Fraunhofer FHR.

For both of these datasets, networks with different sizes and parameters are trained and tested. Further are different regularizations and training methods implemented and tested, which should give an idea of the complexity of these classifiers. Deep neural networks are known to need a lot of training data, which is often not available in radar. One way to overcome this bottleneck is the use of transformations applied to available measurements and create artificial training data. The benefit of these transformations is not only an increased performance, but also an equalization in the distribution of the training data to avoid a bias towards certain overrepresented classes. For both of these applications, examples are given.

The final step is the combination of the CNN feature extraction with support vector machines (SVMs) for classification. This combination gave better results in the past than the CNN alone, but many new improvements in the CNN training methods also improved the test results of the CNNs. Results are given for the CNN alone as well as for the combination with SVMs to verify if this classifier substitution still gives significant improvements.

8.1 Introduction

Radar is used very often to support situational awareness in many applications. These applications of radar are very widespread, from classical applications like battlefield

[1]Cognitive Radar Department, Fraunhofer FHR, Wachtberg, Germany

surveillance in a military context [1] or air traffic control systems [2] to more recent applications like automotive radar [3]. All these applications have in common that they would benefit from an automatic target recognition (ATR) system, which has pushed a lot of research in this direction over the last decades [4,5].

The work in this chapter is limited to the application of imaging radars, namely synthetic aperture radar (SAR) for ground targets and ISAR for air targets. From these imaging techniques, SAR is already very common these days for ground surveillance, and most of the research on image-based ATR is focused on SAR images. ISAR systems are not that common in air surveillance systems, but the potential of these systems for target recognition has been presented several times in the literature [6]. For both of these imaging methods, several classification strategies have been presented over the last years, and a key step was the introduction of the MSTAR database [7,8]. Although this database is already more than 20 years old, it is still the only freely available benchmark dataset and is still frequently used to compare classification approaches.

This comparison of classification performance also showed that the performance for this dataset has increased a lot over the last few years. Especially the use of DL techniques had a significant contribution to this performance gain, and some examples of published results are presented in Section 8.2. The DL structure that is used mostly for ATR applications is the CNN, which possesses an intrinsic combination of feature extraction and classification. In the classical SAR ATR processing chain, which is depicted in Figure 8.1, the actual target classification is either done by a template matching approach or by a feature-based approach with features that are selected experimentally. Both methods have major drawbacks: the template matching is computationally demanding and the generalization capabilities are limited, since only target variations that are in the database can be recognized. The feature-based approach requires an experienced system designer to select appropriate features, which might also be used for inference and to recognize unknown target variations. The DL technique offers an automatic feature selection and extraction algorithm that can be used as classification system, and although it shares the drawback of template matching and requires a large database to learn these features, it gives state-of-the-art results with the available datasets that are described in Section 8.3.

The other steps of the processing chain in Figure 8.1 are assumed to be done in the approach presented here. The pre-screening describes a detection process that gives a list of candidates that might represent targets. The decision if these candidates are actually targets or can be considered as clutter is done by the discrimination block.

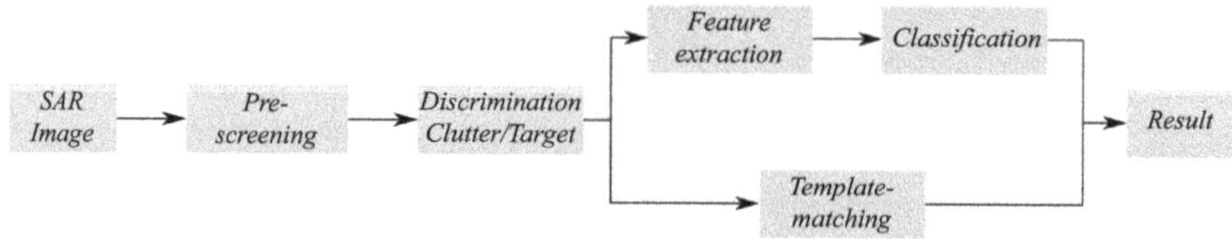

Figure 8.1 Classical ATR structure for SAR images

For this block also, DL approaches were presented in the past, e.g. [9], and thus also a combination of this step and the actual classification is possible. After the discrimination step, the potential targets are presented to the classification system as image chips with a fixed size. For the ISAR scenario used later, the same steps are assumed. The pre-screening refers in this case to a detection by a surveillance radar and the images are produced by a high-resolution imaging radar. The discrimination step is obsolete in this case, since only air targets are considered here and clutter is very unlikely in the airspace.

The classification system of the presented approach is a CNN and the main elements of this system are described in Section 8.4. The achieved results are given in Section 8.5, and Section 8.6 summarizes and concludes the chapter. Although the main elements of the networks are described in this chapter, a more tutorial overview can be found in Chapter 3 on DL theory or in textbooks like [10–12] or in freely available lecture notes like [13]. The theoretical background of SVMs can also be found in [10,12] or in more specialized textbooks like [14,15].

8.2 Deep learning methods used for target recognition

Classical ATR methods are based on the scheme that handcrafted features are extracted and are fed into a classifier [4]. In the last few years, new methods that use DL structures and thus combine feature extraction and classification were presented. To the knowledge of the authors, the first ATR systems using DL structures were published in 2014 independently by [16,17]. Since that time a vast amount of papers appeared on this topic, thus it is not possible to give a full overview on all published methods, instead some selected papers are presented in this section to show the large variety of DL methods that can be used for SAR and ISAR target recognition.

There are mainly two DL structures, i.e. CNNs and autoencoders (AEs), of which CNNs have been used far more often for target recognition purposes than AEs. For a general description of both methods, the reader is referred to the chapter on theoretical foundations of DL or to the literature, e.g. [18] or [19].

The use of AEs for target recognition is limited by the fact that there is no intrinsic classifier as is the case for CNNs. An AE can only be used for feature extraction and an additional classifier or at least a softmax layer must be used for classification. For example in [20], an AE is used for feature extraction and a linear SVM for classification. To improve the performance of linear SVM, an additional Euclidean distance constraint is added to the cost function of the AE. This constraint penalizes code differences of examples of the same class as well as closely spaced codes of examples of different classes. The result should be that the Euclidean distance of different classes should increase in feature space, and therefore, it should be more likely that the SVM can separate the samples linearly.

In [21], an AE is used for the fusion of handcrafted features and a softmax layer as well as an SVM are used as classifier. The main point of this publication is that the AE is not used for feature extraction but to reduce the dimension of the feature space. With the reduction of the dimensionality, the problem becomes computationally less demanding and removes the redundancy in the selected features. The removed redundancy and the efficient representation of the features in the latent space of the AE improved the classification results compared to the baseline methods and also against the classification using the full images. The internal representation of the AE, i.e. the code, was forced to be sparse using an additional sparsity constraint in the cost function during the training. This sparsity constraint allows a very efficient representation using only a few nonzero elements and can, therefore, create distinguishable codes of different targets. The resulting AE is called a sparse AE (SAE), and in this application, two AEs are stacked to further increase the level of abstraction.

A review of several AE structures and some comparative experiments on target recognition using these different AE structures is given in [22]. From the different AE structures, the SAE and the convolutional AE (CAE) should be mentioned, since these structures are also used in other literature examples of this section. The SAE was already introduced previously; it uses an additional sparsity constraint in the cost function to produce a sparse code in the intermediate layer. The CAE is a combination of CNN and AE, and it uses the convolutional feature extraction of the CNN in the encoder and deconvolution methods in the decoder. This kind of AE should be very useful in the representation of images due to the convolutional structure. Additional to the experiments with different AE structures, different parameters of the AEs, e.g. the loss function or the depth of the network, are varied and the influence on the classification result is validated. To perform a classification with the different AEs, a softmax layer is added at the code layer of the AE.

The second mentioned DL structure, the CNN, is already very widespread for ATR applications compared to AE. As already mentioned earlier, the first two publications known to the authors that use DL for ATR are [16,17]. Both of them use CNNs, yet in different ways. In [17], a one-layer CNN is created with convolutional kernels, which are built by using the weights of an SAE that is trained on randomly selected image patches. As a result, an unsupervised feature extraction is created. The output of this feature extraction is then fed into a softmax classifier, which is trained in a supervised manner. Although the results achieved with this classification scheme cannot compete with state-of-the-art results of handcrafted classifiers, the possibilities and benefits of automatic feature extraction are visible.

The CNN in [16] is split into two parts, the feature extraction and the classification, which is the standard separation for CNNs. The feature extraction part consists of the convolution and the pooling layers, while the classification part is a two-layer multilayer perceptron (MLP) [10]. Both elements can be trained together by the backpropagation algorithm but can be treated as independent after the training. This independence is exploited to use SVMs as the actual classification system. The use of SVMs instead of the intrinsic MLP of the CNN is motivated by the superior generalization capability of SVMs, which should give better results with unknown targets.

The separation of CNNs into feature extraction and classification parts is also used in [23]. In this publication, the encoder of a CAE is applied as the feature extraction and a shallow neural network as classifier. Using the CAE structure accelerates the training of the system compared to a standard CNN with backpropagation, while the recognition rate is close to the one in [16].

A problem that arises regularly in DL is overfitting due to a limited amount of training data, and several methods and ideas have been presented during the last years. These methods can be roughly split into two categories: the first one creates additional data from known data by transformations or from models by simulations and the second way is to reduce the capacity of the network. The latter is normally achieved by a reduction of the free parameters and a good example can be found in [24]. In this work, the so-called *all-convolutional network* was introduced, which uses only sparsely connected convolutional layer and no fully connected layers anymore. The classification is performed by a softmax activation function in the last convolutional layer that contains as many feature maps as targets have to be discriminated.

Examples for an increased database can be found in [25,26], where in the first example, the training data is increased using simulated data and the latter uses transformations of real data. The application in [25] is ship classification and the simulations are used to fill the gaps in the dataset for missing azimuth aspect angles. The results show that the simulations improve the results if they are used additionally to the available real data, but a classification based solely on simulations shows a large drop of the classification rate. The work in [26] is applied to ground targets of the MSTAR database, and different transformations are used to simulate effects that happen in radar imaging. For example, a simple rescaling of the range axis is used to compensate the scaling effects of the changing depression angle between training and test data. Another method is elastic distortion, which performs a random shift of scatterers to simulate target variations and scintillation effects. The results show that a larger training set gives better results, and these transformation methods will also be used in the experimental part of this work.

So far, all publications presented in this section are based on image processing, i.e. only magnitude values of the SAR images are exploited for classification. The main reason for this is probably the history of DL methods in computer vision, which uses only real-valued data, albeit sometimes in a multichannel representation, e.g. RGB images. In radar imaging, on the other hand, the standard case is complex-valued data, and a lot of information is contained in the ratio of real and imaginary parts, i.e. the phase. Thus an increased interest in using complex-valued data in DL-based ATR has been observed, for example in [27,28]. In both works, a three-channel CNN is applied and magnitude, real and imaginary parts are used as network channels. The drawback of these channel separations is that real and imaginary parts are handled independently, and the full potential of phase information is still not used. For a full exploitation of this information, a complex-valued network with complex-valued input is necessary. Some attempts to implement such complex-valued networks have been presented in the past, also in the context of radar [29], but to the knowledge of the authors not yet to target recognition in radar imaging. Therefore, it remains as an open point for future research and is also not considered in the experimental part of this chapter.

8.3 Datasets

To demonstrate the classification capabilities of the chosen CNN, several experiments are presented in Section 8.5, which use two different datasets. The first dataset is the public MSTAR dataset, which is also used in most of the publications presented in Section 8.2. This dataset is of very high quality and is commonly used as a benchmark to compare different classification systems. As a counterpart to this SAR image database, an ISAR dataset is included in the experiments here. The raw data of this ISAR dataset is measured with the TIRA of Fraunhofer FHR during controlled flights of the targets. The processing of SAR and ISAR images is not considered here, but details of different imaging algorithms can be found in the literature, for example in [30] or [31].

8.3.1 MSTAR SAR dataset

The MSTAR dataset consists of X-band spotlight-SAR images with 0.3 m resolution in each direction and is available from [32]. A detection step is not necessary, since the database consists of image chips with a size of 128×128 pixels centred on the target. The data collected with 17° depression angle is used for training (3,671 images) and the data with 15° depression angle for testing (3,203 images). The targets have been measured at both depression angles over the full 360° azimuth coverage area and the increment between the imaging varies from 1° to 5°. The distribution of the images over the different targets is given in Table 8.1.

As already mentioned in Section 8.2, the classification system used in this work is limited to real-valued signals, and therefore, the magnitude values are converted to a logarithmic scale. With this step, not just real values are achieved, the compressed dynamic range is also more appropriate for classification.

Another important preprocessing step is the cropping of the image, since it was shown in the past that the clutter within the MSTAR dataset shows such a high

Table 8.1 Numbers of images per class in the MSTAR dataset

Class (target)	Training	Test
Class 1 (BMP-2)	698	587
Class 2 (BTR-70)	233	196
Class 3 (T-72)	691	582
Class 4 (BTR-60)	256	195
Class 5 (2S1)	299	274
Class 6 (BRDM-2)	298	274
Class 7 (D7)	299	274
Class 8 (T-62)	299	273
Class 9 (ZIL-131)	299	274
Class 10 (ZSU-23/4)	299	274
Complete	3,671	3,203

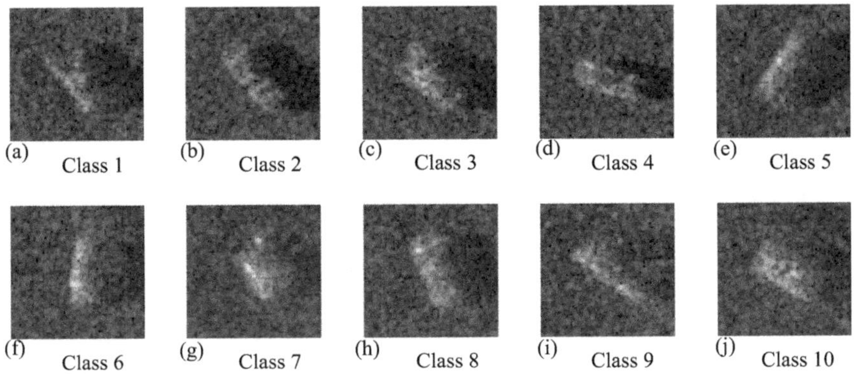

Figure 8.2 Examples of the MSTAR database: (a) class 1, (b) class 2, (c) class 3, (d) class 4, (e) class 5, (f) class 6, (g) class 7, (h) class 8, (i) class 9 and (j) class 10

correlation between training and test data that a classification based solely on the clutter is possible [1,33]. To remove most of the clutter, the image is cropped to a size of 64×64 pixels, which has the side effect of reduced computational demands and thus a reduced training time.

The last preprocessing step is the normalization of the data to a mean value of zero and a standard deviation of one. This rescaling is a very common method to accelerate the training and get more comparable images. Randomly selected examples of the final images for each class are depicted in Figure 8.2. All images in this figure are in the same dynamic range and show a strong background clutter and speckle noise, which results in low contrast images.

8.3.2 TIRA ISAR dataset

The ISAR dataset used in this work was introduced in [34] and contains 1,085 images of five different targets. The targets contain three military fighter aircrafts and two civil airplanes of comparable size. The data was recorded with a bandwidth of 800 MHz, and the processing intervals are chosen according to the estimated rotational speed to achieve a cross-range resolution that matches the resolution in range. The numbers of images per class are shown in Table 8.2, which also shows the separation into training and test data. The dataset is quantized in subsets with a size of 31 images, since for this dataset only front aspect angles from 0° to 30° in azimuth are used and images are processed to have one image for each integer aspect angle. To create independent training and test sets, the data of four classes are separated such that the two subsets contain images from different flights and also from different years. Only for one class, both training and test data are from the same flight, since for this aircraft only one flight was available. To ensure a certain independence of training and test data inside this class, the training and test data were chosen from different circles

Table 8.2 Numbers of images per class in the TIRA dataset

Class (type)	Training	Test
Class 1 (fighter 1)	186	124
Class 2 (fighter 2)	93	62
Class 3 (fighter 3)	155	279
Class 4 (civil jet 1)	62	31
Class 5 (civil jet 2)	62	31
Complete	558	527

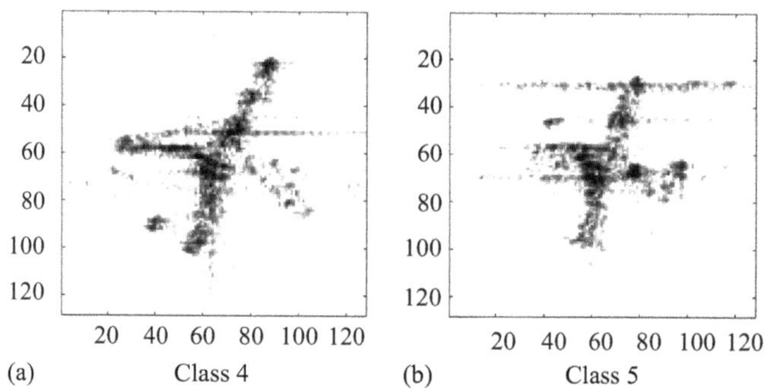

Figure 8.3 Examples of the TIRA database: (a) class 4 and (b) class 5

the target flew during the measurement [35]. Two example images of this dataset are shown in Figure 8.3, in which a threshold has been applied to remove some imaging artefacts.

From Table 8.2, it can be seen that the distribution in this dataset is far from being equally distributed, which can cause problems in the training process. Since the data is presented iteratively during the training to the classifier, an overrepresented class might cause a bias towards this class because the easiest way to reduce the cost function is to assign all images to this class. To avoid this bias, a method to expand the training database artificially is presented in the next part.

8.3.3 Artificial training data

To achieve an equally distributed database and to further robustify the classification system, artificial training data is created from known data. In this work, the techniques presented in [26] are used to increase the size of the database. For a detailed description of these methods, the reader is referred to [26] and the references therein, here only a short description with some examples is given.

The first method is called *elastic distortion*, which generates a small random displacement of the image pixels that are afterwards connected by an interpolation. This method was originally introduced for optical character recognition to simulate

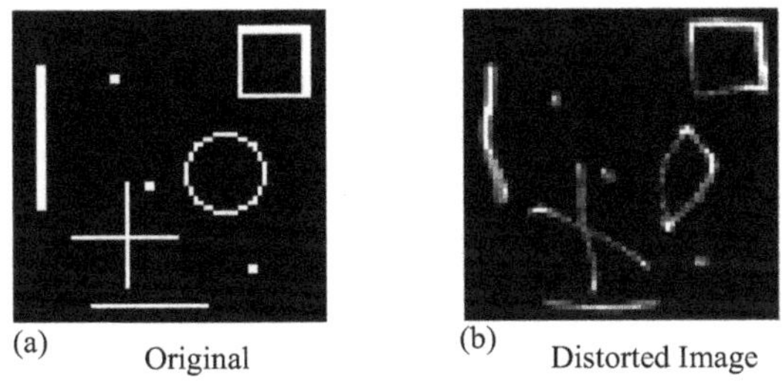

Figure 8.4 Elastic distortion of a synthetic test scene: (a) original image and (b) distorted image. © 2016 IEEE. Reprinted, with permission, from Reference [26].

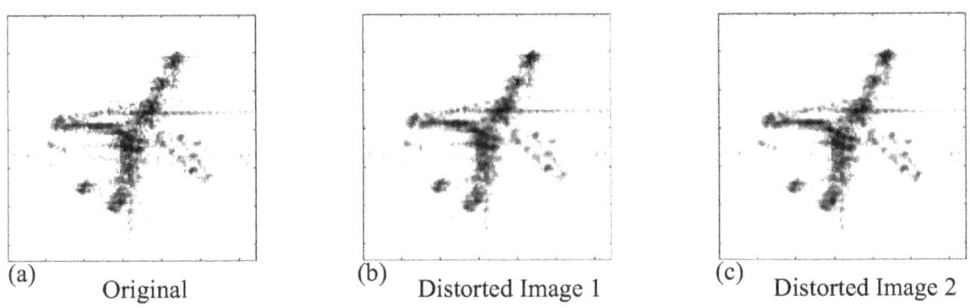

Figure 8.5 Elastic distortion of an ISAR image: (a) original, (b) distorted image 1 and (c) distorted image 2

different handwritings, in terms of radar it can be interpreted as a simulation of small target variations and, especially in ISAR images of air targets, as a minor change of the image projection plane due to a variation in the angular movement of the target. In Figure 8.4, an example with a synthetic test scene is shown to clarify the basic idea of the method, and an example with a real ISAR image of the TIRA dataset is shown in Figure 8.5.

The examples of Figure 8.5 show that the effect of the deformation is not very obvious in the radar images. A small change can be seen on the left wing of the aircraft. A line of strong scattering centres, which is visible in the original image, is varied a little in both shape and amplitude. This effect might also happen if the flaps of the airplane are lifted or changed, which is an example for the minor changes of the target variation or articulation in this case. It should be clearly stated that the main purpose of this image-based data augmentation is the creation of small variations to overcome the small data problem. This method, as well as the second method discussed next, is no replacement for more sophisticated data augmentation methods based on electromagnetic simulations. For more information on this topic, the interested reader is referred to Chapter 6, the highly recommended chapter on data augmentation for SAR ATR in this book.

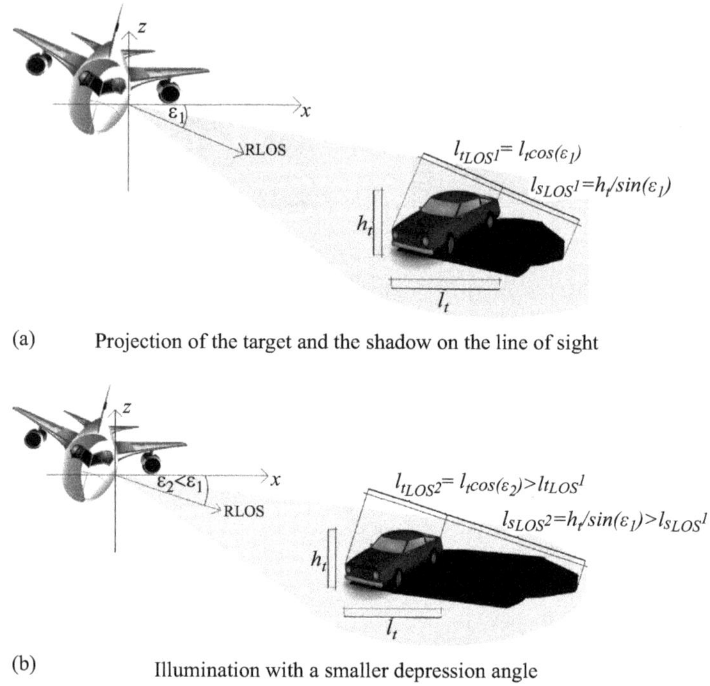

Figure 8.6 Range scaling in SAR imaging: (a) projection of the target and the shadow on the line of sight and (b) illumination with a smaller depression angle. ©2016 IEEE. Reprinted, with permission, from Reference [26].

The second method used to increase the size of the database is a rescaling of the range and cross-range axis by an affine transformation. With this transformation, the uncertainty in ISAR cross-range scaling can be compensated and to some extent also the difference of the projected length of targets on the radar line of sight (RLOS) dependent on the depression angle. The principal of this dependency on the projected length is graphically explained in Figure 8.6, which shows an illustration of SAR measurements with different depression angles. The formulas in the upper picture describe the length of the target and of the shadow dependent on the length and height of the target. The lower picture shows an illumination with a smaller depression angle, and thus the projected lengths of both target and shadow increase.

From the formulas in Figure 8.6, the depression angles of 17° for the training data and 15° for the test data of the MSTAR dataset follow a difference in the shadow size of approximately 13% and about 1% of the projected target size. Though the size of the target is almost the same in both sets, the difference in the shadow size could have a large influence on the classification performance. In the past, it was shown, e.g. in [36], that the use of the shadow area alone is sufficient to achieve

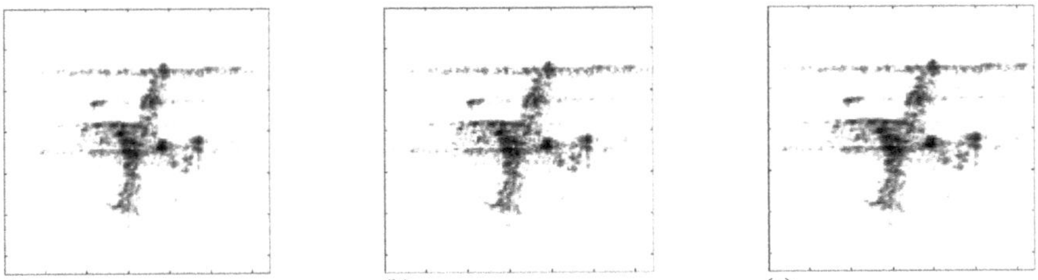

Figure 8.7 Different cross-range scalings of an ISAR image: (a) compressed, (b) original and (c) stretched

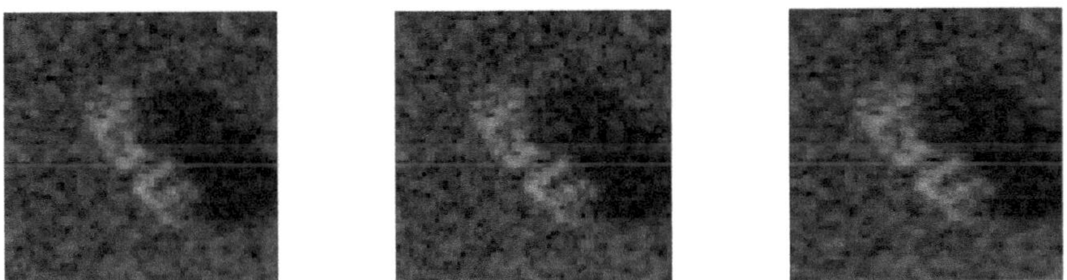

Figure 8.8 Range scaling of an MSTAR image: (a) compressed, (b) original and (c) stretched

a satisfying classification result with the MSTAR dataset. Examples for this affine scaling are shown in Figure 8.7 for an ISAR image with a rescaled cross-range axis and in Figure 8.8 with an example of the MSTAR database and a varying range axis.

Two other options that can be used to increase the size of the database is a shift in position or a rotation around a certain angle. The first method is used to create a system that is more robust against errors in the centering of the targets within the presented image chips. The rotation of the images is used to overcome errors in the aspect angle estimation of targets [26], which is used to present image chips always with the same target orientation. In this work, both these methods are used more generally to create additional examples with small variations to reduce the probability of overfitting to the training data.

8.4 Classification system

The classification system used in this work is a further development of the system presented in [16,26] and was already introduced in [37]. In this chapter, the latest developments of this system, in terms of cost functions and regularization methods, and an investigation on the influence of different parameters are presented. For most of the presented results, the CNN is used for both feature extraction and classification,

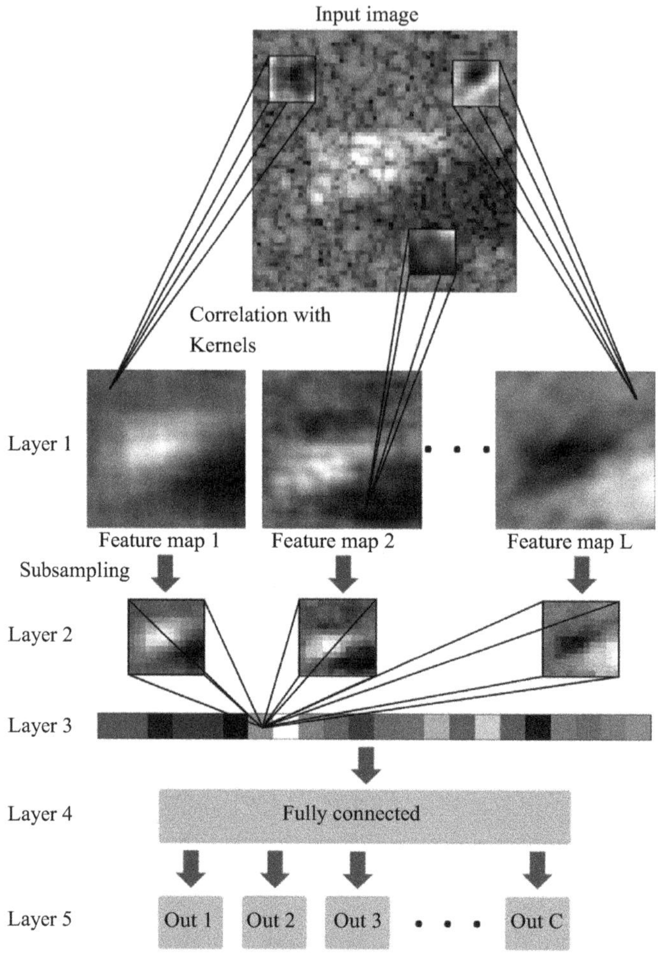

Figure 8.9 Structure of the used ATR system. ©2017 IEEE. Reprinted, with permission, from Reference [37].

only in the last part is the classification done by SVMs, since this classifier has shown to have better generalization capabilities [10]. A graphical representation of a five-layer CNN used here is shown in Figure 8.9, and the different elements of this kind of network are described next.

8.4.1 CNN structure

A CNN, for example the five-layer network in Figure 8.9, can be separated into two parts: the first one is the feature extraction part and the second part is the classification part. The feature extraction part consists of the given example of the first three layers but consists generally of an alternation of convolutional and pooling layers. The neurons in the convolutional layers are organized in so-called feature maps and in each feature map the result of a convolution

$$F(i,j) = (I \times K)(i,j) = \sum_{m}\sum_{n} I(i+m, j+n)K(m,n) \tag{8.1}$$

of the output $I(i,j)$ of the previous layer with one or more learned kernels $K(m,n)$ is stored. This means that all neurons in a single feature map use the same weights, which reduces the number of parameter that must be learned enormously. The first layer of a CNN is always a convolutional layer and each feature map in this layer uses the current image under test as input signal and thus gives the similarity of the input image with a certain spatial feature. In deeper layers, a single feature map is normally connected to several feature maps of the previous layer and thus looks for combinations of different features in the input image. The connections in the deeper layers are set randomly with a probability of 0.5, i.e. each feature map in deeper layer is in average connected to 50% of the feature maps in the previous layer. For each of these connections, an independent kernel is learned to achieve a variation in the weighting of the corresponding features.

Between the convolutional layers, pooling layers that reduce the number of values in the feature maps are placed. The dimensionality can be, for example, reduced by max pooling, which uses the maximum value of a certain area. In this work the area is of the size 2×2 or 4×4, which lead to a reduction by the factor 2 and 4 respectively. Other possibilities to reduce the dimension of feature maps by pooling are the mean or the median value of the mentioned areas. Another approach to achieve feature maps with a reduced dimension, but without pooling, are convolutional layers with a stride different from 1 within the convolution operation [18]. The last mentioned approach is not applied in the work presented here.

Since both kinds of layers use standard model of a neuron, an activation function $\varphi(\cdot)$ is implemented at the output. Over the last years, the so-called rectified linear unit (ReLU)

$$\varphi_{ReLU}(v) = \max\{0, v\} \tag{8.2}$$

has become standard in CNNs and other DL networks. In the networks used here, the ReLU function is used in the convolutional layers, and in the pooling layers either a logistic function of the form [10]

$$\varphi_{Logistic}(v) = \frac{1}{1+\exp(-bv)} \qquad b > 0 \tag{8.3}$$

or a linear function, i.e. no activation, is implemented. The logistic activation function of the pooling layers creates an almost binary image between 0 and 1 as input for the deeper convolutional layers and the saturation values indicate if a feature or a combination of features is present at a specific location or not. The variable b is set to $2/3$ to increase the range of the function before the saturation is reached. This reduced slope should accelerate the training, since the gradient between the saturation areas is much larger and therefore, allows a larger correction in each iteration as long as saturation is not reached. [10]. As part of the experimental results in this chapter, the difference in performance between both activation functions will be investigated. The CNN image of Figure 8.9 was created with a linear pooling. In this case, the output of the pooling layers look like the output of the feature maps in the previous layer, only the dimension is reduced.

The last two layers of all CNNs in this chapter are two fully connected layers, i.e. a standard MLP. The last layer uses a softmax activation function and the maximum value is used to decide which class is present in the image under test. The output values of a softmax classifier are normalized to give a cumulative sum of 1 and thus can the output values be interpreted as likelihoods of the corresponding classes.

To train the network, two different approaches that both are based on the backpropagation algorithm are used and shortly described in the next section. The backpropagation algorithm itself is used to solve the so-called *credit assignment problem*, which described the problem of finding the influence of a certain weight in any layer of the network on the output signal. This means that the training of MLPs became possible with the introduction of this algorithm, since it allows to calculate the gradient of the cost function with respect to any weight of the network. A general description of this algorithm can be found in basic textbooks like [10–12].

8.4.2 Training algorithm

In this work, two training algorithms, namely the *stochastic diagonal Levenberg–Marquardt (SDLM)* method [38,39] and the *Adam* algorithm [40], are used, which are both based on the backpropagation algorithm using a stochastic gradient descent method. This means that the weights $w(k)$ of the network are changed iteratively according to the gradient of a certain cost function $C_k(w)$

$$\Delta w(k) = -\gamma(k)\frac{\partial C_k(w)}{\partial w(k)}, \tag{8.4}$$

where the parameter $\gamma(k)$ represents the learning rate at iteration step k. The variable $\Delta w(k)$ represents the calculated change of the weights or the current iteration k. The weights $w(k)$ are stored as a vector, which contains all free parameters of the network, i.e. the correlation kernels, the bias terms and the weight matrices of the fully connected layers. The weight change $\Delta w(k)$ is a vector of the same size as the weight vector and contains the weight change of the current iteration. The new weight vector for the next iteration is calculated by

$$w(k+1) = w(k) + \Delta w(k). \tag{8.5}$$

The difference between the two mentioned training algorithms is in the determination of the learning rate, which is described next for both methods. The cost function $C_k(w)$ in (8.4) is the instantaneous loss function of the current iteration k calculated with respect to the training image. Normally the weights are not updated after every sample, but after a small batch of K samples to achieve an averaging effect that accelerates the training. Therefore, a new iteration index t is introduced and the weight change is calculated as

$$\Delta w(t) = -\gamma(t)\frac{1}{K}\sum_{k=1}^{K}\frac{\partial C_k(w)}{\partial w(k)}. \tag{8.6}$$

A different method to achieve a stabilized and faster training is the use of *momentum*, which adds a feedback loop to the algorithm. With this technique a certain amount of the weight change of the last iteration is added to the weight change of the current iteration. This amount is determined by the so-called *momentum constant* α with a value between zero and one and leads to the expression

$$\Delta w(t) = \alpha\Delta w(t-1) - \gamma(t)\frac{1}{K}\sum_{k=1}^{K}\frac{\partial C_k(w)}{\partial w(k)}. \tag{8.7}$$

The use of momentum and also the batch update rule should have a positive effect on the behaviour of the training algorithm and may prevent the algorithm from converging to a local minimum of the cost function [10].

8.4.2.1 Stochastic diagonal Levenberg–Marquardt algorithm

This method was introduced already in the first publication on CNNs [38], but the term SDLM algorithm was coined later in [39]. In the original paper, it was stated, 'This algorithm is not believed to bring a tremendous increase in learning speed but it converges reliably without requiring extensive adjustments of the learning parameters' [38]. This reliable convergence is the factor why this algorithm is still in use today and the main reason why no adjustments of the learning parameter are necessary is that the algorithm can be considered as adaptive. It uses not only the gradient but also an estimate of the second derivative h of the cost function with respect to the weights to determine the learning rate

$$\gamma(k) = \frac{\eta}{\mu + h(k)}, \tag{8.8}$$

where η and μ are constants. With this formula, the learning rate is large if the second derivative is small and vice versa. The constant μ determines a minimum value of the learning rate if the second derivative is zero. Since the second derivative can also be negative, an approximation that always gives a positive estimate is used to avoid this situation. This approximation corresponds to the Levenberg–Marquardt approximation from nonlinear regression [38], which is the reason for the naming of the algorithm. The term 'diagonal' is due to the fact that in principal the Hessian with all cross terms of a specific weight must be used to determine the learning rate, but instead only the elements of the main diagonal of the Hessian matrix are used. For a detailed description of the algorithm and the used approximations, the reader is referred to [38,39].

A difference between the implementation in this work and the description above is that the variable η is not fixed over the complete training. It is reduced during the training, but always fixed for several epochs.* This change should avoid large variations towards the end of the training when mostly fine-tuning is done.

8.4.2.2 Adam algorithm

As mentioned earlier, the SDLM algorithm was the first available algorithm to train CNNs. Some years ago, a method coined Adam was introduced [40], which is nowadays used very often and some comparative investigations showed that this algorithm is always among the best results, for example in [22].

The Adam algorithm is a first-order method, i.e. only the gradient is used, which is a major difference compared to the algorithm above. Nevertheless, it is still adaptive, since the first and second moments of the gradient estimates are used to determine the learning rate. This estimation gave the name to the method, since the name is derived from adaptive moment estimation. The details of the algorithm and the derivation can be found in [40], here only the steps and the key points of the explanation are given.

For this algorithm a fixed learning rate α and two hyperparameters β_1 and β_2 are needed. With these hyperparameters biased estimates of the first and second raw moments, the mean $m(t)$ and the uncentred variance $v(t)$, are calculated by

$$m(t) = \beta_1 m(t-1) + (1-\beta_1)\frac{1}{K}\sum_{k=1}^{K}\frac{\partial C_k(w)}{\partial w(k)} \tag{8.9}$$

and

$$v(t) = \beta_2 v(t-1) + (1-\beta_2)\left(\frac{1}{K}\sum_{k=1}^{K}\frac{\partial C_k(w)}{\partial w(k)}\right)^2. \tag{8.10}$$

Both $m(t)$ and $v(t)$ are vectors of the same size as the weight vector $w(t)$ and are initialized as a vector of zeros, which leads to a biased estimation towards zero. In the next step, these biased estimations are corrected to an unbiased estimation by

$$\hat{m}(t) = \frac{m(t)}{1-\beta_1^t} \tag{8.11}$$

and

$$\hat{v}(t) = \frac{v(t)}{1-\beta_2^t}. \tag{8.12}$$

These equations show that the hyperparameters β_1 and β_2 must be smaller than one, but should also be close to one. Default values will be given next, but a lower bound cannot be defined, since both values have an influence on the performance and the optimal value depends on the given problem. With the calculated first and second moment and the learning rate α, the final formula for the actual weight update is given by

$$w(t) = w(t-1) - \alpha\frac{\hat{m}(t)}{\sqrt{\hat{v}(t)}+\epsilon}, \tag{8.13}$$

*During one epoch, all available training samples are presented one time [10].

where ϵ is a very small constant and the calculation is done elementwise, i.e. for each weight individually. In the original papers, default values for the tested machine learning problems are given by $\alpha = 0.001, \beta_1 = 0.9, \beta_2 = 0.999$ and $\epsilon = 10^{-8}$. These values will also be used for the comparative results in this chapter.

The scaling parameter of the learning rate, i.e. the ratio of mean to variance, is interpreted in [40] as signal-to-noise ratio of the gradient. If this value is small, which corresponds to a large variance, the estimated direction of the gradient might differ from the true direction and the step size is reduced. If the value is high, i.e. the estimated gradient is always pointing in the same direction, this value is more trustworthy and thus the step size is increased.

This method together with a batch update of the gradient and momentum, already combines three methods of gradient averaging that should stabilize the training. In the next section, some more methods are introduced that should stabilize the training but are not directly related to the training algorithm.

8.4.3 Regularization methods

Up to this point, the cost function $C_k(w)$ was not explicitly defined. In the regular case, it can be separated into two parts: a basic error function $E(w)$ that controls the ability of the network to classify the training data and the second part should control the complexity of the classifier itself. The error functions considered in this work are the mean square error (MSE)

$$E_k^{MSE}(w) = (d_k - y_k)^2 \tag{8.14}$$

and the cross entropy (CE)

$$E_k^{CE}(w) = -\sum_{i=1}^{C} y_k^i \ln(p_k^i) + \left(1 - y_k^i\right) \ln(1 - p_k^i). \tag{8.15}$$

The variable d_k in (8.14) is the desired output of the network at iteration k and y_k is the current output. This measure gives the squared difference at every neuron in the output layer of the network. The definition of the CE in (8.16) is a sum that runs over all classes C and adds the differences between the ith element of the output layer at iteration k, y_k^i, and the desired probability of this element p_k^i. Since the sum of all output values in a softmax classifier is one, as it was already mentioned at the end of Section 8.4.1, the second term in (8.15) can be ignored. In this special case, the first and the second terms become redundant, since if the first term is reduced to zero, the second term will automatically be zero as well. The simplified CE formula is

$$E_k^{CE}(w) = -\sum_{i=1}^{C} y_k^i \ln(p_k^i). \tag{8.16}$$

The desired output vector for the CE and the MSE error functions contains a one in the element corresponding the class of the input image and zero in all other elements.

The second part of the cost function that should control the complexity of the classifier and also stabilize the training is called *regularizer* $R(w)$. The sum of the error function and the regularizer build the cost function of the network

$$C_k(w) = E_k(w) + \lambda R(w), \tag{8.17}$$

where λ is the weighting of the regularizing function. These regularizers are necessary since the training of neural networks is in general ill-posed and also a random process. The iterative manner and the dependency on the initialization of the network make the training process unstable in the way that small variations in the data might produce a remarkably worse result. To avoid this, several regularization terms were introduced during the last years. One of the first and still very common one is the so-called *weight decay* [41]. The idea of weight decay is to shrink the weights in each iteration by a small amount and in this way the weights that do not contribute to the output value should reduce to almost zero and cannot cause perturbations during the operation of the network. The weight decay is implemented by introducing the ℓ_2-norm of the weights in the cost function

$$C_k(w) = E_k(w) + \frac{1}{2}\lambda||w||_2^2. \tag{8.18}$$

Calculating the gradient of this cost function for the weight update according to (8.4) shows that in each iteration the weights are reduced by λw. Since there is no rule to calculate the optimal weight decay parameter λ, a rule of thumb value of 1% of the initial η (SDLM algorithm) or α (ADAM algorithm) is used in this work.

Other regularization methods considered in this work are applied nonlinearly and cannot be integrated in the cost function in a way according to (8.17). The first one is a constraint comparable to weight decay and prevents the weights from growing to an arbitrary large value. This method was mentioned in [42] and is a max-norm regularizer, i.e. at every iteration the norm of the weight vector is determined and if it is above a certain value c, it is projected back to this value. This means that the norm of the weight vector cannot grow larger than this constant c, $||w||_2 \leq c$. In the implementation used here, this constraint is applied to each convolutional kernel independently, since this method should give feature maps of comparable magnitude and thus should all features be of equal relevance.

Another regularization method that has been used often since its introduction is *dropout* [42]. The idea of this method is to randomly drop (remove) units from the network and train a new network at each epoch. In this way, the result of the network is not the result of a single network, but an averaged result of many networks. However, if dropout is used during training, some aspects must be considered that are given in the appendix of [42]. First of all, the network size should be increased. If for example the dropping probability $1 - p$ is 0.5 in a certain layer, the number of neurons should be doubled in this layer. The general rule of thumb for a layer with n neurons is that the new size is n/p neurons, where p is the probability of keeping a neuron, i.e. $p = 1$ means no dropout. Other points affect the learning rate and the momentum constant. Since dropout introduces a lot of noise in the gradients, the learning rate should be increased 10–100 times and the momentum constant should be between 0.95 and 0.99. The drawback of the large learning rate and momentum is that the step size can become very large, and therefore the above-mentioned max-norm regularization should be used. Unfortunately, there is no fixed rule to find the limiting value c in the max-norm regularization and it must be found experimentally. Typical values for p are given as 0.8 in the input layer and 0.5–0.8 in deeper layers.

The influence of certain parameters of the network structure and the mentioned regularization methods will be shown in Section 8.5.

8.5 Experimental results

In this section, the results achieved with CNNs for both mentioned datasets are presented. The influence of different parameters of the CNN structure is presented in Section 8.5.1, where the most important question is how deep the network should be. The results of several structures with different depths will be shown there. Some experiments regarding the difference between the mentioned training algorithms and cost functions will be presented in Section 8.5.2. The influence of the regularization methods presented earlier are investigated in Section 8.5.3 and the benefit of the artificial training data in Section 8.5.4. Finally, in Section 8.5.5, the use of SVMs as classifier will be shown and some experiments regarding the generalization capabilities of the classification systems will be performed.

8.5.1 CNN structure

In the first experiment, several structures of CNNs with different depths are trained on the MSTAR dataset and the results are compared. The depth of the network ranges from five to nine layers and are all trained with the same training parameters. These are a CE error function and the SDLM training algorithm with a batch size of ten. The momentum constant is set to 0.9 and the max-norm regularization is applied with a value of $c = 0.8$ for each convolutional kernel individually. In the fully connected and pooling layers, no regularization is applied. The training time is dependent on the depth and is either 48 epochs for the five- and seven-layer networks or 96 epochs for the nine-layer networks. The training is stopped after this time or if a CE value of 10^{-3} is reached. The value of μ is set to $5 \cdot 10^{-4}$ for the whole training, while η is initialized with 10^{-4} and halved after every 12th or 24th epoch depending on the total maximum training time. The activation functions in the convolutional layers are all ReLU functions, and in the pooling and first fully connected layer, a sigmoid function is used.

For each network, a 10-fold cross validation is used, this means that for each structure, ten networks are trained and the averaged results are given. This is done to consider the random nature of the training, since the weights are randomly initialized and the gradient descent algorithms are only local optimization methods. Due to this random nature, each training will give a different result and here the average, the minimum and the maximum performances as well as the standard deviation are given as results. The networks in the results are characterized by the number of feature maps in the convolutional layers, the kernel size, the pooling method and stride, and the number of fully connected neurons in the layers previous to the last layer. The output

Table 8.3 Network structures for the MSTAR dataset

Layer	Five-layer CNN	Seven-layer CNN	Nine-layer CNN
1	Convolutional 13 × 13	Convolutional 7 × 7	Convolutional 5 × 5
2	Pooling, stride 4	Pooling, stride 2	Pooling, stride 2
3	Convolutional 13 × 13	Convolutional 8 × 8	Convolutional 3 × 3
4	Fully connected	Pooling, stride 2	Pooling, stride 2
5	Softmax	Convolutional 11 × 11	Convolutional 3 × 3
6	–	Fully connected	Pooling, stride 2
7	–	Softmax	Convolutional 6 × 6
8	–	–	Fully connected
9	–	–	Softmax

layer is a softmax classifier and consists always of as many neurons as classes that have to be separated. For each depth, all parameters except the number of feature maps and the number of fully connected neurons are fixed.

The kernel size, the pooling method and the pooling stride are fixed for each structure, and they are chosen to result in feature maps of size 1 × 1 pixel for input images of size 64 × 64 pixels. This choice is motivated by the interpretation of the last convolutional layer as the output of the feature extraction part of the network. In this way, the values of the feature maps are the elements of the feature vector built by all feature maps of the last convolutional layer. The architectures of the different structures are summarized in Table 8.3, and the number of feature maps and fully connected neurons is varied in the later experiments.

The results of the five-layer CNNs are summarized in the upper rows of Table 8.4 and the different structures are described by three numbers. The first number gives the number of feature maps in the first layer, the second number the size of the third layer and the last one the number of neurons in the first fully connected layer of the network. The results show that the classification rates are stable as soon as a certain size of the network is reached and after that point only small increases in performance are achieved by increasing the number of neurons in the single layers.

Although the results of the test data are slightly increasing with an increasing number of neurons in the layers, this increase comes with the cost of a higher complexity and this has two major drawbacks. The first one is the high computational demand, i.e. the increased training and classification time, and the second one is related to the generalization capability of the classifiers. Generally speaking, the higher the complexity of a classifier, the worse the generalization capability. This relationship is the main point of the so-called *Vapnik–Chervonenkis dimension* (VC dimension) of a classifier. The VC dimension measures the ability of a classifier to separate a training set in arbitrary ways. If the VC dimension is high, the better a classifier can adapt to the training data, but this should have a negative influence on the generalization. More details on this theory can be found in textbooks like [15] or [10].

Table 8.4 MSTAR test data results of the different CNN structures

CNN structure	$\bar{P}_{CC}$ (%)	$P_{CC}^{\max}$ (%)	$P_{CC}^{\min}$ (%)	$\sigma_{P_{CC}}$ (%)
Five-layer				
4-8-14	70.89	74.15	67.34	2.80
6-14-18	82.42	86.61	76.87	2.96
8-18-22	91.60	92.85	90.45	0.84
10-22-28	93.62	94.58	92.88	0.57
12-26-32	93.59	94.01	92.35	0.48
14-28-38	93.90	94.69	92.98	0.48
16-32-42	94.61	95.22	93.19	0.69
18-36-48	94.16	95.82	93.13	0.72
20-40-52	94.64	95.44	94.10	0.51
22-44-58	94.11	96.00	92.16	1.14
Seven-layer				
4-8-14-20	83.25	87.23	78.49	2.82
6-12-20-28	90.23	92.29	87.29	1.60
8-16-26-36	90.48	92.44	88.70	1.46
10-20-32-44	92.78	94.63	88.26	1.73
12-24-38-52	93.28	94.47	91.07	0.97
14-28-44-60	92.38	93.51	91.07	0.88
16-32-50-66	95.11	96.28	93.72	0.78
18-36-56-72	94.57	95.19	92.88	0.74
Nine-layer				
6-10-16-24-32	80.07	90.76	37.37	17.43
8-14-22-32-40	88.75	93.79	81.61	3.55
10-18-26-36-46	87.42	95.50	39.40	16.96
12-22-30-42-54	91.05	94.32	88.57	1.80
14-26-34-48-60	85.01	94.51	46.27	15.47
16-30-38-54-68	94.44	95.94	93.13	0.96

The results for the seven- and nine-layer structures are also shown in Table 8.4. The description scheme is the same as for the five-layer case, the last number gives the number of neurons in the first fully connected layer and the remaining numbers give the number of feature maps in the order of the convolutional layers. The results of the seven-layer networks are comparable to the five-layer networks, a steady increase with an increasing complexity of the networks. Also the absolute numbers are in a comparable range, although the maximum correct classification rate is achieved with a seven-layer structure, namely the structure with 16 correlation kernels in the first layer.

The results of the nine-layer structures already show a problem of the high complexity. From the minimum classification rates, it can be seen that some of these networks did not converge and thus the classification rates are very low. On the other hand, some of these networks give excellent results, which results in very high standard deviations for these structures.

Table 8.5 MSTAR test data results with linear pooling layers

CNN structure	$\bar{P}_{CC}$ (%)	P_{CC}^{max} (%)	P_{CC}^{min} (%)	$\sigma_{P_{CC}}$(%)
10-22-28	90.81	94.01	85.51	2.36
16-32-42	91.19	94.07	88.07	1.49
16-32-50-66	42.82	62.97	21.57	14.26

The results on the MSTAR dataset lead to the conclusion that depth is not the most important criteria in neural network classification systems for this dataset. It is more important that the classifier possess a certain complexity to learn the data. But this complexity should also be not too high to avoid problems with overfitting.

As it was already mentioned in the description of the system in Section 8.4.1, two different activation functions are used in the pooling layers. Therefore, the training is repeated for some of the network structures of Table 8.4 with a linear, which means no activation, in the pooling layer. For this experiment, only three of the structures trained above are used, the 10-22-28, the 16-32-42 and the 16-32-50-66 structures. These were chosen since they are good examples for the used complexity levels, which achieved good results. The results of this experiment are shown in Table 8.5.

The results with the linear activation function show a reduced correct classification rate compared to the results in Table 8.4. Although the maximum values of the five-layer structures are close to the results with the nonlinear activation, the mean values are reduced by some per cents due to the higher variations in the results. For the seven-layer structure, the mean result is reduced by more than 50%, which means that the training has failed completely. These results indicate that the nonlinear activation function increases the capacity of the classifier.

In the next step, these experiments are repeated with the TIRA ISAR dataset and slightly changed parameters due to the smaller amount of data. The training time is reduced to 24 epochs, but the early stopping with a value of 10^{-3} for the CE error function is still used. As already mentioned at the end of Section 8.3.2, this dataset is always expanded with the methods of Section 8.3.3 to avoid a bias due to the unbalanced dataset. Due to the larger image size of this dataset, the kernel sizes and pooling parameter of the different structures must be changed and the new values are printed in Table 8.6.

The results for the TIRA dataset with the different structures are shown in Table 8.7. The numbers in the structure determine also here the number of feature maps and number of fully connected neurons. The minimum and maximum network sizes are, compared to the networks used with the MSTAR data, reduced here. This reduction is due to a halved number of classes and thus the problem is of a smaller complexity, but the data itself is more challenging due to the higher amount of imaging artefacts in ISAR imaging in comparison to the MSTAR dataset. Therefore, the reduction is not as large as one would expect for a comparable SAR dataset.

The results of the TIRA data show a behaviour that is comparable to the results of the MSTAR data. The first point is that an increasing depth is not a guarantee for better

Table 8.6 Network structures for the TIRA dataset

Layer	Five-layer CNN	Seven-layer CNN	Nine-layer CNN	Eleven-layer CNN
1	Convolutional 29 × 29	Convolutional 21 × 21	Convolutional 9 × 9	Convolutional 5 × 5
2	Pooling, stride 4	Pooling, stride 2	Pooling, stride 2	Pooling, stride 2
3	Convolutional 25 × 25	Convolutional 19 × 19	Convolutional 9 × 9	Convolutional 5 × 5
4	Fully connected	Pooling, stride 2	Pooling, stride 2	Pooling, stride 2
5	Softmax	Convolutional 18 × 18	Convolutional 9 × 9	Convolutional 6 × 6
6	–	Fully connected	Pooling, stride 2	Pooling, stride 2
7	–	Softmax	Convolutional 9 × 9	Convolutional 5 × 5
8	–	–	Fully connected	Pooling, stride 2
9	–	–	Softmax	Convolutional 4 × 4
10	–	–	–	Fully connected
11	–	–	–	Softmax

Table 8.7 TIRA test data results of the different CNN structures

CNN Structure	$\bar{P}_{CC}$ (%)	P_{CC}^{max} (%)	P_{CC}^{min} (%)	$\sigma_{P_{CC}}$ (%)
Five-layer				
2-8-16	90.46	95.26	75.14	5.72
4-12-20	89.35	96.02	74.95	8.01
6-16-24	94.04	95.45	91.27	1.56
8-20-28	95.16	96.77	92.41	1.43
10-24-32	94.61	98.86	90.51	2.23
12-28-36	95.24	97.15	91.84	1.74
14-30-38	95.05	97.53	92.03	1.74
16-34-42	94.00	96.20	91.65	1.57
18-38-46	94.00	96.39	92.41	1.28
Seven-layer				
2-6-10-16	82.47	94.69	68.50	9.07
4-10-14-22	93.47	94.88	91.46	1.18
6-12-18-26	83.15	94.50	72.49	8.81
8-16-22-32	91.04	95.83	83.11	3.71
10-20-24-36	92.13	92.60	91.65	0.67
Nine-layer				
6-14-20-24-30	78.52	92.60	68.88	7.01
8-18-24-28-36	88.61	91.65	83.68	2.32
10-22-28-32-40	80.70	87.86	72.49	6.02
12-26-32-38-46	81.69	88.05	72.49	5.70
Eleven-layer				
4-8-14-20-24-30	80.53	90.89	69.26	6.64
6-12-18-26-30-38	79.35	90.51	56.74	10.22

Table 8.8 TIRA test data results with linear pooling layers

CNN structure	$\bar{P}_{CC}$ (%)	P_{CC}^{max} (%)	P_{CC}^{min} (%)	$\sigma_{P_{CC}}$ (%)
10-24-32	91.48	92.98	88.99	1.33
12-28-36	86.55	92.03	81.78	3.43
4-10-14-22	74.67	88.43	38.71	17.95

results; in fact, the best results are obtained with the five-layer structure. The best mean performance is achieved with the 12-28-36 structure and the best maximum correct classification rate of 98.86% is achieved with the 10-24-32 structure. The deeper structures cannot compete with the results of the five-layer network, which is also an indicator that this problem is not that complex due to the small number of classes.

A problem that arises with both datasets is the convergence problem of deeper structures, which is tackled by the regularization methods presented earlier, and results of these methods are shown in the next sections.

For the TIRA data, the experiment with the linear pooling activation is also repeated for three selected structures. In this case, the 10-24-32, the 12-28-36 and the 4-10-14-22 structures are used. The results of this experiment are shown in Table 8.8.

The results of the experiment with the TIRA data are comparable to the result of the MSTAR experiment. The correct classification rates of the five-layer networks are some per cents below the results of sigmoid activation function. The seven-layer network has already strong convergence problems, and the correct classification rate is not comparable to the network with the nonlinear activation. Compared to the MSTAR dataset, the performance of the seven-layer network is still much better, especially the maximum value. One reason for this could be the reduced complexity of the classification problem.

8.5.2 Training method and cost function

The experiments in this section are supposed to show the difference between the presented training methods of Sections 8.4.2.1 and 8.4.2.2, namely the SDLM method, which was used so far in the experiments of the previous section, and the ADAM method. Further is the influence of the applied cost function, i.e. the MSE or the CE, investigated with regard to the used training algorithm. The results in this section are again the classification rates of the test data for the different combinations of training method and cost function. The structures chosen in this section are the same as have been selected in Section 8.5.1 for the experiments with the linear pooling activation. In addition to the correct classification rates, the value of the cost function during the training is shown to compare the convergence properties of the different combinations. For both cost functions, the same minimum value of 0.001 is used as stopping criteria.

Table 8.9 MSTAR test data results with different combinations of cost function and training algorithm

CNN structure	$\bar{P}_{CC}$ (%)	P_{CC}^{max} (%)	P_{CC}^{min} (%)	$\sigma_{P_{CC}}$ (%)
SDLM-CE				
10-22-28	93.62	94.58	92.88	0.57
16-32-42	94.61	95.22	93.19	0.69
16-32-50-66	95.11	96.28	93.72	0.78
SDLM-MSE				
10-22-28	38.46	51.83	26.97	8.04
16-32-42	54.38	61.94	48.02	4.99
16-32-50-66	75.49	78.93	70.28	2.56
ADAM-CE				
10-22-28	89.36	94.41	77.14	5.73
16-32-42	91.15	95.00	82.76	4.45
16-32-50-66	89.96	95.00	71.90	6.77
ADAM-MSE				
10-22-28	87.63	91.41	79.36	4.71
16-32-42	83.81	89.54	74.15	5.34
16-32-50-66	88.38	93.88	79.43	4.08

The results are separated as earlier for the MSTAR and TIRA data and are shown in Tables 8.9 and 8.10, respectively. The results for the combination of SDLM and CE are taken from Tables 8.4 and 8.7 but are included here as well for a better overview.

Table 8.9 shows that the choice of the cost function is more important than the choice of the training algorithm. Especially in case of the SDLM, the CE cost function gives much better results than the MSE, although the bad results of the MSE cost function might be due to convergence problems for this network size. Different network sizes might give better results, since this combination of training algorithm and cost function was also used in the past by the authors [26] with a performance comparable to the CE results. A comparison between the training algorithms is not that obvious. If the CE cost function is applied, the SDLM gives better results, but with the MSE cost function, the performance of the ADAM algorithm is much better. Although the variation of the results in the latter case is still quite high, the stabilizing behaviour of the ADAM might have a positive effect on the output and achieves a stable convergence for the MSE cost function. One possible reason for the better results of the second-order SDLM method with the CE cost function might be a more stable gradient than with the MSE cost function.

The TIRA results of Table 8.10 confirm the results of the MSTAR data, since also for this dataset the choice of the cost function possesses a higher influence on the classification rates than the choice of the training method. With both training methods, the CE cost functions gives better classification rates. The performance of the training algorithms is also comparable to Table 8.9, i.e. with the MSE cost function, the ADAM algorithm performs a little bit better than the SDLM, at least

Table 8.10 TIRA test data results with different combinations of cost function and training algorithm

CNN structure	$\bar{P}_{CC}$ (%)	P_{CC}^{max} (%)	P_{CC}^{min} (%)	$\sigma_{P_{CC}}$ (%)
SDLM-CE				
10-24-32	94.61	98.86	90.51	2.23
12-28-36	95.24	97.15	91.84	1.74
4-10-14-22	93.47	94.88	91.46	1.18
SDLM-MSE				
10-24-32	90.66	92.98	83.87	2.94
12-28-36	91.33	95.07	85.96	3.03
4-10-14-22	89.09	93.17	81.78	3.44
ADAM-CE				
10-24-32	94.06	95.45	91.46	1.10
12-28-36	94.52	95.64	92.60	1.13
4-10-14-22	88.05	96.77	77.61	6.12
ADAM-MSE				
10-24-32	93.66	96.58	90.89	2.03
12-28-36	91.97	96.20	82.54	5.00
4-10-14-22	86.77	95.45	73.06	6.34

for the five-layer structures. Nevertheless, the best results are achieved by the SDLM with the CE cost function, as in the case of the MSTAR data. An interesting point is the standard deviation of the test results, which is in three out of four cases for the five-layer structures minimized by the ADAM algorithm. This is definitely a benefit of this algorithm, since in a practical application a stable training result is sometimes more important than 1% or 2% higher classification rates. In case of the seven-layer network, the ADAM algorithm gives with both cost functions the highest maximum classification rate, but due to the high variation of the results, the mean value is below the SDLM results.

An explanation for the contradictive points between both datasets, like the smallest standard deviation, could be the difference between the two datasets. Although the MSTAR is the more complex classification problem, i.e. ten classes instead of five, the ISAR imaging artefacts of the TIRA dataset make the dataset itself more challenging.

8.5.3 Regularization methods

To validate the usefulness of the regularization methods of Section 8.4.3, several experiments with and without these methods are presented in this section. The reference ($\lambda = 0$) are again the results of Tables 8.4 and 8.7, and comparative results are shown for different values of the regularization parameter.

The first method mentioned in the description of Section 8.4.3 is the *weight decay* method, which reduces the weights of the network in each iteration by a small amount. In this way, the network should be reduced to the important weights and should give

Table 8.11 MSTAR test data results with different weight decay parameters

Structure	$\bar{P}_{CC}$ (%)	P_{CC}^{max} (%)	P_{CC}^{min} (%)	$\sigma_{P_{CC}}$ (%)
10-22-28				
$\lambda = 0$	93.62	94.58	92.88	0.57
$\lambda = 10^{-6}$	90.91	93.16	88.23	1.46
$\lambda = 10^{-4}$	90.31	92.88	87.82	1.83
$\lambda = 10^{-2}$	27.32	31.03	24.66	1.82
16-32-42				
$\lambda = 0$	94.61	95.22	93.19	0.69
$\lambda = 10^{-6}$	88.57	91.20	86.82	1.54
$\lambda = 10^{-4}$	91.84	93.78	88.10	2.18
$\lambda = 10^{-2}$	28.38	30.00	26.57	1.09
16-32-50-66				
$\lambda = 0$	95.11	96.28	93.72	0.78
$\lambda = 10^{-6}$	84.05	89.07	75.93	4.35
$\lambda = 10^{-4}$	83.18	90.42	67.22	6.39
$\lambda = 10^{-2}$	20.53	26.91	18.17	3.75

better results, especially in terms of generalization. Results for the MSTAR and the TIRA dataset are shown in Tables 8.11 and 8.12, respectively.

The results of the MSTAR dataset in Table 8.11 show that no improvement with the test data is achieved. The reason for this might be the fact that the targets of the MSTAR dataset are exactly the same in both training and test data. Therefore, a classifier that memorizes the data might be better for this particular dataset than a classifier with good generalization. This and other problems of the MSTAR dataset was already discussed in [1], where the clutter correlation between training and test data was shown. Another point visible in the results is that the decay parameter should not be too large, since if a certain value is overcome the decay is larger than the learning rate, which makes a proper training process impossible.

For the TIRA dataset, the independence of the data was achieved by using, if possible, measurements of different flights and in case of the fighter aircrafts also different configurations of the targets. These target variations are a further factor that makes this dataset more challenging than the MSTAR dataset and a part of the results of the experiment with the weight decay confirms in this case that a better generalization improves the performance of the classifier. Although the maximum value is not reached in case of the 10-24-32 structure without weight decay, the mean value increases and the significantly lower standard deviation indicates a more stable training process. With the 12-28-36 structure, the mean value and also the maximum value decrease, but the standard deviation decreases with the smallest decay parameter. With the medium decay parameter, a single outlier increases the standard deviation a lot. If this outlier of 80.83% is removed, the minimum value is 92.60%, which increases the mean value to 93.84% and the standard deviation is reduced to 0.91%.

Table 8.12 TIRA test data results with different weight decay parameters

Structure	$\bar{P}_{CC}$ (%)	P_{CC}^{max} (%)	P_{CC}^{min} (%)	$\sigma_{P_{CC}}$ (%)
10-24-32				
$\lambda = 0$	94.61	98.86	90.51	2.23
$\lambda = 10^{-6}$	94.74	95.64	92.60	1.02
$\lambda = 10^{-4}$	95.26	96.39	92.60	1.20
$\lambda = 10^{-2}$	73.81	85.20	62.43	6.71
12-28-36				
$\lambda = 0$	95.24	97.15	91.84	1.74
$\lambda = 10^{-6}$	93.36	95.26	90.32	1.56
$\lambda = 10^{-4}$	92.54	95.45	80.83	4.20
$\lambda = 10^{-2}$	74.27	82.35	69.26	4.13
4-10-14-22				
$\lambda = 0$	93.47	94.88	91.46	1.18
$\lambda = 10^{-6}$	82.26	92.22	51.99	12.30
$\lambda = 10^{-4}$	85.18	94.88	62.62	9.85
$\lambda = 10^{-2}$	57.97	77.80	31.88	12.84

For large decay parameter and the larger network, the same effect as with the MSTAR dataset appears and the classification rates decrease.

8.5.4 Artificial training data

In this part, the benefit of the artificial training data of Section 8.3.3 is presented. The results of this section are limited to the MSTAR dataset, since the TIRA dataset is always extended with these methods due to the unbalanced amount of training data.

In [26], the dataset was also extended with most of the transformations applied here and the results were presented separated for different combinations of the extended dataset to find the influence of different transformations. In this work, only results with the maximum amount of artificial training data are presented, since also in [26] the best results were achieved with the maximum amount of training data. The transformations used for the MSTAR dataset are affine rotation and scaling of the range axis, elastic distortion and a random shift of the target position. For each transformation, two realizations of each training sample are generated and added to the original data.

The training is done for the same network structures as in the previous sections and the results are shown in Table 8.13.

From Table 8.13, it can be seen that the smaller one of the five-layer networks is not able to handle this huge amount of training data. Although the maximum classification rate is over 90%, the mean value is below 50% since many of the ten networks did not converge. The extreme case happened for the network with the minimum classification rate, all test images were assigned to the same class. The results of the larger networks are more as one would expect, in both cases the mean value is increased compared to the reference results of Table 8.4. Further is

Table 8.13 MSTAR test data results with artificial training data

Structure	$\bar{P}_{CC}$ (%)	P_{CC}^{max} (%)	P_{CC}^{min} (%)	$\sigma_{P_{CC}}$ (%)
10-22-28	49.89	90.85	6.09	40.23
16-32-42	95.88	96.63	94.97	0.48
16-32-50-66	96.65	96.78	94.47	0.67

the maximum value of this dataset achieved with the 16-32-42 structure, although the improvement and the absolute number are not comparable to the result of [26]. This is due to the fact that the knowledge of the aspect angle is not used here, which is the more realistic situation. Another benefit of the artificial training data is the stabilizing effect that can be seen in the reduced standard deviation of the results for the two larger networks in Table 8.13.

8.5.5 Use of SVMs as classifier

In our previous work, the classification part of the CNN was often replaced by SVMs to improve the classification results [16,26]. This experiment is repeated here to verify if this step is still useful with the improved network implementation compared to our previous publications. For this experiment, the same network structures as in the previous sections are selected to illustrate the results. The results are shown for the TIRA data and for the MSTAR data. The structure of this classification system can be seen in Figure 8.10 with an example of the MSTAR data.

The last two layers of the network structure have been removed and replaced by SVMs, which use a one vs. all strategy to classify the data. This means that in a ten class problem ten SVMs are created and each of them is trained to recognize a specific class. Since SVMs produce a separating hyperplane in a high-dimensional feature space, the distance to this hyperplane can be used as confidence measure of the classification. If two or more SVMs give a positive result, the class with the highest confidence is chosen as the final result in a forced decision scheme. If a rejection class is used, a minimum confidence measure can be defined that must be overcome to get a classification result, see for example [43].

For the implementation of the SVMs, polynomial kernels up to the order of three (linear, quadratic and poly3) and radial basis functions (RBF) are used. For the RBF kernels, several σ values are tried and only the best ones are shown in Tables 8.14 and 8.15. The result of the original CNN structure is included in the tables as well for comparison.

The results of the MSTAR data in Table 8.14 show that the use of SVMs as classifier improves the performance in most cases. In the given examples, only the linear SVM is always below the CNN results. The performances of the remaining polynomial SVMs are in the vicinity of the CNN results, some are above the CNN and some are below. Only the results of the SVMs with RBF kernels are always above the CNN alone, and in the case of the seven-layer structure, the maximum value is the

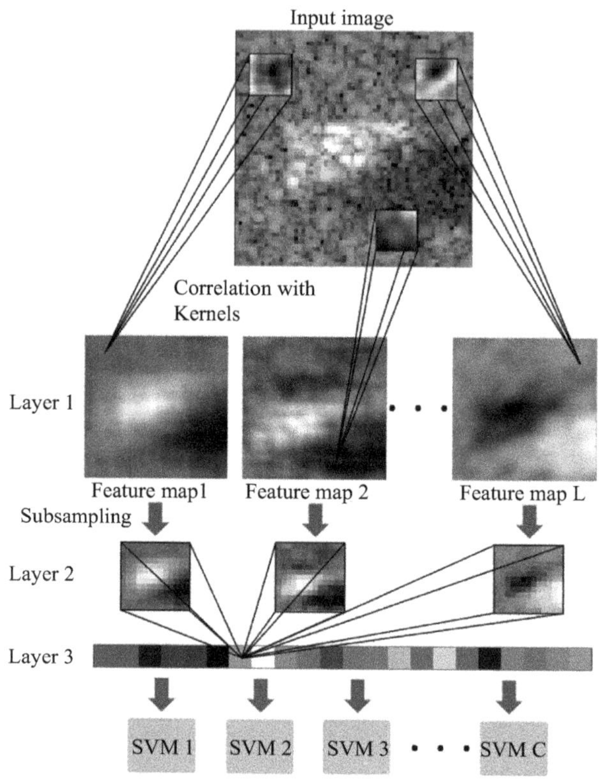

Figure 8.10 Structure of a CNN in combination with SVMs

highest value achieved for this dataset within this chapter, although the mean value of the CNN trained with the extended dataset is higher.

The results of the TIRA dataset in Table 8.15 principally confirm the results of the MSTAR data. Although not as clearly as earlier, with the 12-28-36 structure for example, the highest average result is achieved with the original CNN, whereas the maximum value is increased in three of four cases. With the smaller of the five-layer structures, the results are clearly increased and give the best mean and maximum values of all experiments in this chapter for this dataset. Compared to the MSTAR dataset, the RBF kernel is not the best kernel, except for the 12-28-36 structure CNN. Although the maximum, minimum and standard deviation numbers seem to be better for the SVM classifier in that case, the averaged result is still below the original CNN, if only by a very small amount. For the other two chosen network structures, the polynomial kernels give the best results.

The overall result of this section shows that the use of SVMs as classifier improves the classification results in most cases. The drawback of this combination is the increased training time and the further increased number of parameters to choose.

Table 8.14 MSTAR test data results with support vector machines as classifiers

Structure	$\bar{P}_{CC}$ (%)	P_{CC}^{max} (%)	P_{CC}^{min} (%)	$\sigma_{P_{CC}}$ (%)
10-22-28				
CNN	93.62	94.58	92.88	0.57
SVM linear	92.14	92.88	91.26	0.53
SVM quadratic	93.67	94.66	92.98	0.60
SVM poly3	93.74	94.94	92.98	0.59
SVM RBF	94.66	95.38	94.04	0.49
16-32-42				
CNN	94.61	95.22	93.19	0.69
SVM linear	92.87	93.79	90.98	0.86
SVM quadratic	94.36	95.04	92.91	0.78
SVM poly3	94.59	95.35	92.88	0.80
SVM RBF	95.32	96.35	93.72	0.80
16-32-50-66				
CNN	95.11	96.28	93.72	0.78
SVM linear	93.30	95.35	91.38	1.18
SVM quadratic	95.01	96.13	93.63	0.85
SVM poly3	95.54	96.44	94.13	0.84
SVM RBF	95.98	97.16	94.69	0.81

Table 8.15 TIRA test data results with support vector machines as classifiers

Structure	$\bar{P}_{CC}$ (%)	P_{CC}^{max} (%)	P_{CC}^{min} (%)	$\sigma_{P_{CC}}$ (%)
10-24-32				
CNN	94.61	98.86	90.51	2.23
linear	95.37	97.53	92.79	1.39
quadratic	95.67	98.48	93.93	1.48
poly3	95.45	99.24	91.65	2.07
RBF	95.29	97.53	93.17	1.41
12-28-36				
CNN	95.24	97.15	91.84	1.74
linear	94.78	97.72	92.60	1.74
quadratic	95.01	97.72	93.55	1.37
poly3	93.59	96.96	88.61	2.72
RBF	95.22	98.10	93.17	1.51
4-10-14-22				
CNN	93.47	94.88	91.46	1.18
linear	94.38	96.20	92.41	1.06
quadratic	94.02	95.64	91.08	1.34
poly3	93.98	97.53	92.22	1.56
RBF	93.91	95.64	91.84	1.08

8.6 Summary and conclusion

The analysis done in this chapter shows that CNNs are a state-of-the-art method for image-based radar target recognition. The results achieved here give classification rates of more than 96% for the MSTAR dataset without the use of any prior knowledge, which is a very challenging situation. The aspect angle was assumed to be completely unknown, which is more realistic than the exact knowledge of it, but in operational scenarios, the aspect angle might be known with a few degrees uncertainty and the 180° ambiguity. This ambiguity comes from the commonly used bounding box estimation of the aspect angle [44].

From the presented optimization techniques, the one with the highest benefit was the use of artificial training data. Although the smaller tested network did not converge in this case due to the higher complexity, the results of the larger networks have improved and gave the best averaged result achieved in this paper. The highest maximum value and also an increased averaged classification rate were achieved with the CNN-SVM combination in most cases. The benefit of the weight decay parameter could be shown with the TIRA dataset, but not with the MSTAR dataset. The reason for this is probably the high correlation between the training and the test data of the MSTAR dataset, while the TIRA dataset is created from independent measurements with different target configurations in some cases.

In general, it is hard to say *a priori* what network size with which parameters should be used, since depending on the data and the complexity, the results could vary a lot. This could be seen, for example, in the comparison of the training algorithms with the different cost function for the MSTAR and TIRA datasets. Although both datasets showed roughly the same behaviour, there were large differences in the convergence properties and the stability of the training. Therefore, and for other reasons, the search for an operational CNN classifier is still going and there is still a long way to an *off-the-shelf-classifier* solution with DL methods. Yet the results achieved so far are a good motivation to continue with the research on this topic.

References

[1] Schumacher R and Schiller J. Non-cooperative target identification of battlefield targets – classification results based on SAR images. In: *Radar Conference, 2005 IEEE International*, IEEE, Arlington, VA, USA; 2005. p. 167–172.

[2] Mahboubi Z and Kochenderfer MJ. Learning Traffic Patterns at Small Airports From Flight Tracks. *IEEE Transactions on Intelligent Transportation Systems.* 2017;18(4):917–926.

[3] Villeval S, Bilik I, and Gürbüz SZ. Application of a 24 GHz FMCW automotive radar for urban target classification. In: *2014 IEEE Radar Conference*, IEEE, Cincinnati, OH, USA; 2014. p. 1237–1240.

[4] Blacknell D and Griffiths H, editors. *Radar Automatic Target Recognition (ATR) and Non-Cooperative Target Recognition (NCTR).* Radar, Sonar

and Navigation Series. Stevenage, UK: The Institution of Engineering and Technology; 2013.
[5] El-Darymli K, Gill EW, Mcguire P, *et al.* Automatic Target Recognition in Synthetic Aperture Radar Imagery: A State-of-the-Art Review. *IEEE Access.* 2016;4:6014–6058.
[6] Rihaczek AW and Hershkowitz SJ. *Theory and Practice of Radar Target Identification.* Norwood, MA: Artech House; 2000.
[7] Ross TD, Worrell SW, Velten VJ, *et al.* Standard SAR ATR Evaluation Experiments Using the MSTAR Public Release Data Set. *Proc SPIE.* 1998; 3370:566–573.
[8] Diemunsch JR and Wissinger J. Moving and stationary target acquisition and recognition (MSTAR) model-based automatic target recognition: search technology for a robust ATR. In: *Proc. SPIE*, Orlando, FL, USA. vol. 3370; 1998. p. 481–492.
[9] Blacknell D. A comparison of compressive sensing and Fourier reconstructions for radar target recognition. In: *4th Int. Workshop on Compressed Sensing on Radar, Sonar, and Remote Sensing (CoSeRa 2016).* Aachen, Germany; 2016. p. 217 – 221.
[10] Haykin S. Neural *Networks and Learning Machines.* 3rd ed. Upper Saddle River, NJ: Pearson Education Inc.; 2009.
[11] Bishop CM. *Neural Networks for Pattern Recognition.* Oxford University Press, Oxford, UK; 1995.
[12] Duda RO, Hart PF, and Stork DG. *Pattern Classification.* 2nd ed. Wiley, Hoboken, NJ, USA; 2000.
[13] Ng A. Sparse Autoencoder; 2011. CS294A Lecture Notes, Stanford University, Stanford, CA, USA.
[14] Schölkopf B, Smola AJ. *Learning with Kernels.* Cambridge, Massachusetts: MIT Press; 2002.
[15] Vapnik VN. *The Nature of Statistical Learning Theory. 2nd ed. Statistics for Engineering and Information Science.* New York, NY: Springer-Verlag; 2000.
[16] Wagner S. Combination of convolutional feature extraction and support vector machines for radar ATR. In: *Information Fusion (FUSION), 2014 17th International Conference on*, IEEE, Salamanca, Spain; 2014.
[17] Chen S and Wang H. SAR target recognition based on deep learning. In: *Data Science and Advanced Analytics (DSAA), 2014 International Conference on*, IEEE, Shanghai, China; 2014. p. 541–547.
[18] Goodfellow I, Bengio Y, and Courville A. *Deep Learning.* The MIT Press, Cambridge, MA, USA; 2016.
[19] Bengio Y. Learning Deep Architectures for AI. Foundations and Trends® in *Machine Learning.* 2009;2(1):1–127.
[20] Deng S, Du L, Li C, *et al.* SAR Automatic Target Recognition Based on Euclidean Distance Restricted Autoencoder. *IEEE Journal of Selected Topics in Applied Earth Observations and Remote Sensing.* 2017;10(7):3323–3333.
[21] Kang M, Ji K, Leng X, *et al.* Synthetic Aperture Radar Target Recognition with Feature Fusion Based on a Stacked Autoencoder. *Sensors.* 2017;17(1).

[22] Dong G, Liao G, Liu H, *et al.* A Review of the Autoencoder and Its Variants: A Comparative Perspective from Target Recognition in Synthetic-Aperture Radar Images. *IEEE Geoscience and Remote Sensing Magazine.* 2018;6(3): 44–68.

[23] Li X, Li C, Wang P, *et al.* SAR ATR based on dividing CNN into CAE and SNN. In: *Synthetic Aperture Radar (APSAR), 2015 IEEE 5th Asia-Pacific Conference on*, IEEE, Singapore, Singapore; 2015. p. 676–679.

[24] Chen S, Wang H, Xu F, *et al.* Target Classification Using the Deep Convolutional Networks for SAR Images. *IEEE Transactions on Geoscience and Remote Sensing.* 2016;54(8):4806–4817.

[25] Ødegaard N, Knapskog AO, Cochin C, *et al.* Classification of ships using real and simulated data in a convolutional neural network. In: *Radar Conference (RadarConf)*. IEEE, Philadelphia, PA, USA, 2016 IEEE; 2016.

[26] Wagner SA. SAR ATR by a Combination of Convolutional Neural Network and Support Vector Machines. *IEEE Transactions on Aerospace and Electronic Systems.* 2016;52(6):2861–2872.

[27] Burak Alver M, Atito S, and Çetin M. SAR ATR in the phase history domain using deep convolutional neural networks. In: *Proc. SPIE*, Berlin, Germany. vol. 10789; 2018. p. 1–10.

[28] Coman C and Thaens R. A deep learning SAR target classification experiment on MSTAR dataset. In: *2018 19th International Radar Symposium (IRS)*, IEEE, Berlin, Germany; 2018. p. 1–6.

[29] Hänsch R and Hellwich O. Complex-valued convolutional neural networks for object detection in PolSAR data. In: *8th European Conference on Synthetic Aperture Radar*; 2010. p. 1–4.

[30] Wehner DR. *High-Resolution Radar.* 2nd ed. Artech House, Norwood, MA, USA; 1995.

[31] Chen VC and Martorella M. *Inverse Synthetic Aperture Radar Imaging.* Edison, NJ: SciTech Publishing; 2014.

[32] Sensor Data Management System. MSTAR Overview; 1995. Available from: www.sdms.afrl.af.mil/index.php?collection=mstar.

[33] Zhou F, Wang L, Bai X, *et al.* SAR ATR of Ground Vehicles Based on LM-BN-CNN. *IEEE Transactions on Geoscience and Remote Sensing.* 2018;56(12):7282–7293.

[34] Ligori D, Wagner S, Fabbrini L, *et al.* Nonparametric ISAR Autofocusing Via Entropy- Based Doppler Centroid Search. *IEEE Geoscience and Remote Sensing Letters.* 2018;15(11):1725–1729.

[35] Schiller J and Rosenbach K. Comparing classifier effectiveness. In: Blacknell D, Griffiths H, editors. *Radar Automatic Target Recognition (ATR) and Non-Cooperative Target Recognition (NCTR).* Radar, Sonar and Navigation Series. Stevenage, UK: The Institution of Engineering and Technology; 2013. p. 157–175.

[36] Papson S and Narayanan RM. Classification via the Shadow Region in SAR Imagery. *Aerospace and Electronic Systems, IEEE Transactions on.* 2012;48(2):969–980.

[37] Wagner S, Barth K, and Brüggenwirth S. A deep learning SAR ATR system using regularization and prioritized classes. In: *2017 IEEE Radar Conference (RadarConf)*; 2017. p. 0772–0777.

[38] le Cun Y. *Generalization and Network Design Strategies.* Department of Computer Science, University of Toronto; 1989. CRG-TR-89-4.

[39] LeCun Y, Bottou L, Bengio Y, *et al.* Gradient-Based Learning Applied to Document Recognition. *Proceedings of the IEEE.* 1998;86(11):2278–2324.

[40] Kingma DP and Ba J. Adam: A Method for Stochastic Optimization, 3rd International Conference on Learning Representations, ICLR 2015, San Diego, CA, USA. arXiv:1412.698. May 7–9, 2015.

[41] Plaut DC, Nowlan SJ, and Hinton GE. *Experiments on Learning by Back Propagation.* Pittsburgh, PA: Carnegie–Mellon University; 1986. CMU-CS-86-126.

[42] Srivastava N, Hinton G, Krizhevsky A, *et al.* Dropout: A Simple Way to Prevent Neural Networks From Overfitting. *Journal of Machine Learning Research.* 2014;15(1):1929–1958.

[43] Wagner S. Morphological component analysis in SAR images to improve the generalization of ATR systems. In: *Compressed Sensing Theory and its Applications to Radar, Sonar and Remote Sensing (CoSeRa), 2015 3rd International Workshop on*, IEEE, Pisa, Italy; 2015. p. 46–50.

[44] Sun Y, Liu Z, Todorovic S, *et al.* Adaptive Boosting for SAR Automatic Target Recognition. *IEEE Transactions on Aerospace and Electronic Systems.* 2007;43(1):112–125.

Chapter 9
Deep learning for passive synthetic aperture radar imaging

Samia Kazemi[1], Eric Mason[2], Bariscan Yonel[1], and Birsen Yazici[1]

9.1 Introduction

9.1.1 Passive radar

This chapter presents a data-driven learning and model-based approach to passive synthetic aperture radar (SAR) imaging within a deep learning (DL) framework. This approach enables simultaneous estimation of unknown model parameters, effective incorporation of prior knowledge, lower computational cost and superior performance than other state-of-the-art methods.

Unlike an active SAR, passive SAR does not have a dedicated transmitter, and instead it utilizes existing sources of electromagnetic signals, commonly referred to as "sources of opportunity" or "illuminator of opportunity." It has been an active area of research due to the proliferation of transmitters of opportunity, such as broadcast systems and communication systems [1], and other advantages over active radar. These include efficient use of electromagnetic spectrum, increased stealth and reduced cost among others [2–9]. A passive radar may use single or multiple sources of opportunity as well as a single receiver or multiple receivers [1,10,11]. In the passive setting, the transmitter-related terms may be entirely or partially unknown. This presents a major challenge for passive imaging, because it poses image reconstruction as a bilinear inverse problem, in which the underlying forward model is only partially known. Several fundamental approaches were proposed in literature to address this problem. These include passive coherent localization (PCL) [12–16], and interferometric imaging techniques [2,17–25].

There has been growing interest in PCL both in academic setting and military air surveillance in recent years due to its robust technological basis accompanied by less maintenance requirements compared to an active radar [1]. In addition to

[1]Department of Electrical, Computer, and Systems Engineering, Rensselaer Polytechnic Institute, Troy, NY, USA
[2]US Naval Research Laboratory, Washington, DC, United States

measuring the reflected signal from the scene of interest, PCL receives the signal from a transmitter of opportunity as well and uses it for matched filtering. This method therefore requires two antennas at the receiver location, the highly directional reference receiver and the surveillance receiver, direct line-of-sight to a transmitter of opportunity and high signal-to-noise ratio (SNR) for the signal received directly from the transmitter. Flow diagram for analog and digital signal processing for PCL is shown in Figure 9.1.

In the disturbance cancellation step, the reference signal is removed from the combined reflected signal measured at the surveillance receiver. The reference signal measured at the reference receiver can be removed either spatially or in the time domain [1]. In the case of digital PCL, the original signal can be reconstructed from the surveillance receiver measurement given that the transmission standard is known. In the range-Doppler cross correlation stage, the target reflection obtained by disturbance cancellation step is used for target detection and tracking.

Interferometric imaging approaches use two or more sufficiently far apart receivers deployed on the same or different locations. The signals at different receiver locations are correlated and backprojected based on time or frequency difference of arrival to form an image of the scene. These methods do not require direct-line-of-sight to a transmitter, high SNR or knowledge of transmitter location. However, they are limited to imaging widely separated point scatterers. More recently, alternative interferometric imaging methods based on low-rank matrix recovery (LRMR) and Wirtinger flow have been developed. However, computational requirements of LRMR [26–28] preclude its applicability in practice. A recent advancement over LRMR has been the generalization of Wirtinger flow (GWF) to interferometric inversion problem [29–33], which offers exact recovery and computational efficiency in practical imaging applications.

Both LRMR and GWF assume full knowledge of transmitted waveforms and make small scene and far-field approximations to eliminate the need for transmitter location information. Furthermore, these methods do not explicitly incorporate

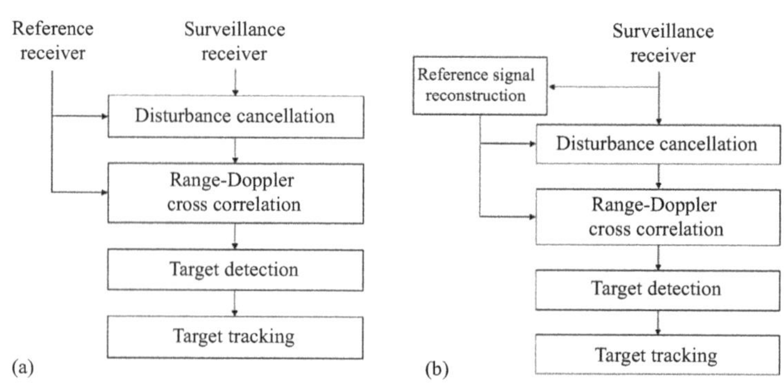

Figure 9.1 (a) and (b) are the signal processing stages for analog and digital PCL, respectively [1]

data-driven a priori information into image reconstruction. Our objective is to combine data-driven and model-based approaches within a DL framework to reconstruct SAR images without the a priori knowledge of transmitter location or transmitted waveforms.

9.1.2 DL and its application to SAR

DL has advanced the state of the art in a wide range of science and engineering problems in recent years [34–37]. DL falls into a subgroup of machine learning (ML) techniques, referred to as representation learning, characterized by multilevel structures. The multilevel structure of DL networks enables task-specific feature extraction mechanism and eliminates the need for hand-crafted feature engineering. Inherent nonlinearity enables deep networks to approximate complex mappings between the input and output domains for seemingly complex tasks, such as target classification from images [38–41] and translation between two natural languages [42,43]. In general, DL provides an effective framework to combine model-based and data-driven learning-based approaches. Due to the complexity associated with nonlinearity and large network dimensions, convergence guarantees are often not available for DL, and its performance is usually interpreted and justified in terms of the universal approximation theorem [44,45].

In recent years, there has been growing interest in adapting DL methods to a wide range of problems in SAR. An important category of DL applications in SAR imaging includes automatic target recognition (ATR) [46–52]. The DL methods adapted to ATR are often motivated by their state-of-the-art performance in the area of visual object classification. Unlike visual image classification problem, for SAR ATR tasks, it is often difficult to collect large amounts of labeled data. To overcome this problem, a transfer-learning-based approach is presented in [53], where a convolutional neural network (NN) (CNN)-based architecture is utilized for learning from a limited number of labeled data as well as a larger set of unlabeled data. In [54], a multi-view SAR dataset is used to train a CNN with a parallel network architecture. Inspired by [55,56], a framework for exploring DL for radar signal processing and imaging is introduced in [57–60]. A recurrent NN (RNN) architecture in which the network parameters are related to underlying forward model is introduced for blind deconvolution and specifically for SAR image reconstruction problems. In [59], the advantages and applicability of RNN-based architecture to passive SAR with unknown transmitter locations are presented. Similar networks with a different parameterization to enable unknown transmitted waveform estimation in passive SAR image reconstruction problem are introduced in [60,61].

DL is an effective framework to combine model-based and data-driven approaches. It allows for the incorporation of functional priors such as those related to imaging geometry-based and physics-based models. The learning-based aspect of DL can be utilized to estimate unknown network parameters from training data. Additionally, the DL method eliminates the need for solving an optimization problem during the imaging process and requires only a fixed number of nonlinear transformations for image formation. Furthermore, RNN framework is also suitable in addressing other blind deconvolution problems in radar, such as waveform design for active SAR and autofocus.

Since the radar-received signal is complex valued, one distinctive feature of the DL-based image reconstruction approach is the necessity for the network to process complex-valued signals. This poses the question of whether to use real- or complex-valued network parameters. For real-valued parameters, there are further choices to be made regarding how to transform complex input into a corresponding multichannel signal consisting of real and imaginary parts; magnitude, phase or any combination of these quantities. These decisions guide the model structures and impact task outcomes, computational costs and training times. The model presented in this chapter is designed so that the input and output signals of each stage, as well as the associated parameters, are all complex valued. This choice is motivated by the underlying model, optimization problem and an iterative algorithm.

9.1.3 Organization of the chapter

The rest of this chapter is organized as follows: in Section 9.2, we present a brief discussion on adapting DL for general inverse problems; in Section 9.3, we present the received signal model for passive SAR and define various parameters; in Section 9.4, we present a brief discussion on DL-based approach for radar image reconstruction; we discuss joint passive SAR imaging and waveform estimation in Section 9.5; in Section 9.6, we present a deep model for image reconstruction when transmitted waveform and transmitter location are unknown; Section 9.7 presents numerical results; and finally, Section 9.8 concludes this chapter.

9.2 DL for inverse problem

One of the simplest ML networks is the perceptron that is a linear binary classifier for an input vector. A perceptron with arbitrary activation function, referred to a as *neuron* or *unit,* is the basic building block of NN. A multilayer NN, with the same activation function $f(.)$ at each layer, is shown in Figure 9.2. According to universal approximation theorem [44,45], any continuous measurable function can be approximated by standard three-layer feedforward networks. In DL, the network includes many more hidden layers than standard NN, which increases the expressive power and nonlinearity of the network. RNNs are multilevel NN where all the stages share the same weight matrix and bias vector. This architecture reduces the number of unknown parameters to be learned and thereby minimizes computational cost, and training data requirement.

In this section, we briefly discuss the adaptation of DL approach for the inverse problem. Inverse problems in the context of radar imaging include forward modeling, inversion and computationally efficient forward and inverse solver algorithm design. DL framework can be utilized in all aspects of inverse problem in radar imaging. The network architecture that estimates the inversion can be derived from the underlying algorithm, while the forward model can be designed and estimated during the learning

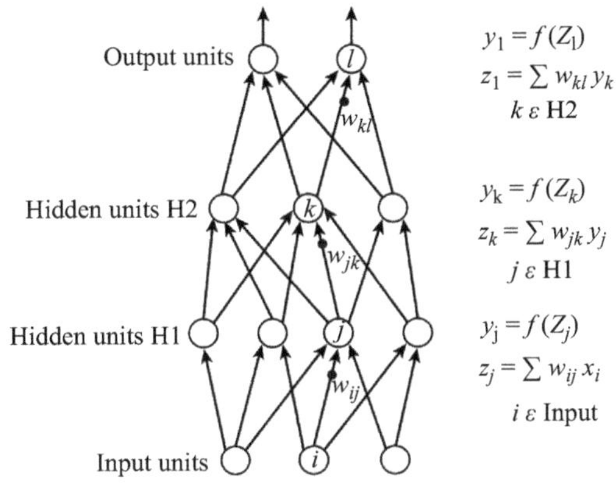

Figure 9.2 Multilayer NN [62]

process. In the next subsections, we discuss the adaptation of DL in addressing forward mapping and inversion problems [59]. The overall DL network design and learning will be discussed in later sections in the context of passive SAR imaging.

9.2.1 DL for forward modeling

Forward modeling refers to the task of learning a mapping operator from the quantities of interest to the physical measurements. This mapping often involves uncertainties and relies on simplifying assumptions, such as single scattering assumption for electromagnetic wave measurements in radar imaging problem that results in an approximately linear mapping [63]. Learning-based approach enables the estimation of a nonlinear mapping, in which case, the network is required to be parameterized by the functions of the unknown measurement operator. Let $\boldsymbol{\rho} : \mathbb{R}^2 \to \mathbb{R}^N$ be the quantity of interest, $\mathbf{d} \in \mathbb{C}^M$ be the corresponding measured value, $\mathbf{n}$ denotes the measurement noise and $\mathscr{M}$ the measurement or forward operator. Then,

$$\mathbf{d} = \mathscr{M}[\boldsymbol{\rho}] + \mathbf{n}. \tag{9.1}$$

The goal is to estimate mapping $\mathscr{M}$ using a deep network where the lth layer, for $l \in [1, L]$, involves an affine transformation $\mathscr{W}_l$ followed by a nonlinear operator σ. $\boldsymbol{\rho}$ is applied as the network input and the output is treated as an approximation of the measurement value $\mathbf{d}$. If $\hat{\mathbf{d}}$ denotes the network output, then

$$\hat{\mathbf{d}} = \sigma\left(\mathscr{W}_L \sigma\left(\mathscr{W}_{L-1} \cdots \sigma\left(\mathscr{W}_1[\boldsymbol{\rho}]\right)\right)\right). \tag{9.2}$$

Suppose, the training data consists of $\{\boldsymbol{\rho}_1, \boldsymbol{\rho}_2, \ldots, \boldsymbol{\rho}_T\}$ and $\{\mathbf{d}_1, \mathbf{d}_2, \ldots, \mathbf{d}_T\}$ that are two sets containing quantity of interest and corresponding measurement value, respectively. Comparing (9.1) and (9.2), it is evident that the approximation of $\mathscr{M}$ within the network is parameterized by $\mathscr{W}_l$s. Therefore, learning the measurement operator requires learning the equivalent set of operators $\mathscr{W}_l$s from the training data.

9.2.2 DL for inversion

Inversion refers to the estimation of the quantity of interest from the measurement value. It can be implemented as the forward propagation of a deep network. With the conventional filtered-backprojection method where $\mathscr{H}$ and $\mathscr{M}^{\dagger}$ are the filter and the backprojection operators, respectively, estimated quantity of interest $\hat{\boldsymbol{\rho}}$ from measurement $\mathbf{d}$ is $\mathscr{M}^{\dagger}\mathscr{H}[\mathbf{d}]$. Solving this inversion problem with deep network involves starting with an initial estimation of the quantity of interest and gradually updating this value [57,59], as a function of the measurement, as it propagates forward through the network. With linear weight operators $\mathscr{W}_l$s for $l \in [1, L]$ and nonlinear operator σ, estimated inversion outcome $\hat{\boldsymbol{\rho}}$ can be expressed as

$$\hat{\boldsymbol{\rho}} = \sigma\left(\mathscr{W}_L \sigma\left(\mathscr{W}_{L-1} \cdots \sigma\left(\mathscr{W}_1[\mathbf{d}]\right)\right)\right). \tag{9.3}$$

For the same training set specified in the previous subsection, learning aspect of the DL model allows for estimating the combined operator $\mathscr{M}^{\dagger}\mathscr{H}$ as a function of W_l's whose values are optimized during training. Nonlinear operator σ can be known completely or only its functional form may be known with unknown parameters learned along with the affine transformation operators during training. If the deep network is an RNN, the role of this function is to impose prior knowledge about the quantity of interest if the RNN is designed from an iterative solution for the underlying optimization problem [57,59]. This is explained further in the context of passive SAR imaging later in this chapter.

9.3 Problem statement

We assume a stationary transmitter located at $\boldsymbol{\gamma}_T \in \mathbb{R}^3$ and a moving receiver traversing a trajectory $\boldsymbol{\gamma}_R : \mathbb{R} \rightarrow \mathbb{R}^3$. Let $\mathbf{x} \in \mathbb{R}^3$ and $\boldsymbol{x} \in \mathbb{R}^2$. Without the loss of generality, we assume flat topography for the scene that we intend to image. Thus, the location of a scatter on the ground is $\mathbf{x} = [\boldsymbol{x}, \boldsymbol{0}]$. Let $\rho : \mathbb{R}^2 \rightarrow \mathbb{C}$ represents the surface reflectivity function.

Under the single scattering and start–stop approximations [63], the received signal can be modeled by the following forward map:

$$d(\omega, s) = \mathscr{F}[\rho](\omega, s) := \int e^{-i((\omega R(s,\mathbf{x}))/c_0)} A(\omega, s, \mathbf{x}) \rho(\boldsymbol{x}) d\boldsymbol{x}, \tag{9.4}$$

where $\omega \in [\omega_1, \omega_2]$ is the fast-time frequency, $s \in [s_1, s_2]$ is the slow-time parameterizing the position of the receiver, $i = \sqrt{-1}$, and c_0 is the speed of light in free-space. $R(s, \boldsymbol{x})$ is the bistatic range given by

$$R(s, \mathbf{x}) = |\boldsymbol{\gamma}_T - \mathbf{x}| + |\mathbf{x} - \boldsymbol{\gamma}_R(s)|. \tag{9.5}$$

It is the total distance traveled by a signal sent from the transmitter of opportunity at $\boldsymbol{\gamma}_T$ to a point in the scene $\mathbf{x}$ and then back to the receiver location at $\boldsymbol{\gamma}_R(s)$. $A(\omega, s, \mathbf{x})$ is a slowly varying amplitude function of ω and includes the transmitted waveforms and geometric spreading factors. Geometry of the SAR system is depicted in Figure 9.3.

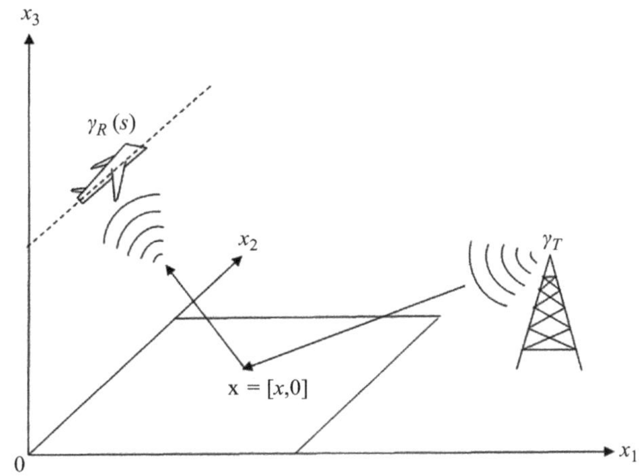

Figure 9.3 SAR data-collection geometry

We assume here that the transmitted waveform and the fixed transmitter location $\boldsymbol{\gamma}_T$ are unknown. Under these assumptions, bistatic range and the amplitude function are both unknown. Suppose, the 3D environment encompassing the scene, the transmitter location and the receiver trajectory are discretized and the scene is sampled into N discrete points resulting in the set $\{\boldsymbol{x}_n\}_{n=1}^{N}$. The quantity $\boldsymbol{\rho}$ is an $N \times 1$-dimensional complex-valued vector, and its nth component is equal to $\rho(\boldsymbol{x}_n)$. The variable pair (ω, s), corresponding to fast-time and slow-time values, is discretized into M sample where K and L are the numbers of fast-time and slow-time points, respectively, and $M = K \times L$. The data vector $\mathbf{d} \in \mathbb{C}^{M\times 1}$ is formed by stacking the received signal value at each fast-time slow-time pair with mth component being equal to $d((\omega, s)_m)$. Thus, we write

$$\mathbf{d} = \mathbf{F}\boldsymbol{\rho}, \tag{9.6}$$

where $\mathbf{F}$ is the discretized forward map $\mathscr{F}$. Our goal is to design a data-driven learning and model-based method for image reconstruction, under the condition that the transmitted waveform or transmitter locations are unknown, and estimate these quantities from a set of training data.

9.4 Bayesian and optimization-inspired DL for radar imaging

In this section, we present some background on radar image reconstruction formulated in the framework of an RNN, its relation to the iterations of an optimization problem and how it leads to the interpretation of the network parameters in terms of the underlying forward model [57,59]. One of the main advantages of RNN-based implementation is the reduced number of parameters to be learned. This is especially beneficial to radar imaging applications where acquiring large amount of radar data for the training set is often unrealistic. Assuming additive Gaussian noise model for the measured received signal, the Bayesian estimation of the scene reflectivity vector $\boldsymbol{\rho}$ in (9.6) is given as the solution of the following optimization problem:

$$\boldsymbol{\rho}^* = \underset{\rho}{\operatorname{argmin}} f(\boldsymbol{\rho}) + g(\boldsymbol{\rho}) = \underset{\rho}{\operatorname{argmin}} \|\mathbf{d} - \mathbf{F}\boldsymbol{\rho}\|_2^2 + \|\boldsymbol{\rho}\|_0, \tag{9.7}$$

where

$$f(\boldsymbol{\rho}) = \|\mathbf{d} - \mathbf{F}\boldsymbol{\rho}\|_2^2, \tag{9.8}$$

$$g(\boldsymbol{\rho}) \quad = \|\boldsymbol{\rho}\|_0. \tag{9.9}$$

Here (9.7)–(9.9) is a non-convex optimization problem where the quadratic term $f(\boldsymbol{\rho})$ imposes consistency with the collected data vector, and the l_0 norm of $\boldsymbol{\rho}$ imposes sparsity on the reconstructed scene. Using the forward–backward splitting algorithm [64], a solution of (9.7) is obtained by an iterative algorithm involving a gradient descent and a projection step during each iteration. The gradient descent update at the lth iteration involves one step of the l_2 norm minimization over $\boldsymbol{\rho}$ to get an updated estimation $\tilde{\boldsymbol{\rho}}^{(l)}$:

$$\tilde{\boldsymbol{\rho}}^{(l)} = \underset{\rho}{\operatorname{argmin}} f(\boldsymbol{\rho}) = \underset{\rho}{argmin} \|\mathbf{d} - \mathbf{F}\boldsymbol{\rho}\|_2^2. \tag{9.10}$$

Resultant output $\tilde{\boldsymbol{\rho}}^{(l)}$ is then projected onto the solution space to impose the regularization condition:

$$\boldsymbol{\rho}^{(l)} = \underset{\rho}{\operatorname{argmin}} \|\tilde{\boldsymbol{\rho}}^{(l)} - \boldsymbol{\rho}\|_2^2 + i_C(\boldsymbol{\rho}). \tag{9.11}$$

Here C denotes a set of $N \times 1$ dimensional sparse vectors. An RNN-based implementation of radar image reconstruction involves a repetitive architecture where each stage simulates the nonlinear transformation within each iteration step. The recurrent model is derived by unfolding fixed number of iteration steps for solving the original optimization problem, which leads to direct interpretation of network parameters unlike using generic deep networks. For a fixed $\mathbf{F}$ and received signal vector $\mathbf{d}$, derivative of the objective function $f(\boldsymbol{\rho})$ in the optimization problem in (9.10) with respect to scene reflectivity vector $\boldsymbol{\rho}$ is

$$\frac{\partial f(\boldsymbol{\rho})}{\partial \boldsymbol{\rho}} = \mathbf{F}^H(\mathbf{d} - \mathbf{F}\boldsymbol{\rho}) = \mathbf{F}^H\mathbf{d} - \mathbf{F}^H\mathbf{F}\boldsymbol{\rho}. \tag{9.12}$$

Thus, the updated value of $\boldsymbol{\rho}$ at lth iteration step is

$$\tilde{\boldsymbol{\rho}}^{(l)} = \boldsymbol{\rho}^{(l-1)} - \alpha \frac{\partial f(\boldsymbol{\rho})}{\partial \boldsymbol{\rho}}\Big|_{\boldsymbol{\rho}=\boldsymbol{\rho}^{(l-1)}}. \tag{9.13}$$

Learning rate $\alpha \in \mathbb{R}^+$, where $\mathbb{R}^+$ is the set of positive real numbers. The projection step is implemented using a nonlinear proximity function $\mathscr{P}_{\alpha g}(.)$ for regularization term $g(.)$ and depends on the prior condition on the reconstructed image. For example, for sparse scenes, $g(.)$ is modeled by a rectified linear function with parameter τ, which refers to an arbitrary threshold value [59], i.e.,

$$\boldsymbol{\rho}^{(l)} = \mathscr{P}_{\alpha g}(\tilde{\boldsymbol{\rho}}^{(l)}) = \mathscr{P}_{\alpha g}(\mathbf{Q}\boldsymbol{\rho}^{(l-1)} + \mathbf{b}), \tag{9.14}$$

$$= \begin{cases} |\mathbf{Q}\boldsymbol{\rho}^{(l-1)} + \mathbf{b}| - \tau; & |\mathbf{Q}\boldsymbol{\rho}^{(l-1)} + \mathbf{b}| > \tau, \\ 0; & \text{otherwise.} \end{cases} \tag{9.15}$$

$\mathbf{Q}$ is the weight matrix associated with each step and related to forward mapping matrix $\mathbf{F}$ and learning rate α by $\mathbf{Q} = \mathbf{I} - \alpha\mathbf{F}^H\mathbf{F}$. Similarly, common bias vector for the internal layers is a function of $\mathbf{F}$, α and $\mathbf{d}$ expressed as $\mathbf{b} = \alpha\mathbf{F}^H\mathbf{d}$. Image reconstruction network designed by unfolding L iteration steps, therefore, has L stages with network parameters $\mathbf{Q}$ and $\mathbf{b}$ relating to the data collection geometry, sensor data and prior knowledge about the scenes to the network parameters and nonlinearity.

For simplification, $A(\omega, s, \mathbf{x})$ in (9.4) is typically modeled by a flat spectrum and set equal to a constant value. If we have prior knowledge about the location of the transmitter, which is the case for active SAR, then the forward mapping matrix $\mathbf{F}$ is fully known. During the learning process, the goal is then to find an optimum step size α such that for all the received signals in the sample set, it is possible to reconstruct corresponding images with high accuracy using L iteration steps.

Forward propagation for an RNN refers to the process of calculating hierarchical feature representation and the final output in a step-by-step manner starting from the input vector. Let $\mathscr{L} : \Omega_{\mathbf{d}} \rightarrow \Omega_{\rho}$ denote the mapping from the input space $\Omega_{\mathbf{d}}$ to the output space Ω_{ρ} of the forward propagation. Then, the output of the lth layer is given by

$$\boldsymbol{\rho}^{(l)} = \mathscr{P}_{\alpha g}\left(\mathbf{Q}\boldsymbol{\rho}^{(l-1)} + \mathbf{b}\right) = \mathscr{P}_{\alpha g}\left(\begin{bmatrix}\mathbf{Q} & \mathbf{b}\end{bmatrix}\begin{bmatrix}\boldsymbol{\rho}^{(l-1)} \\ 1\end{bmatrix}\right), \tag{9.16}$$

$$= \mathscr{P}_{\alpha g}\left(\mathbf{Q}_1\boldsymbol{\rho}_1^{(l-1)}\right), \tag{9.17}$$

where $\mathbf{Q}' = \begin{bmatrix}\mathbf{Q} & \mathbf{b}\end{bmatrix}$ and $\boldsymbol{\rho}'^{(l-1)} = \begin{bmatrix}\boldsymbol{\rho}^{(l-1)} & 1\end{bmatrix}^T$. If all the L stages in the RNN implement the same activation function $\mathscr{P}_{\alpha g}(.)$, output vector from the last stage $\boldsymbol{\rho}^{(L)}$ is expressed as

$$\boldsymbol{\rho}^L = \mathscr{P}_{\alpha g}(\mathbf{Q}'\mathscr{P}_{\alpha g}(\mathbf{Q}'\mathscr{P}_{\alpha g}(\cdots\mathbf{Q}'\mathscr{P}_{\alpha g}(\mathbf{Q}'\boldsymbol{\rho}'^{(0)})))). \tag{9.18}$$

As stated in [44,45], as long as sufficient number of hidden layer neurons are available, multilayer feedforward networks can adequately approximate input–output mapping within $l_p(\boldsymbol{\rho})$ performance criteria, where $1 \leq p < \infty$, for finite input measure $\boldsymbol{\rho}$ with even a single hidden layer and arbitrary bounded and nonconstant activation functions. With a deep L-layer RNN architecture with N neurons per network stage, the same as input vector length, and deterministic relationship between the input and output space, the network successfully models the inverse mapping for image reconstruction task if L is large.

9.4.1 Parameter learning via backpropagation

Backpropagation refers to the process by which the values of the network parameters are updated using gradient descent steps with the goal of minimizing a task-specific cost function. To enable unsupervised training, the RNN is implemented as part of an autoencoder [65,66] that first generates an image using the RNN and then maps this vector back to the data domain. Using the RNN forward mapping operator $\mathscr{L}$, parameterized by $\mathbf{W}_1$ and τ, the output of the Lth layer becomes

$$\mathscr{L}[\mathbf{d}] = \boldsymbol{\rho}^L. \tag{9.19}$$

Commonly used cost function for deep classification network is cross-entropy loss function. For the inverse problem, a suitable function is the l_2 norm of the difference between the projected received-signal estimate $\mathbf{F}\boldsymbol{\rho}^L$ and its given value $\mathbf{d}$. For a training set $\mathscr{D}$ with T samples, average loss $\mathscr{C}[\mathscr{D}]$ is

$$\mathscr{C}[\mathscr{D}] = \frac{1}{T}\sum_{t=1}^{T} \|\mathbf{F}\mathscr{L}[\mathbf{d}_t] - \mathbf{d}_t\|_2^2. \tag{9.20}$$

Regardless of whether the cost function is based on l_2 norm of estimation error, cross-entropy etc. and the associated non-convexity, stochastic gradient descent approach turns out to be adequate for parameter optimization for most tasks. Derivative of the cost function with respect to the weight matrix and scalar threshold value can be calculated using chain rule for partial derivative as follows:

$$\frac{\partial \mathscr{C}[\mathscr{D}]}{\partial \mathbf{Q}'} = \frac{\partial \boldsymbol{\rho}^{(1)}}{\partial \mathbf{Q}'} \frac{\partial \boldsymbol{\rho}^{(2)}}{\partial \boldsymbol{\rho}^{(1)}} \cdots \frac{\partial \boldsymbol{\rho}^{(L)}}{\partial \boldsymbol{\rho}^{(L-1)}} \frac{\partial \mathscr{C}[\mathscr{D}]}{\partial \boldsymbol{\rho}^{(L)}}, \tag{9.21}$$

$$\frac{\partial \mathscr{C}[\mathscr{D}]}{\partial \tau} = \frac{\partial \boldsymbol{\rho}^{(1)}}{\partial \tau} \frac{\partial \boldsymbol{\rho}^{(2)}}{\partial \boldsymbol{\rho}^{(1)}} \cdots \frac{\partial \boldsymbol{\rho}^{(L)}}{\partial \boldsymbol{\rho}^{(L-1)}} \frac{\partial \mathscr{C}[\mathscr{D}]}{\partial \boldsymbol{\rho}^{(L)}}. \tag{9.22}$$

For a learning rate $\lambda > 0$, the updated parameter values after the kth training step are given by

$$\mathbf{Q}'^{k+1} = \mathbf{Q}'^{k} - \lambda \frac{\partial \mathscr{C}[\mathscr{D}]}{\partial \mathbf{Q}'}, \quad \tau^{k+1} = \tau^{k} - \lambda \frac{\partial \mathscr{C}[\mathscr{D}]}{\partial \tau}. \tag{9.23}$$

9.5 Passive SAR imaging for unknown transmission waveform

In this section, we summarize the RNN-based image reconstruction and waveform estimation process from passive SAR-received signal discussed in [60,61,67]. For passive SAR imaging unknown transmitted waveform results in partially known forward mapping matrix $\mathbf{F}$. It is assumed that for all the samples in the training set, as well as for any new sample in the test set, the transmitted signal from an arbitrary transmitter of opportunity and the location of the transmitter remain fixed. Additionally, the transmitter location is assumed to be known during training.

For active radars, electromagnetic pulses are transmitted by the radar toward the scene, and reflected signal from the illuminated surface is detected at the receiver and then processed for image reconstruction. This requires the SAR to perform transmission operation that entails increased operational cost. Additionally, in military applications, this signal can be detected and the radar can be jammed by adversarial entity. Passive radar takes advantage of the proliferation of communication and broadcasting signals, such as radio or television transmissions, to deliver increased stealth in a cost-effective manner. Passive radar operation under different types of illuminators of opportunity has been investigated extensively in the literature, including using FM [68–70], GSM base stations [71], WiMAX and DVB-T [8,72–74]. WiMAX and DVB-T standards, both of which implement OFDM waveform, are widely implemented across the world.

For the estimation of the unknown transmitted waveform in addition to imaging, the deep network is parameterized using the unknown waveform, which leads to the estimation of its value during the learning process. Since $A(\omega, s, \mathbf{x})$ in (9.4) is a function of the waveform, it can be factored and the received signal can be modeled as follows:

$$d(\omega, s) = W(\omega, s) \int e^{-i(\omega/c_0)R(s,\mathbf{x})} a_1(\omega, s, \mathbf{x}) \rho(\boldsymbol{x}) d\boldsymbol{x}. \tag{9.24}$$

$W(\omega, s)$ is the transmitted waveform value at slow-time and fast-time points s and ω, respectively, and $a_1(\omega, s, \mathbf{x})$ models the antenna beam pattern and geometric spreading factors and is assumed to be a constant equal to 1. Let

$$\tilde{\mathscr{F}}[\rho](\omega, s) := \int e^{-i(\omega/c_0)R(s,\mathbf{x})} \rho(\boldsymbol{x}) d\boldsymbol{x}. \tag{9.25}$$

Then, we can express the received signal as

$$d(\omega, s) \approx W(\omega, s) \tilde{\mathscr{F}}[\rho](\omega, s). \tag{9.26}$$

Let $\boldsymbol{\rho} \in \mathbb{C}^{N \times 1}$, $\mathbf{d} \in \mathbb{C}^{M \times 1}$ and $\tilde{\mathbf{F}} \in \mathbb{C}^{M \times N}$ denote discretized ρ, d and $\tilde{\mathscr{F}}$, respectively. From (9.26), we have,

$$\mathbf{d} \approx diag(\mathbf{w}) \tilde{\mathbf{F}} \boldsymbol{\rho}, \tag{9.27}$$

where $diag(\mathbf{w})$ is an $M \times M$-dimensional diagonal matrix with $\mathbf{w}$ along its diagonal. A data-driven approach for image reconstruction for this problem involves designing a network that has unknown waveform vector $\mathbf{w}$ as parameters and $\boldsymbol{\rho}$ as output. As in the previous case when the waveform is known, the optimum reconstructed image is the solution of the following minimization problem [61],

$$\boldsymbol{\rho}^* = \underset{\rho}{\operatorname{argmin}} \, \|\mathbf{d} - diag(\mathbf{w}) \tilde{\mathbf{F}} \boldsymbol{\rho}\|_2^2 + \Phi(\boldsymbol{\rho}), \tag{9.28}$$

where $\Phi(\boldsymbol{\rho})$ is the regularization term. Similar to the previous case, the optimum value can be found by iterative proximal gradient descent approach with each iteration comprising a gradient descent step and a projection function $\mathscr{P}_{\alpha\Phi}(.)$ onto a feasible set to enforce the regularization term. The updated value of the lth step is

$$\boldsymbol{\rho}^{(l)} = \mathscr{P}_{\alpha\Phi} \left((\mathbf{I} - \alpha \tilde{\mathbf{F}}^H diag(|\mathbf{w}|^2) \tilde{\mathbf{F}}) \boldsymbol{\rho}^{(l-1)} + \alpha \tilde{\mathbf{F}}^H diag(\mathbf{w})^H \mathbf{d} \right), \tag{9.29}$$

$$= \mathscr{P}_{\alpha\Phi} \left(\mathbf{Q} \boldsymbol{\rho}^{(l-1)} + \alpha \mathbf{F}^H \mathbf{d} \right). \tag{9.30}$$

Here $\mathbf{Q} = \mathbf{I} - \alpha\tilde{\mathbf{F}}^H diag(|\mathbf{w}|^2)\tilde{\mathbf{F}}$ and $\mathbf{F} = diag(\mathbf{w})\tilde{\mathbf{F}}$. An L stage RNN with activation function $\mathscr{P}_{\alpha\phi}(.)$ at each stage is then used for image reconstruction. For sparsity, inducing prior, for example, when the regularization term is an l_1 norm, activation function can be effectively implemented by a soft-thresholding function with threshold τ. Reconstructed image from the last layer $\boldsymbol{\rho}^{(L)}$, which is function of unknown waveform vector $\mathbf{w}$, is then projected back to the SAR data domain using the relation $\hat{\mathbf{d}} = diag(\mathbf{w})\tilde{\mathbf{F}}\boldsymbol{\rho}^{(L)}$ to construct the autoencoder model to enable unsupervised learning.

9.6 Passive SAR imaging for unknown transmission waveform and transmitter location

In this section, we present an extended model for image reconstruction for passive SAR for the case when transmission waveform and transmitter locations are unknown. Assuming that the antennas are sufficiently broadband, the same region stays in the main lobe of the antennas in a spotlight configuration for high-resolution SAR imaging, and the amplitude remains approximately constant over the aperture. Of the commonly used illuminators for opportunity, FM, DVB-T and digital audio broadcast (DAB) operate in frequency bands 88–108, 470–862 and 174–240 MHz with single-channel bandwidths equal to 150 kHz, 7.612 MHz and 1.536 MHz, respectively [1]. Under these considerations, the antenna beam pattern and geometric spreading are approximately constant and without the loss of generality we set $a_1(\omega, s, \mathbf{x})$ in (9.24) to 1.

We consider two models. Different than our previous works in[59,61], we merge the unknown components of the bistatic SAR forward model and parameterize the image reconstruction algorithm by a single structured unknown. In the first model, we assume that the transmitter look direction is known. Invoking the far field, and small scene approximations, we write

$$d(\omega, s) \approx W(\omega, s)e^{-i(\omega/c_0)(|\boldsymbol{\gamma}_T|)} \int e^{-i(\omega/c_0)\,(|\boldsymbol{\gamma}_R(s)-\mathbf{x}|-\hat{\boldsymbol{\gamma}}_T\cdot\mathbf{x})}\rho(\boldsymbol{x})d\boldsymbol{x} \tag{9.31}$$

$$:= \mathscr{G}[\rho](\omega, s). \tag{9.32}$$

Observe that the integral operator acting on the reflectivities is fully known in (9.31), and the unknown terms are grouped into a complex scalar multiplier at each (ω, s) pair on the received signal. Hence, letting

$$\tilde{\mathscr{G}}[\rho](\omega, s) := \int e^{-i(\omega/c_0)(|\boldsymbol{\gamma}_R(s)-\mathbf{x}|-\hat{\boldsymbol{\gamma}}_T\cdot\mathbf{x})}\rho(\boldsymbol{x})d\boldsymbol{x} \tag{9.33}$$

denote the known component of the forward mapping, we write (9.31) as

$$d(\omega, s) = \tilde{W}(\omega, s)\tilde{\mathscr{G}}[\rho](\omega, s), \tag{9.34}$$

where $\tilde{W}$ captures all the structured uncertainty in the measurement map.

In the second model, we assume that there is no knowledge on any transmitter-related term. This yields a received signal model in which the kernel of the integral operator has a multiplicative unknown at each $\boldsymbol{x}$, as follows:

$$d(\omega, s) \approx \int W(\omega, s) e^{-i(\omega/c_0)\,(|\boldsymbol{\gamma}_T - \mathbf{x}|)} e^{-i(\omega/c_0)\,(|\boldsymbol{\gamma}_R(s) - \mathbf{x}|)} \rho(\boldsymbol{x}) d\boldsymbol{x} := \mathscr{K}[\rho](\omega, \mathbf{s}). \tag{9.35}$$

We see that the kernel of $\mathscr{K}$ yields a *Hadamard* product of the fully known receiver-related component $\tilde{\mathscr{K}}_R$,

$$\tilde{\mathscr{K}}_R(\o, s, \mathbf{x}) = e^{-i(\o/c_0)\,(|\boldsymbol{\gamma}_R(s) - \mathbf{x}|)}, \tag{9.36}$$

and the unknown transmitter-related component $\tilde{\mathscr{K}}_T$,

$$\tilde{\mathscr{K}}_T(\o, s, \mathbf{x}) = W(\o, s) e^{-i(\o/c_0)\,(|\boldsymbol{\gamma}_T - \mathbf{x}|)}. \tag{9.37}$$

With the same sampling scheme as in Section 9.3, we stack the sampled waveform elements $\tilde{W}(\omega, s)|_{(\omega,s)_m}$ into a vector $\tilde{\mathbf{w}} \in \mathbb{C}^M$. Thus, for the first model, we obtain

$$\mathbf{d} \approx \mathrm{diag}(\tilde{\mathbf{w}})\tilde{\mathbf{G}}\boldsymbol{\rho} = \mathbf{G}\boldsymbol{\rho}, \tag{9.38}$$

where $\mathbf{G} \in \mathbb{C}^{M\times N}$ and $\tilde{\mathbf{G}} \in \mathbb{C}^{M\times N}$ are the matrices corresponding to the finite dimensional representation of $\mathscr{G}$ in (9.32) and $\tilde{\mathscr{G}}$ in (9.33), respectively. For the second model, we obtain

$$\mathbf{d} \approx (\tilde{\mathbf{K}}_R \odot \tilde{\mathbf{K}}_T)\boldsymbol{\rho} = \mathbf{K}\boldsymbol{\rho}, \tag{9.39}$$

where $\odot$ refers to the Hadamard product between $\tilde{\mathbf{K}}_R$ and $\tilde{\mathbf{K}}_T$, and $\mathbf{K}$, $\tilde{\mathbf{K}}_R$ and $\tilde{\mathbf{K}}_T$ are discretized versions of the operators with kernels given in (9.35), (9.36) and (9.37), respectively. Hadamard product is the element-wise product of two matrices with same dimensions, i.e., the element at the ith row and jth column of $\mathbf{C} \odot \mathbf{D}$, denoted by $(\mathbf{C} \odot \mathbf{D})[i,j]$ equals to $\mathbf{C}[i,j]\mathbf{D}[i,j]$.

9.6.1 Network design for imaging

Similar to the two approaches discussed in Sections 9.4 and 9.5, unfolding the iterations, we write an RNN that applies a general proximal gradient descent algorithm for a fixed number of iterations:

$$\boldsymbol{\rho}^{(k+1)} = \mathscr{P}_{\alpha\Phi}(\mathbf{Q}\boldsymbol{\rho}^{(k)} + \alpha\mathbf{b}). \tag{9.40}$$

For the first model in (9.38) where $\mathbf{G}$ denotes the full forward model and the unknown components are initialized within a feasible set, $\mathbf{Q}$ is the image domain filter, initialized as $\mathbf{Q} = \mathbf{I} - \alpha\mathbf{G}^H\mathbf{G}$ and $\mathbf{b} = \mathbf{G}^H\mathbf{d}$. α is the step size of the iterative algorithm and upper bounded by the reciprocal of the Lipschitz constant of $\mathbf{G}$. Similarly for the second model in (9.39), $\mathbf{Q} = \mathbf{I} - \alpha\mathbf{K}^H\mathbf{K}$ and $\mathbf{b} = \mathbf{K}^H\mathbf{d}$. The image domain filter $\mathbf{Q}$ becomes the network weight matrix, whereas the backprojection images $\mathbf{G}^H\mathbf{d}$ and $\mathbf{K}^H\mathbf{d}$ are the bias elements. Observe that if allowed setting $\alpha = 1$, the layer simply interpolates between the result of the filter and the backprojection image. For flat spectrum waveforms, in the model expressed by (9.38), filter $\mathbf{Q}$ is fully known as

$$\mathbf{Q} = \mathbf{I} - \alpha \tilde{\mathbf{G}}^H \mathrm{diag}(|\tilde{\mathbf{w}}|^2)\tilde{\mathbf{G}} = \mathbf{I} - \alpha \tilde{\mathbf{G}}^H \tilde{\mathbf{G}}, \tag{9.41}$$

since $\tilde{\mathbf{w}}$ has all complex exponential entries. Then the network can be parameterized by only $\tilde{\mathbf{w}}$ in the bias terms, as $\mathbf{b} = \alpha \tilde{\mathbf{G}}^H \mathrm{diag}(\tilde{\mathbf{w}})^H \mathbf{d}$.

For the second model in (9.39), the transmitter-related terms can still be merged into a single unknown with unit modulo constraint under the flat spectrum assumption over transmitted waveforms. However, this model requires the backpropagation over a full matrix $\tilde{\mathbf{K}}_T$ defined in the Hadamard product with the known component, along with learning the image domain filter $\mathbf{Q}$, which is no longer known fully. We consider the more challenging formulation of the second model and formulate the training as

$$\mathscr{C}_2[\mathscr{D}] = \frac{1}{T}\sum_{t=1}^{T} \ell(\mathbf{K}\boldsymbol{\rho}_t^*(\tilde{\mathbf{K}}_T, \mathbf{Q}, \tau), \mathbf{d}_t) + \boldsymbol{i}_{C_{\tilde{K}_T}}(\tilde{\mathbf{K}}_T) + \boldsymbol{i}_{C_Q}(\mathbf{Q}) + \boldsymbol{i}_+(\tau), \tag{9.42}$$

where $\boldsymbol{\rho}_t^*$ is the image estimate produced by the forward propagation using the training data $\{\mathbf{d}_t\}_{t=1}^T$ and ℓ is an appropriately chosen mismatch function, hence forming a recurrent autoencoder. In the objective function of backpropagation, we insert the indicator function $\boldsymbol{i}$ for the network parameters, in which $C_{\tilde{K}_T}$ denotes the set of matrices $\mathbb{C}^{M\times N}$ with unit modulo entries, C_Q denotes the set of Hermitian symmetric matrices in $\mathbb{C}^{N\times N}$ and $+$ denotes the nonnegative real numbers $\mathbb{R}^+$ for the threshold parameter τ.

We consider the formulation in which $\mathbf{Q}$ and $\mathbf{K}$ are *untied*, i.e., are learned as different network parameters. This allows $\mathbf{Q}$ to obtain a meaning beyond the characterization of the forward model, one that not only governs image quality through model accuracy but also improves the convergence speed [55]. The remaining parameterization depends on the model selection, and functional constraints of the unknowns.

This parameterization allows for learning process to estimate the unknown transmission location and transmission signal term in the measurement matrix. Separately estimating both terms is challenging at least without some prior condition on the transmission signal. Figure 9.4 shows an L stage RNN network for the first model where the transmitter look direction is known. The normalization block at the end is used under the assumption that maximum scene reflectivity value is 1. $\|\boldsymbol{\rho}^{(L)}\|_\infty$ here denotes the maximum value of the image vector $\boldsymbol{\rho}$.

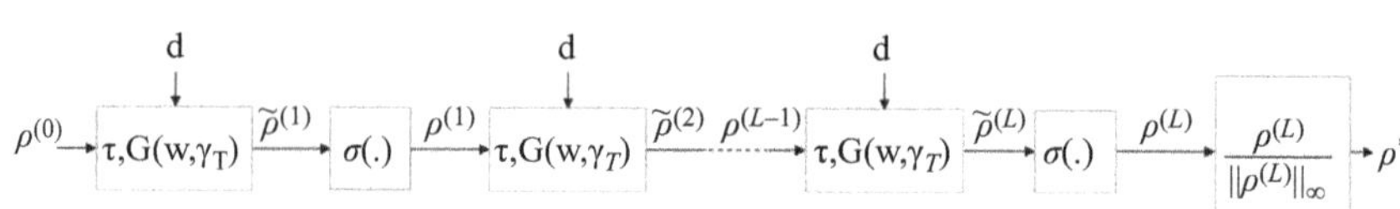

Figure 9.4 Deep network for passive SAR imaging when the distance of the transmitter from the origin and transmitted waveform is unknown, but direction of transmission is known

For the second network where neither the transmitter location or the transmission direction are known, deep network model for imaging from passive SAR is shown in Figure 9.5.

9.6.2 Network training and partial derivatives

As in the previous models discussed in this chapter, network training is performed in an unsupervised manner. The main advantage of unsupervised approach is the fact that the performance of the network is not going to be upper bounded by the quality of the training images. The reconstructed images are back-projected onto the data domain and compared with the received signal in the autoencoder model. Loss function $\mathscr{C}_1[\mathscr{D}]$ for the first network in (9.38) is defined as

$$\mathscr{C}_1[\mathscr{D}] = \sum_{t=1}^{T} \ell(\mathbf{G}(\boldsymbol{\gamma}_T, \mathbf{w})\boldsymbol{\rho}_t^*(\mathbf{G}, \tau), \mathbf{d}_t) + \boldsymbol{i}_{C_{\mathbf{w}}}(\mathbf{w}) + \boldsymbol{i}_+(\tau), \tag{9.43}$$

where $C_{\mathbf{w}}$ refers to the set of matrices with unit modulo components, and indicator function $\boldsymbol{i}_+$ is infinity when τ is zero or negative and has a small value otherwise. Here, $\ell(.)$ is defined as follows:

$$\ell(\mathbf{G}(\boldsymbol{\gamma}_T, \mathbf{w})\boldsymbol{\rho}_t^*(\mathbf{G}, \tau), \mathbf{d}_t) = \frac{1}{2}\|\mathbf{d}_t - \mathbf{G}(\boldsymbol{\gamma}_T, \mathbf{w})\boldsymbol{\rho}_t^*(\mathbf{G}, \tau)\|_2^2. \tag{9.44}$$

The initial values for $\mathbf{w}$ are set equal to a complex vector of length M with each element equals to $(1 + 1i)/\sqrt{2}$. During training, values of $\boldsymbol{\gamma}_T$, $\mathbf{w}$ and τ are updated so that $\ell(\mathbf{G}\boldsymbol{\rho}_t^*, \mathbf{d}_t)$ is minimized by gaining information from the training samples. At stage k of the training process, updated values of $\boldsymbol{\gamma}_T^{(k)}$, $\mathbf{w}^{(k)}$ and $\tau^{(k)}$ for a single training sample are

$$\boldsymbol{\gamma}_T^{(k)} = \boldsymbol{\gamma}_T^{(k-1)} - \lambda_{\gamma_T} \overline{\left(\frac{\partial \ell}{\partial \boldsymbol{\gamma}_T}\right)}_{\boldsymbol{\gamma}_T = \boldsymbol{\gamma}_T^{(k-1)}}, \tag{9.45}$$

$$\mathbf{w}^{(k)} = \mathscr{P}_{C_w}\left[\mathbf{w}_T^{(k-1)} - \lambda_w \overline{\left(\frac{\partial \ell}{\partial \mathbf{w}}\right)}_{\mathbf{w} = \mathbf{w}_T^{(k-1)}}\right], \tag{9.46}$$

$$\tau^{(k)} = \mathscr{P}_+\left[\tau^{(k-1)} - \lambda_\tau \overline{\left(\frac{\partial \ell}{\partial \tau}\right)}_{\tau = \tau^{(k-1)}}\right]. \tag{9.47}$$

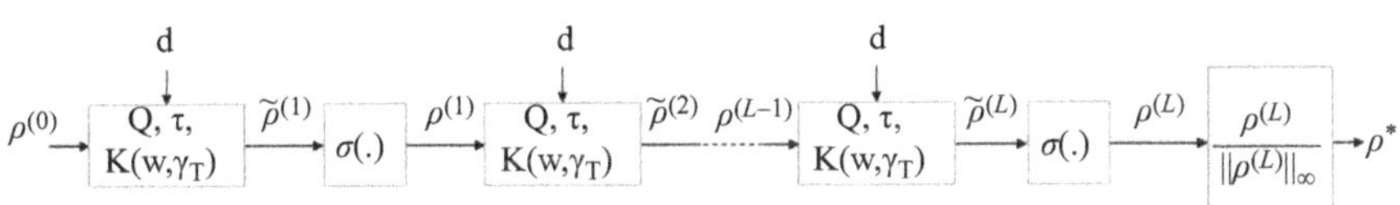

Figure 9.5 Deep network for passive SAR imaging when transmitter location and transmitted waveform are unknown

Partial derivatives $(\partial\ell/\partial\boldsymbol{\gamma}_T)$, $(\partial\ell/\partial\mathbf{w})$ and $(\partial\ell/\partial\tau)$ can be expressed as

$$\frac{\partial\ell}{\partial\boldsymbol{\gamma}_T} = \left(\frac{-\mathrm{i}}{c_0}\right)\sum_{m=1}^{M}\omega_{m_1(m)}e^{-i(\omega_{m1(m)}/c_0)|\boldsymbol{\gamma}_T|}\mathbf{w}[m]\hat{\boldsymbol{\gamma}}_T\sum_{n=1}^{N}\mathbf{A}[m,n]\tilde{\mathbf{G}}[m,n], \quad (9.48)$$

$$\frac{\partial\ell}{\partial\mathbf{w}} = diag(\mathbf{v})\left[\sum_{n=1}^{N}\mathbf{A}[1,n]\tilde{\mathbf{G}}[1,n] \quad \cdots \quad \sum_{n=1}^{N}\mathbf{A}[M,n]\tilde{\mathbf{G}}[M,n]\right]^T, \quad (9.49)$$

$$\frac{\partial\ell}{\partial\tau} = -\frac{1}{2}\sum_{n=1}^{N}\mu(|\tilde{\boldsymbol{\rho}}^{(1)}[n]| > \tau)$$
$$\times\prod_{l=1}^{L-1}\left((\mathbf{i}_1(n) - \alpha(\mathbf{G}[\,:,n])^T\bar{\mathbf{G}})\mathbf{Y}(\boldsymbol{\rho}^{(l+1)})\right)\mathbf{Z}(\boldsymbol{\rho}^*,\boldsymbol{\rho}^{(L)})\mathbf{G}^T(\bar{\mathbf{d}} - \bar{\mathbf{G}}\bar{\boldsymbol{\rho}}^*). \quad (9.50)$$

$\mathbf{A}$ in (9.48) and (9.49) is a complex-valued matrix of dimension $M \times N$, and its component at the ith row and jth column is

$$\mathbf{A}[i,j] = -\tfrac{1}{2}M\boldsymbol{\rho}^*[j] + \tfrac{1}{2}(\alpha\boldsymbol{\rho}^{(0)}[j]\textstyle\sum_{m=1}^{M}\sum_{n=1}^{N}\mathbf{F}[m,n]\mathbf{F}^H[i,:\,]$$
$$\times\textstyle\prod_{l=1}^{L-1}\left(\mathbf{Y}(\boldsymbol{\rho}^{(l)})(\mathbf{I} - \alpha\mathbf{F}^T\bar{\mathbf{F}})\right)\mathbf{Y}(\boldsymbol{\rho}^{(L)})\mathbf{z}(\boldsymbol{\rho}^*[n],\boldsymbol{\rho}^{(L)})). \quad (9.51)$$

Here, $m_1(m) = \lfloor\frac{m}{L}\rfloor + 1$ for $m \in [1,M]$. $\mathbf{Y}(\boldsymbol{\rho}^{(l)}) = diag(\mathbf{y}(\boldsymbol{\rho}^{(l)}))$ and $\mathbf{y}(\boldsymbol{\rho}^{(l)}) \in \mathbb{C}^{N\times 1}$ is expressed in (A.11). $\mathbf{z} \in \mathbb{R}^{N\times 1}$ and $\mathbf{v} \in \mathbb{C}^{M\times 1}$ are expressed in (A.12) and (A.6), respectively. $\mu(.)$ is an indicator function with a value of 1 when the logical expression in its argument is true, otherwise it is 0. $\mathbf{Z} \in \mathbb{R}^{N\times N}$ in (9.50) is defined in (A.18). $\mathbf{i}_1$ is an N-dimensional vector of all 0's except for the nth component where its value is 1.

The second network is parameterized by four variables $\mathbf{Q} \in \mathbb{C}^{N\times N}$, $\mathbf{w} \in \mathbb{C}^{M\times 1}$, $\boldsymbol{\gamma}_T \in \mathbb{R}^{3\times 1}$ and $\tau \in \mathbb{R}^+$, optimum values of which are learned during training. Loss function $\mathscr{C}_2[\mathscr{D}]$ defined in (9.42) can be rewritten as follows:

$$\mathscr{C}_2[\mathscr{D}] = \sum_{t=1}^{T}\ell(\mathbf{K}(\boldsymbol{\gamma}_T,\mathbf{w}),\boldsymbol{\rho}_t^*(\mathbf{Q},\boldsymbol{\gamma}_T,\mathbf{w},\tau),\mathbf{d}_t) + \boldsymbol{i}_{C_{\tilde{K}_T}}(\tilde{\mathbf{K}}_T) + \boldsymbol{i}_{C_Q}(\mathbf{Q}) + \boldsymbol{i}_+(\tau), \quad (9.52)$$

where $\ell(\mathbf{K}(\boldsymbol{\gamma}_T,\mathbf{w})\boldsymbol{\rho}_t^*(\mathbf{Q},\boldsymbol{\gamma}_T,\mathbf{w},\tau),\mathbf{d}_t)$ has the same definition as in (9.44) except the forward mapping matrix is $\mathbf{K}$ in this case. Partial derivative $(\partial\ell/\partial\tau)$ for this model also has the same expression presented in (9.50) with $\mathbf{G}$ replaced by $\tilde{\mathbf{K}}_T \odot \tilde{\mathbf{K}}_R$. Partial derivative $(\partial\ell/\partial\mathbf{Q}) = \mathbf{B} \in \mathbb{C}^{N\times N}$ where

$$\mathbf{B}[n_1,n_2]$$
$$= \boldsymbol{\rho}^{(0)}[n_2]\prod_{l=1}^{L-1}\left(\mathbf{y}(\boldsymbol{\rho}^{(l)})[n_2](\mathbf{i}_1[n_1] - \alpha(\mathbf{K}[n_1,:\,])^T\bar{\mathbf{K}})\right)\mathbf{Y}(\boldsymbol{\rho}^{(L)})\mathbf{Z}(\boldsymbol{\rho}^*,\boldsymbol{\rho}^{(L)}), \quad (9.53)$$

for $n_1, n_2 \in [1, N]$. Partial derivatives of ℓ with respect to $\boldsymbol{\gamma}_T$ is a vector with three components where the cth component is

$$\left(\frac{\partial \ell}{\partial \boldsymbol{\gamma}_T}\right)[c] = \sum_{n=1}^{N}\sum_{m=1}^{M}(\bar{\mathbf{K}}[m,n]\frac{\mathrm{i}(\boldsymbol{\gamma}_T[c] - \boldsymbol{x}_n[c])}{c_0|\boldsymbol{\gamma}_T - \boldsymbol{x}_n|}\omega_{m_1(m)}\alpha\mathbf{d}[m]$$
$$\times \prod_{l=1}^{L-1}\left(\mathbf{y}(\boldsymbol{\rho}^{(l)})[n](\mathbf{i}_i[n] - \alpha(\mathbf{K}[n,:])^T\bar{\mathbf{K}})\right)\mathbf{Y}(\boldsymbol{\rho}^{(L)})\mathbf{Z}(\boldsymbol{\rho}^*, \boldsymbol{\rho}^{(L)})). \tag{9.54}$$

Finally, partial derivative $(\partial\ell/\partial\mathbf{w})$ is vector of length M and its mth component is

$$\left(\frac{\partial \ell}{\partial \mathbf{w}}\right)[m] = \frac{1}{2}\alpha\sum_{n=1}^{N}(\frac{\tilde{\boldsymbol{\rho}}^{(1)}[n]}{|\tilde{\boldsymbol{\rho}}^{(1)}[n]|}\mu(|\tilde{\boldsymbol{\rho}}^{(1)}[n]| > \tau)\tilde{\mathbf{K}}_R[m,n]\bar{\mathbf{d}}[m]e^{-i(\omega_{m_1(m)}/c_0)|\boldsymbol{\gamma}_T - \boldsymbol{x}_n|}$$
$$\times \prod_{l=1}^{L-1}\left((\mathbf{i}_1[n] - \alpha\mathbf{K}^T\bar{\mathbf{K}})\mathbf{Y}(\boldsymbol{\rho}^{(l+1)})\right)\mathbf{Z}(\boldsymbol{\rho}^*, \boldsymbol{\rho}^{(L)})\mathbf{K}^T(\bar{\mathbf{d}} - \bar{\mathbf{K}}\bar{\boldsymbol{\rho}}^*)). \tag{9.55}$$

Detailed calculations for the first and the second models are presented in Appendices A.1 and B.1, respectively. For both the networks, all parameter estimations are updated using stochastic gradient descent, each corresponds to a different parameter, until convergence of the loss function to a small value. Even though in the discussion presented in Appendices A.1 and B.1, we take into account the estimation of the transmitter location $\boldsymbol{\gamma}_T$ separately from the transmitted waveform $\mathbf{w}$, estimating the correct value of transmitter location may be difficult due to the fact that it appears only in the angle of a complex number. Due to the periodic nature of the sinusoidal function, multiple transmitter locations can result in the same complex value corresponding to the sensor data in combination with the parameter value of the waveform variable. For example, in the first network this combined term is $\tilde{w}(\omega, s) = W(\omega, s)e^{-i(\omega/c_0)|\boldsymbol{\gamma}_T|}$, which indicates that separate estimation of this term from the waveform may not result in correct transmitter location. However, this does not affect the reconstructed image since as long as the combined term $\tilde{\mathbf{w}}$ is learned correctly, reconstructed image will still be an estimate of the underlying scene reflectivity function.

9.7 Numerical results

In this section, we verify the proposed network model by presenting preliminary imaging results using a simulated dataset. The information available during training is a set of passive SAR sensor data collected from a variety of different scenes. Due to the manner in which both the networks are conceived and parameterized, it is a requirement that the transmitted waveform and transmitter location to be the same for all samples during training as well as for imaging from any new sample using the trained model. This is because the network parameters are directly related to these unknowns and having training samples with widely varying transmission signal or transmitter location and transmission direction is likely to cause oscillation in loss function value during training, and it may not converge. However, both the networks are adaptable for the case when transmitter location varies during data collection interval as long as this variation is common for all data samples.

Both the image reconstruction networks in Figures 9.4 and 9.5 work for any arbitrary geometry. Here, we assume that the SAR operates in spotlight mode with a circular receiver trajectory with origin at the central point of the scene. We use simulated scenes with a single rectangular object with arbitrary shape and location. The SAR receiver is mounted on a moving platform at a fixed height of 1 km and circulates around the scene at a radius of 7 km. The transmitter is at an unknown location such that the SAR operates under small scene and far-field condition. The transmitted waveform has a center frequency of 760 MHz and bandwidth of 8 MHz. The waveform is modeled by complex-valued signal with uniformly distributed real and imaginary parts and unit magnitude. No prior assumption is made on whether the signal transmitted at each slow-time point is required to be identical. Received signal is contaminated by complex-valued zero mean Gaussian-distributed noise.

Original scene and corresponding reconstructed images for SNR values equal to ∞, 20, 10 and 0 dB are shown in Figures 9.6 and 9.7. The numbers of samples in the training set and the test set are 900 and 100, respectively. Training process with stochastic gradient descent steps is carried out for 15,000 epochs for each SNR value after which training is terminated regardless of the loss function values. Received signals are sampled at 100 slow-time points and 10 fast-time frequencies equal to the center frequency. From Figures 9.6 and 9.7, it can be observed that as the SNR value decreases, quality of the reconstructed image degrades characterized by defocused and noisy images. To quantify reconstructed image quality, we compare the following three matrices: average data error (η_d), average image error (η_ρ) and average waveform and transmitter location error (η_w) for different SNR levels. For a test set with S samples, η_d, η_ρ and η_w are defined as

$$\eta_d = \sum_{s=1}^{S} \frac{\|\mathbf{F}\boldsymbol{\rho}_s^* - \mathbf{d}_s\|_2^2}{\|\mathbf{d}_s\|_2^2}, \quad \eta_\rho = \sum_{s=1}^{S} \frac{\|\boldsymbol{\rho}_s^* - \boldsymbol{\rho}_s\|_2^2}{\|\boldsymbol{\rho}_s\|_2^2}, \quad \eta_w = \frac{\|\tilde{\mathbf{w}}^* - \tilde{\mathbf{w}}\|_2^2}{\|\tilde{\mathbf{w}}\|_2^2}. \tag{9.56}$$

Here $\mathbf{d}_s$ is the sth received signal vector in the test set. $\boldsymbol{\rho}_s$ and $\boldsymbol{\rho}_s^*$ are the real image vector and reconstructed image vector, respectively, corresponding to the received signal vector $\mathbf{d}_s$. $\tilde{\mathbf{w}}$ is the actual waveform and transmitter-location-related term, and $\tilde{\mathbf{w}}^*$ refers to the corresponding estimated value learned via training. We evaluate these parameters on four sets of training and test data with SNR $= 0$, 10, 20 dB and ∞ additive Gaussian noise cases. Plots with η_d, η_ρ and η_w values vs. epochs for different SNR levels are shown in Figures 9.8–9.10, respectively. As expected, average errors fall as training continues, and at the same level of training, all three errors are generally higher for higher noise level. For example, after 4,000 training epochs, average data error η_d in Figure 9.8 are -22.4, -19.3, -13.3 and -7.8 dB for SNR values ∞, 20, 10 and 0 dB, respectively. After 15,000 epochs, η_d values for different SNR levels are

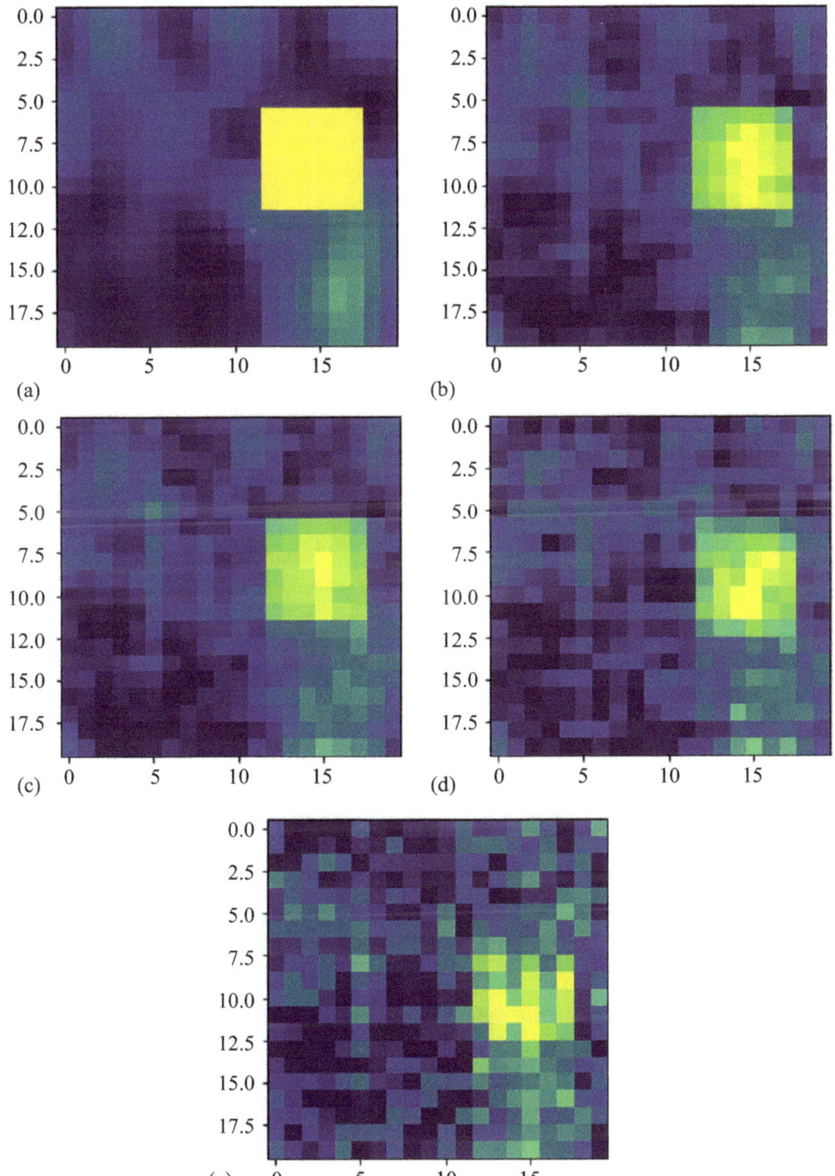

Figure 9.6 *(a) Original scene and reconstructed scenes when (b) SNR = ∞, (c) SNR = 20 dB, (d) SNR = 10 dB and (e) SNR = 0 dB for 20 × 20-pixel scene, 100 slow-time samples and 10 fast-time frequencies*

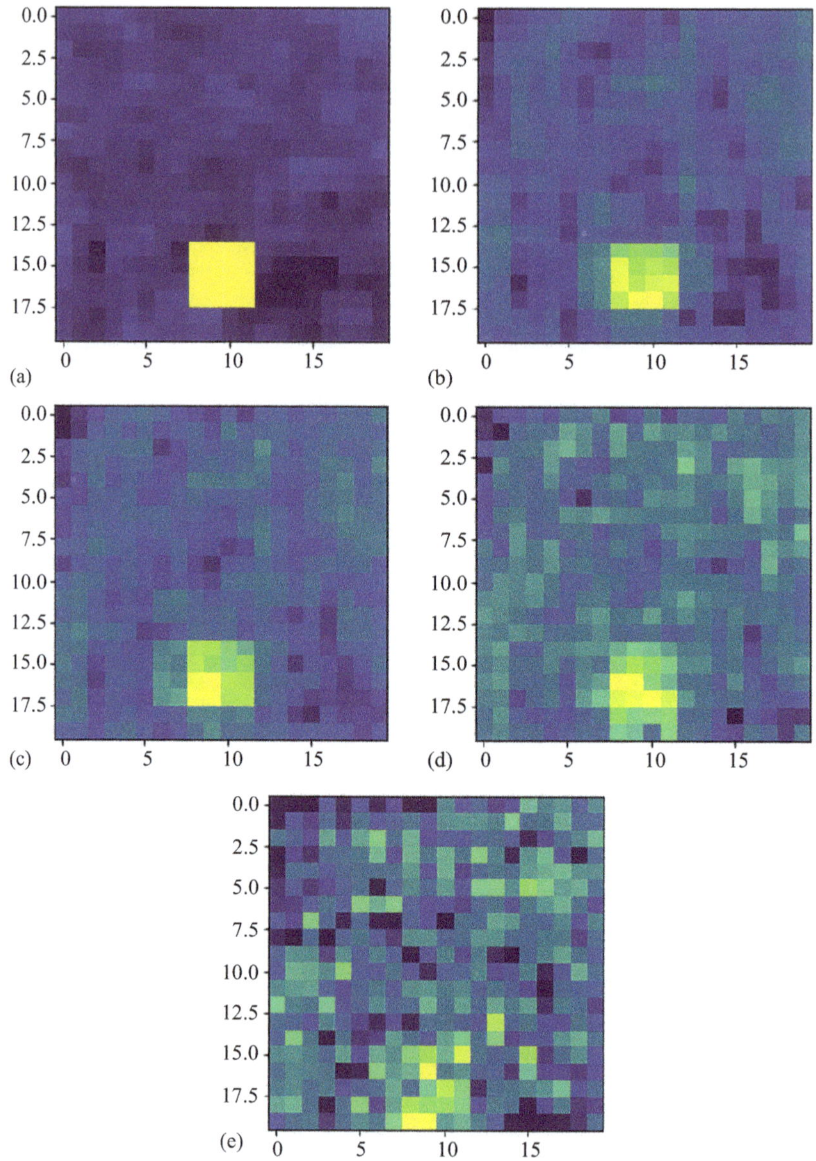

Figure 9.7 (a) Original scene and reconstructed scenes when (b) SNR = ∞, (c) SNR = 20 dB, (d) SNR = 10 dB and (e) SNR = 0 dB for 20 × 20-pixel scene, 100 slow-time samples and 10 fast-time frequencies

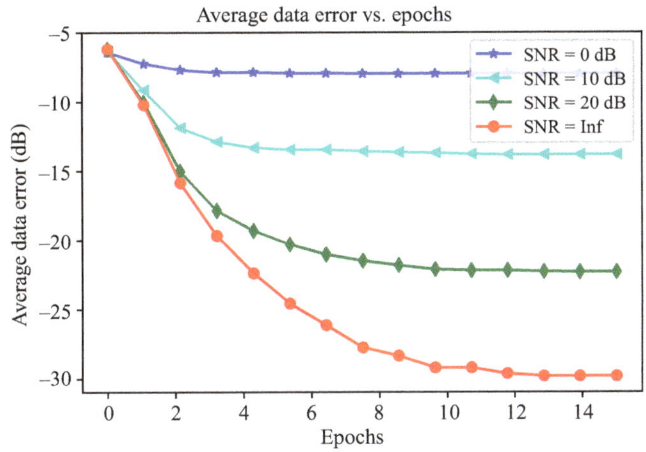

Figure 9.8 Data error (dB) vs. epochs (×1,000) for different SNR, 15 RNN stages and received signal with 100 slow-time samples and 10 fast-time frequencies

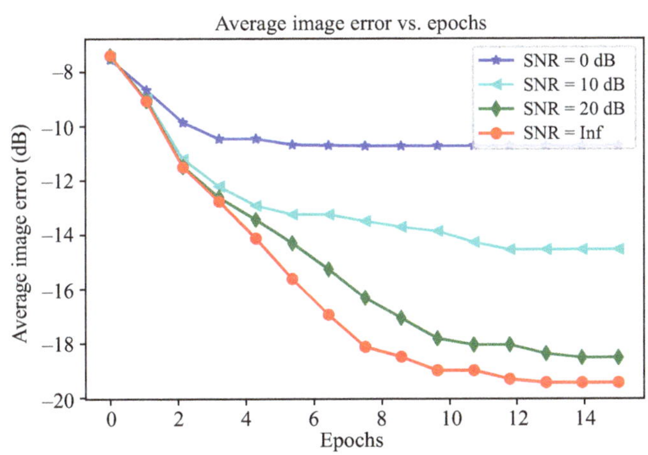

Figure 9.9 Image error (dB) vs. epochs (×1,000) for different SNR, 15 RNN stages and received signal with 100 slow-time samples and 10 fast-time frequencies

further apart with the lowest error −29.8 dB for infinite SNR case. Error levels for consecutively increasing SNR values decrease by approximately 6–9 dB indicating the effect of noise on reconstruction performance. Similar effects are observed for η_ρ and η_w as well where lower noise level is associated with better performance.

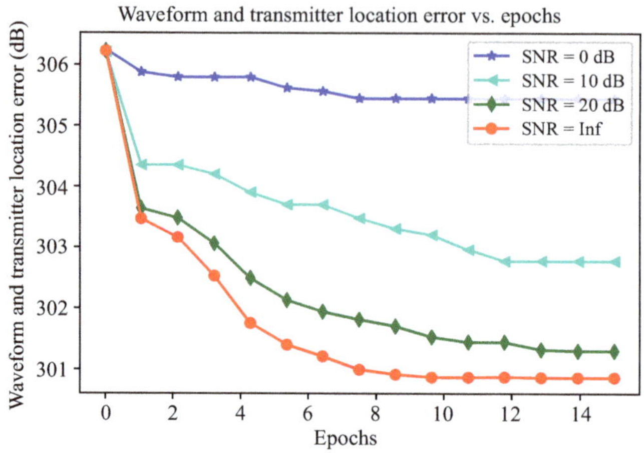

Figure 9.10 Waveform and transmitter location error (dB) vs. epochs (×1,000) for different SNR, 15 RNN stages and received signal with 100 slow-time samples and 10 fast-time frequencies

9.8 Conclusion

In this chapter, we have presented an expansion of the previous RNN-based image reconstruction techniques proposed for passive SAR imaging for operating conditions where the forward mapping operator is either fully or partially known with information about the transmission signal. We have considered two scenarios, one where there is complete absence of transmitter location information and the other case where transmission direction is known as prior information. The discussion presented here indicates that it is relatively simple to adapt RNN-based models to cases with increasing level of uncertainty, and as long as we have an iterative formulation for the underlying optimization, it is possible to relate unknown parameters of the image reconstruction problem to the parameters of the RNN. Unlike iterative methods, learning-based aspect provides a trade-off between computational cost of iterative imaging steps during test phase and the requirement of dataset and computational cost during training. Numerical results presented here show validity of this reconstruction approach for different noise levels as well as in estimating a joint term that relates the unknown transmission signal and the transmitter location, during the training process.

Acknowledgement

This work was supported by the Air Force Office of Scientific Research (AFOSR) under the agreement FA9550-16-1-0234 and FA9550-19-1-0284, Office of Naval Research (ONR) under the agreement N0001418-1-2068 and by the National Science Foundation (NSF) under Grant No ECCS-1809234.

Appendix

A.1 Partial derivatives when transmission direction is known

For the first model presented in (9.38), and corresponding loss function $\mathscr{L}(\mathscr{D})$ in (9.43), the gradients of ℓ with respect to $\boldsymbol{\gamma}_T$, $\mathbf{w}$ and τ can be calculated using the chain rule for partial derivation as follow:

$$\left(\frac{\partial \ell}{\partial \boldsymbol{\gamma}_T}\right)^{3\times 1} = \left(\frac{\partial \tilde{\mathbf{w}}}{\partial \boldsymbol{\gamma}_T}\right)^{3\times M} \left(\frac{\partial \mathbf{G}}{\partial \tilde{\mathbf{w}}}\right)^{M\times M\times N} \left(\frac{\partial \ell}{\partial \mathbf{G}}\right)^{M\times N\times 1}, \tag{A.1}$$

$$\left(\frac{\partial \ell}{\partial \mathbf{w}}\right)^{M\times 1} = \left(\frac{\partial \tilde{\mathbf{w}}}{\partial \mathbf{w}}\right)^{M\times M} \left(\frac{\partial \mathbf{G}}{\partial \tilde{\mathbf{w}}}\right)^{M\times M\times N} \left(\frac{\partial \ell}{\partial \mathbf{G}}\right)^{M\times N\times 1}, \tag{A.2}$$

$$\left(\frac{\partial \ell}{\partial \tau}\right)^{1\times 1} = \left(\frac{\partial \boldsymbol{\rho}^{(1)}}{\partial \tau}\right)^{1\times M} \left(\frac{\partial \ell}{\partial \boldsymbol{\rho}^{(1)}}\right)^{M\times 1}. \tag{A.3}$$

The partial derivative term $\partial\tilde{\mathbf{w}}/\partial\boldsymbol{\gamma}_T$ in (A.1) is the following $3 \times M$ dimensional matrix,

$$\frac{\partial \tilde{\mathbf{w}}}{\partial \boldsymbol{\gamma}_T} = -\frac{i}{c_0}\left[\omega_1 e^{-ij(\omega_1/c_0)|\boldsymbol{\gamma}_T|}\mathbf{w}[1]\hat{\boldsymbol{\gamma}}_T \ \cdots \ \omega_K e^{-i(\omega_K/c_0)|\boldsymbol{\gamma}_T|}\mathbf{w}[M]\hat{\boldsymbol{\gamma}}_T\right]. \tag{A.4}$$

Partial derivative $(\partial\tilde{\mathbf{w}}/\partial\mathbf{w})$ in (A.2) is the following $M \times M$ diagonal matrix:

$$\frac{\partial \tilde{\mathbf{w}}}{\partial \mathbf{w}} = diag(\mathbf{v}), \tag{A.5}$$

where

$$\mathbf{v} = \left[e^{-i|\boldsymbol{\gamma}_T|(\omega_1/c_0)} \ \ldots \ e^{-i|\boldsymbol{\gamma}_T|(\omega_1/c_0)} \ \ldots \ e^{-i|\boldsymbol{\gamma}_T|(\omega_K/c_0)} \ \ldots \ e^{-i|\boldsymbol{\gamma}_T|(\omega_K/c_0)}\right]. \tag{A.6}$$

$\partial\mathbf{G}/\partial\tilde{\mathbf{w}}$ in (A.1) and (A.2) is the following $M \times M \times N$ dimensional tensor,

$$\frac{\partial \mathbf{G}}{\partial \tilde{\mathbf{w}}} = \left[\left(\frac{\partial \mathbf{G}}{\partial \tilde{\mathbf{w}}_1}\right)^{M\times N} \ \left(\frac{\partial \mathbf{G}}{\partial \tilde{\mathbf{w}}_2}\right)^{M\times N} \ \cdots \ \left(\frac{\partial \mathbf{G}}{\partial \tilde{\mathbf{w}}_M}\right)^{M\times N}\right], \tag{A.7}$$

where $(\partial\mathbf{G}/\partial\tilde{\mathbf{w}}_n)^{M\times N} = \left[\mathbf{0} \ \cdots \ \tilde{\mathbf{G}}[n,:] \ \cdots \ \mathbf{0}\right]^T$ for $n \in [1, N]$, and $\mathbf{0}$ is an $N \times 1$ dimensional vector of all zeros. For the partial derivative term $\partial\ell/\partial\mathbf{G}$ in (A.1), and (A.2),

$$\frac{\partial \ell}{\partial \mathbf{G}} = -\frac{1}{2}\frac{\partial(\mathbf{G}\boldsymbol{\rho}^*)}{\partial \mathbf{G}}(\bar{\mathbf{d}} - \bar{\mathbf{G}}\bar{\boldsymbol{\rho}}^*) - \frac{1}{2}\frac{\partial(\bar{\mathbf{G}}\bar{\boldsymbol{\rho}}^*)}{\partial \mathbf{G}}(\mathbf{d} - \mathbf{G}\boldsymbol{\rho}^*). \tag{A.8}$$

Now, for $i \in [1, M]$ and $j \in [1, N]$,

$$\frac{\partial \boldsymbol{\rho}^*[n]}{\partial \mathbf{G}[i,j]} = \frac{\partial \tilde{\boldsymbol{\rho}}^{(1)}}{\partial \mathbf{G}[i,j]} \prod_{l=1}^{L-1}\left(\frac{\partial \boldsymbol{\rho}^{(l)}}{\partial \tilde{\boldsymbol{\rho}}^{(l)}}\frac{\partial \tilde{\boldsymbol{\rho}}^{(l+1)}}{\partial \boldsymbol{\rho}^{(l)}}\right)\frac{\partial \boldsymbol{\rho}^{(L)}}{\partial \tilde{\boldsymbol{\rho}}^{(L)}}\frac{\partial \boldsymbol{\rho}^*[n]}{\partial \boldsymbol{\rho}^{(L)}} \tag{A.9}$$

$$= -\alpha\boldsymbol{\rho}^{(0)}[j]\mathbf{F}^H[i,:]\prod_{l=1}^{L-1}\left(\mathbf{Y}(\boldsymbol{\rho}^{(l)})(\mathbf{I} - \alpha\mathbf{G}^T\bar{\mathbf{F}})\right)\mathbf{Y}(\boldsymbol{\rho}^{(L)})\mathbf{z}(\boldsymbol{\rho}^*[n], \boldsymbol{\rho}^{(L)}), \tag{A.10}$$

where, $\mathbf{Y}(\boldsymbol{\rho}^{(l)}) = diag(\mathbf{y}(\boldsymbol{\rho}^{(l)}))$ and

$$\mathbf{y}(\boldsymbol{\rho}^{(l)}) = \left[\begin{array}{ccc} \frac{\tilde{\boldsymbol{\rho}}^{(l)}[1]}{2|\tilde{\boldsymbol{\rho}}^{(l)}[1]|}\mu(|\tilde{\boldsymbol{\rho}}^{(l)}[1]| > \tau) & \cdots & \frac{\tilde{\boldsymbol{\rho}}^{(l)}[N]}{2|\tilde{\boldsymbol{\rho}}^{(l)}[N]|}\mu(|\tilde{\boldsymbol{\rho}}^{(l)}[N]| > \tau) \end{array}\right]^T. \quad \text{(A.11)}$$

$\mathbf{z}(\boldsymbol{\rho}^*[n], \boldsymbol{\rho}^{(L)})$ is an $N \times 1$ vector with kth component $\mathbf{z}[k]$ equals to

$$\mathbf{z}[k] = \frac{\mu(\boldsymbol{\rho}^{(L)}[n] \neq \|\boldsymbol{\rho}^{(L)}\|_\infty \wedge (n = k))}{\|\boldsymbol{\rho}^{(L)}\|_\infty} - \frac{\boldsymbol{\rho}^{(L)}[n]\mu(\boldsymbol{\rho}^{(L)}[k] = \|\boldsymbol{\rho}^{(L)}\|_\infty \wedge (n \neq k))}{\|\boldsymbol{\rho}^{(L)}\|_\infty^2}. \quad \text{(A.12)}$$

$(\partial(\mathbf{G}\boldsymbol{\rho}^*)/\partial\mathbf{G})$ in (A.8) is the following $M \times N \times M$ tensor,

$$\frac{\partial(\mathbf{G}\boldsymbol{\rho}^*)}{\partial\mathbf{G}} = \begin{bmatrix} \left(\frac{\partial(\mathbf{G}\boldsymbol{\rho}^*)}{\partial\mathbf{G}[1,1]}\right)^{M\times 1} & \cdots & \left(\frac{\partial(\mathbf{G}\boldsymbol{\rho}^*)}{\partial\mathbf{G}[1,N]}\right)^{M\times 1} \\ \vdots & & \vdots \\ \left(\frac{\partial(\mathbf{G}\boldsymbol{\rho}^*)}{\partial\mathbf{G}[M,1]}\right)^{M\times 1} & \cdots & \left(\frac{\partial(\mathbf{G}\boldsymbol{\rho}^*)}{\partial\mathbf{G}[M,N]}\right)^{M\times 1} \end{bmatrix}, \quad \text{(A.13)}$$

$$\frac{\partial(\mathbf{G}\boldsymbol{\rho}^*)}{\partial\mathbf{G}[i,j]} = \left[\begin{array}{ccc} \sum_{n=1}^{N} \frac{\partial(\mathbf{G}[1,n]\boldsymbol{\rho}^*[n])}{\partial\mathbf{G}[i,j]} & \cdots & \sum_{n=1}^{N} \frac{\partial(\mathbf{G}[M,n]\boldsymbol{\rho}^*[n])}{\partial\mathbf{G}[i,j]} \end{array}\right]^T, \quad \text{(A.14)}$$

where for $(i,j) \neq (m,n)$, $((\partial(\mathbf{G}[m,n]\boldsymbol{\rho}^*[n]))/\partial\mathbf{G}[i,j])$ equals to $\mathbf{G}[m,n](\partial\boldsymbol{\rho}^*[n]/\partial\mathbf{G}[i,j])$, and for $(i,j) = (m,n)$, it equals to $\boldsymbol{\rho}^*[n] + \mathbf{G}[m,n](\partial\boldsymbol{\rho}^*[n]/\partial\mathbf{G}[m,n])$. $(\partial\boldsymbol{\rho}^*[n]/\partial\mathbf{G}[i,j])$ is expressed in (A.10). $(\partial(\bar{\mathbf{G}}\bar{\boldsymbol{\rho}}^*)/\partial\mathbf{G})$ in (A.8) is a tensor of all zeros which implies that $\partial\ell/\partial\mathbf{G}$ in (A.8) is equal to $-(1/2)((\partial(\mathbf{G}\boldsymbol{\rho}^*))/\partial\mathbf{G})(\bar{\mathbf{d}} - \bar{\mathbf{G}}\bar{\boldsymbol{\rho}}^*)$. Combining the expressions in (A.4), (A.7) and (A.8), we get the expression for $\partial\ell/\partial\boldsymbol{\gamma}_T$ in (9.48). Similarly, $\partial\ell/\partial\mathbf{w}$ in (9.49) can be derived by combining the partial derivative expressions in (A.5), (A.7) and (A.8). Finally, the partial derivative term $\partial\ell/\partial\tau$ in (A.3) can be expanded as follows:

$$\frac{\partial\ell}{\partial\tau} = \frac{\partial\boldsymbol{\rho}^{(1)}}{\partial\tau}\prod_{l=1}^{L-1}\left(\frac{\partial\tilde{\boldsymbol{\rho}}^{(l+1)}}{\partial\boldsymbol{\rho}^{(l)}}\frac{\partial\boldsymbol{\rho}^{(l+1)}}{\partial\tilde{\boldsymbol{\rho}}^{(l+1)}}\right)\frac{\partial\boldsymbol{\rho}^*}{\partial\boldsymbol{\rho}^{(L)}}\frac{\partial\ell}{\partial\boldsymbol{\rho}^*}, \quad \text{(A.15)}$$

$$= -\frac{1}{2}\frac{\partial\tilde{\boldsymbol{\rho}}^{(1)}}{\partial\tau}\prod_{l=1}^{L-1}((\mathbf{I} - \alpha\mathbf{G}^T\bar{\mathbf{G}})\mathbf{Y}(\boldsymbol{\rho}^{(l+1)}))\frac{\partial\boldsymbol{\rho}^*}{\partial\boldsymbol{\rho}^{(L)}}\mathbf{G}^T(\bar{\mathbf{d}} - \bar{\mathbf{G}}\bar{\boldsymbol{\rho}}^*). \quad \text{(A.16)}$$

Partial derivatives $\partial\boldsymbol{\rho}^{(1)}/\partial\tau$ and $\partial\boldsymbol{\rho}^*/\partial\boldsymbol{\rho}^{(L)}$ in (A.16) can be expressed by

$$\frac{\partial\boldsymbol{\rho}^{(1)}}{\partial\tau} = -\left[\begin{array}{ccc} \mu(|\tilde{\boldsymbol{\rho}}^{(1)}[1]| > \tau) & \cdots & \mu(|\tilde{\boldsymbol{\rho}}^{(1)}[N]| > \tau) \end{array}\right]^T, \quad \text{(A.17)}$$

$$\frac{\partial\boldsymbol{\rho}^*}{\partial\boldsymbol{\rho}^{(L)}} := \mathbf{Z}(\boldsymbol{\rho}^*, \boldsymbol{\rho}^{(L)}) = \left[\begin{array}{ccc} \left(\frac{\partial\boldsymbol{\rho}^*[1]}{\partial\boldsymbol{\rho}^{(L)}}\right) & \cdots & \left(\frac{\partial\boldsymbol{\rho}^*[N]}{\partial\boldsymbol{\rho}^{(L)}}\right) \end{array}\right] \quad \text{(A.18)}$$

$$= \left[\mathbf{z}(\boldsymbol{\rho}^*[1], \boldsymbol{\rho}^{(L)}) \cdots \mathbf{z}(\boldsymbol{\rho}^*[N], \boldsymbol{\rho}^{(L)})\right]. \quad \text{(A.19)}$$

Using the expressions in (A.17) and (A.18) in (A.16), we get the expression for $(\partial\ell/\partial\tau)$ in (9.50).

B.1 Partial derivatives when transmission direction is unknown

For the second network in (9.39), parameter updates require the calculation of the partial derivatives $\partial\ell/\partial\mathbf{Q}$, $\partial\ell/\partial\mathbf{w}$, $\partial\ell/\partial\boldsymbol{\gamma}_T$ and $\partial\ell/\partial\tau$. Using chain rule,

$$\left(\frac{\partial\ell}{\partial\mathbf{Q}}\right)^{N\times N} = \frac{\partial\tilde{\boldsymbol{\rho}}^{(1)}}{\partial\mathbf{Q}}\prod_{l=1}^{L-1}\left(\frac{\partial\boldsymbol{\rho}^{(l)}}{\partial\tilde{\boldsymbol{\rho}}^{(l)}}\frac{\partial\tilde{\boldsymbol{\rho}}^{(l+1)}}{\partial\boldsymbol{\rho}^{(l)}}\right)\frac{\partial\boldsymbol{\rho}^{(L)}}{\partial\tilde{\boldsymbol{\rho}}_{(L)}}\frac{\partial\boldsymbol{\rho}^{*}}{\partial\boldsymbol{\rho}^{(L)}}\frac{\partial\ell}{\partial\boldsymbol{\rho}^{*}}, \tag{B.1}$$

$$\left(\frac{\partial\ell}{\partial\boldsymbol{\gamma}_T}\right)^{3\times 1} = \frac{\partial\mathbf{K}^H}{\partial\boldsymbol{\gamma}_T}\frac{\partial\tilde{\boldsymbol{\rho}}^{(1)}}{\partial\mathbf{K}^H}\prod_{l=1}^{L-1}\left(\frac{\partial\boldsymbol{\rho}^{(l)}}{\partial\tilde{\boldsymbol{\rho}}^{(l)}}\frac{\partial\tilde{\boldsymbol{\rho}}^{(l+1)}}{\partial\boldsymbol{\rho}^{(l)}}\right)\frac{\partial\boldsymbol{\rho}^{(L)}}{\partial\tilde{\boldsymbol{\rho}}^{(L)}}\frac{\partial\boldsymbol{\rho}^{*}}{\partial\boldsymbol{\rho}^{(L)}}\frac{\partial\ell}{\partial\boldsymbol{\rho}^{*}}, \tag{B.2}$$

$$\left(\frac{\partial\ell}{\partial\mathbf{w}}\right)^{M\times 1} = \frac{\partial\bar{\tilde{\boldsymbol{\rho}}}^{(1)}}{\partial\mathbf{w}}\frac{\partial\boldsymbol{\rho}^{(1)}}{\partial\bar{\tilde{\boldsymbol{\rho}}}^{(1)}}\prod_{l=1}^{L-1}\left(\frac{\partial\tilde{\boldsymbol{\rho}}^{(l+1)}}{\partial\boldsymbol{\rho}^{(l)}}\frac{\partial\boldsymbol{\rho}^{(l+1)}}{\partial\tilde{\boldsymbol{\rho}}^{(l+1)}}\right)\frac{\partial\boldsymbol{\rho}^{*}}{\partial\boldsymbol{\rho}^{(L)}}\frac{\partial\ell}{\partial\boldsymbol{\rho}^{*}}, \tag{B.3}$$

$$\left(\frac{\partial\ell}{\partial\tau}\right)^{1\times 1} = \frac{\partial\boldsymbol{\rho}^{(1)}}{\partial\tau}\prod_{l=1}^{L-1}\left(\frac{\partial\tilde{\boldsymbol{\rho}}^{(l+1)}}{\partial\boldsymbol{\rho}^{(l)}}\frac{\partial\boldsymbol{\rho}^{(l+1)}}{\partial\tilde{\boldsymbol{\rho}}^{(l+1)}}\right)\frac{\partial\boldsymbol{\rho}^{*}}{\partial\boldsymbol{\rho}^{(L)}}\frac{\partial\ell}{\partial\boldsymbol{\rho}^{*}}. \tag{B.4}$$

From (B.4), the expression for $\partial\ell/\partial\tau$ for this model can be calculated using (A.16), (A.17) and (A.18) with $\mathbf{G}$ replaced by $\tilde{\mathbf{K}}_R \odot \tilde{\mathbf{K}}_T$. $\partial\ell/\partial\boldsymbol{\rho}^*$ in (B.1), (B.2), (B.3) and (B.4) can be expressed as

$$\frac{\partial\ell}{\partial\boldsymbol{\rho}^*} = \mathbf{K}^T(\bar{\mathbf{d}} - \bar{\mathbf{K}}\bar{\boldsymbol{\rho}}^*). \tag{B.5}$$

Product term $p_1 = \prod_{l=1}^{L-1}\left((\partial\boldsymbol{\rho}^{(l)}/\partial\tilde{\boldsymbol{\rho}}^{(l)})(\partial\tilde{\boldsymbol{\rho}}^{(l+1)}/\partial\boldsymbol{\rho}^{(l)})\right)(\partial\boldsymbol{\rho}^{(L)}/\partial\tilde{\boldsymbol{\rho}}^{(L)})(\partial\boldsymbol{\rho}^*/\partial\boldsymbol{\rho}^{(L)})$ in (B.1) and (B.2) has the following expression:

$$p_1 = \prod_{l=1}^{L-1}(\mathbf{Y}(\boldsymbol{\rho}^{(l)})(\mathbf{I} - \alpha\mathbf{F}^T\bar{\mathbf{F}}))\mathbf{Y}(\boldsymbol{\rho}^{(L)})\mathbf{Z}(\boldsymbol{\rho}^*, \boldsymbol{\rho}^{(L)}), \tag{B.6}$$

where $\mathbf{y}$, which is the diagonal of $\mathbf{Y}$, and $\mathbf{Z}$ are defined in (A.11) and (A.18), respectively. Similarly, the term $p_2 = \prod_{l=1}^{L-1}\left((\partial\tilde{\boldsymbol{\rho}}^{(l+1)}/\partial\boldsymbol{\rho}^{(l)})(\partial\boldsymbol{\rho}^{(l+1)}/\partial\tilde{\boldsymbol{\rho}}^{(l+1)})\right)(\partial\boldsymbol{\rho}^*/\partial\boldsymbol{\rho}^{(L)})$ in (B.3) can be expressed as

$$p_2 = \prod_{l=1}^{L-1}((\mathbf{I} - \alpha\mathbf{K}^T\bar{\mathbf{K}})\mathbf{Y}(\boldsymbol{\rho}^{(l+1)}))\mathbf{Z}(\boldsymbol{\rho}^*, \boldsymbol{\rho}^{(L)}). \tag{B.7}$$

Partial derivative $\partial\tilde{\boldsymbol{\rho}}^{(1)}/\partial\mathbf{Q}$ in (B.1) is an $N \times N \times N$ dimensional tensor,

$$\frac{\partial\tilde{\boldsymbol{\rho}}^{(1)}}{\partial\mathbf{Q}} = \begin{bmatrix} \frac{\partial\tilde{\boldsymbol{\rho}}^{(1)}}{\partial\mathbf{Q}[1,1]} & \cdots & \frac{\partial\tilde{\boldsymbol{\rho}}^{(1)}}{\partial\mathbf{Q}[1,N]} \\ \vdots & & \vdots \\ \frac{\partial\tilde{\boldsymbol{\rho}}^{(1)}}{\partial\mathbf{Q}[N,1]} & \cdots & \frac{\partial\tilde{\boldsymbol{\rho}}^{(1)}}{\partial\mathbf{Q}[N,N]} \end{bmatrix}, \tag{B.8}$$

where $\partial\tilde{\boldsymbol{\rho}}^{(1)}/\partial\mathbf{Q}[n_1, n_2]$ is an $N \times 1$ vector of all zeros except at the n_1^{th} index where it has a value of $\boldsymbol{\rho}^{(0)}[n_2]$ for all $n_1, n_2 \in [1, N]$. Using this expression, along with the expressions in (B.5) and (B.6), we have $\partial\ell/\partial\mathbf{Q} = \mathbf{B}$ with components presented in (9.53).

Partial derivative $\partial\tilde{\boldsymbol{\rho}}^{(1)}/\partial\mathbf{K}^H$ in (B.2) is an $N \times M \times N$ dimensional tensor. Similar to the expression for $\partial\tilde{\boldsymbol{\rho}}^{(1)}/\partial\mathbf{Q}$ in (B.8), $\partial\tilde{\boldsymbol{\rho}}^{(1)}/\partial\mathbf{K}^H$ consists of $N \times 1$ component vectors $\partial\tilde{\boldsymbol{\rho}}^{(1)}/\partial\mathbf{K}^H[i,j]$ for all $i \in [1,N]$ and $j \in [1,M]$. Here, the ith component of $\partial\tilde{\boldsymbol{\rho}}^{(1)}/\partial\mathbf{K}^H[i,j]$ is equal to $\alpha\mathbf{d}[j]$ and the remaining elements are all 0's. Let operator $\mathscr{K}_T(\omega, s, \mathbf{x}) = e^{-i(\omega/c_0)\,(|\boldsymbol{\gamma}_T-\mathbf{x}|)}$ and $\mathbf{K}_T$ is its discretized version. In that case, $\mathbf{K} = diag(\mathbf{w})\mathbf{K}_T(\boldsymbol{\gamma}_T) \odot \tilde{\mathbf{K}}_R$, and the partial derivative $\partial\mathbf{K}^H/\partial\boldsymbol{\gamma}_T \in \mathbb{C}^{3\times N\times M}$ in (B.2) can be expressed by $\left[(\partial\mathbf{K}^H/\partial\boldsymbol{\gamma}_T[1])\ (\partial\mathbf{K}^H/\partial\boldsymbol{\gamma}_T[2])\ (\partial\mathbf{K}^H/\partial\boldsymbol{\gamma}_T[3])\right]$ where $(\partial\mathbf{K}^H/\partial\boldsymbol{\gamma}_T[c]) \in \mathbb{C}^{N\times M}$ for $c \in [1,3]$ has the following expression:

$$\frac{\partial\mathbf{K}^H}{\partial\boldsymbol{\gamma}_T[c]} = \mathbf{K}^H \odot \begin{bmatrix} \frac{i(\boldsymbol{\gamma}_T[c]-\boldsymbol{x}_1[c])}{c_0|\boldsymbol{\gamma}_T-\boldsymbol{x}_1|}\begin{bmatrix}\omega_1 & \cdots & \omega_K\end{bmatrix}^{1\times M} \\ \vdots \\ \frac{i(\boldsymbol{\gamma}_T[c]-\boldsymbol{x}_N[c])}{c_0|\boldsymbol{\gamma}_T-\boldsymbol{x}_N|}\begin{bmatrix}\omega_1 & \cdots & \omega_K\end{bmatrix}^{1\times M} \end{bmatrix}. \tag{B.9}$$

Using the expressions in (B.5), (B.6) and (B.9) in (B.2), we get the expression for the components of $\partial\ell/\partial\boldsymbol{\gamma}_T$ in (9.54).

Finally, the partial derivatives $\partial\bar{\tilde{\boldsymbol{\rho}}}^{(1)}/\partial\mathbf{w}$ and $\partial\boldsymbol{\rho}^{(1)}/\partial\bar{\tilde{\boldsymbol{\rho}}}^{(1)}$ in (B.3) are $M \times N$ and $N \times N$ dimensional matrices with complex-valued components and can be expressed as follows:

$$\frac{\partial\bar{\tilde{\boldsymbol{\rho}}}^{(1)}}{\partial\mathbf{w}} = \alpha\tilde{\mathbf{K}}_R \odot \begin{bmatrix} \bar{\mathbf{d}}[1]\begin{bmatrix}e^{-i(\omega_1/c_0)|\boldsymbol{\gamma}_T-\boldsymbol{x}_1|} & \ldots & e^{-i(\omega_1/c_0)|\boldsymbol{\gamma}_T-\boldsymbol{x}_N|}\end{bmatrix}^{1\times N} \\ \vdots \\ \bar{\mathbf{d}}[M]\begin{bmatrix}e^{-i(\omega_K/c_0)|\boldsymbol{\gamma}_T-\boldsymbol{x}_1|} & \ldots & e^{-i(\omega_K/c_0)|\boldsymbol{\gamma}_T-\boldsymbol{x}_N|}\end{bmatrix}^{1\times N} \end{bmatrix}, \tag{B.10}$$

$$\frac{\partial\boldsymbol{\rho}^{(1)}}{\partial\bar{\boldsymbol{\rho}}_1'} = \frac{1}{2}diag\left(\left[\frac{\tilde{\boldsymbol{\rho}}^{(1)}[1]}{|\tilde{\boldsymbol{\rho}}^{(1)}[1]|}\mu(|\tilde{\boldsymbol{\rho}}^{(1)}[1]| > \tau)\ \cdots\ \frac{\tilde{\boldsymbol{\rho}}^{(1)}[N]}{|\tilde{\boldsymbol{\rho}}^{(1)}[N]|}\mu(|\tilde{\boldsymbol{\rho}}^{(1)}[N]| > \tau)\right]\right). \tag{B.11}$$

Using the expressions in (B.5), (B.7), (B.10) and (B.11), we get the expression for the components of $\partial\ell/\partial\mathbf{w}$ in (9.55).

References

[1] Kuschel H, Cristallini D, and Olsen KE. Tutorial: passive radar tutorial. *IEEE Aerospace and Electronic Systems Magazine.* 2019;34(2):2–19.

[2] Yarman CE, Wang L, and Yazici B. Doppler synthetic aperture hitchhiker imaging. *Inverse Problems.* 2010;26(6):065006.

[3] Malanowski M, Kulpa K, Dryjański M, *et al.* Fixed WiMAX (IEEE 802.16 d) base station signal analysis for passive radar applications. In: Photonics Applications in Astronomy, Communications, Industry, and High-Energy Physics Experiments 2009, *International Society for Optics and Photonics. SPIE*; 2009;7502:270–275.

[4] Dawidowicz B, Samczynski P, Malanowski M, *et al.* Detection of moving targets with multichannel airborne passive radar. *IEEE Aerospace and Electronic Systems Magazine.* 2012;27(11):42–49.

[5] Rapson M. Passive Multistatic Radar Imaging using an OFDM Based Signal of Opportunity [thesis]. Wright-Patterson AFB, OH: Air Force Institute of Technology; 2012.

[6] Stinco P, Greco MS, Gini F, *et al.* Posterior CramRao lower bounds for passive bistatic radar tracking with uncertain target measurements. *Signal Processing.* 2013;93(12):3528–3540.

[7] Palmer JE, Harms HA, Searle SJ, *et al.* DVB-T passive radar signal processing. *IEEE Transactions on Signal Processing.* 2013;61(8):2116–2126.

[8] Gutierrez del Arroyo Jr., JA. WiMAX OFDM for passive SAR ground imaging. *IEEE Transactions on Aerospace and Electronic Systems.* 2013;49(2): 945–959.

[9] Malanowski M, Haugen R, Greco MS, *et al.* Land and sea clutter from FM-based passive bistatic radars. *IET Radar, Sonar Navigation.* 2014;8(2): 160–166.

[10] Son IY and Yazici B. Passive imaging with multistatic polarimetric radar. In: *2015 IEEE Radar Conference*, IEEE; 2015. p. 1584–1589.

[11] Son IY and Yazici B. Passive polarimetrie multistatic radar for ground moving target. In: *2016 IEEE Radar Conference*, IEEE; 2016. p. 1–6.

[12] Griffiths HD and Baker CJ. Passive coherent location radar systems. Part 1: Performance prediction. *IEEE Proceedings – Radar, Sonar and Navigation.* 2005;152(3):153–159.

[13] Baker CJ, Griffiths HD, and Papoutsis I. Passive coherent location radar systems. Part 2: Waveform properties. *IEEE Proceedings – Radar, Sonar and Navigation.* 2005;152(3):160–168.

[14] Wu Q, Zhang YD, Amin MG, *et al.* High-resolution passive SAR imaging exploiting structured Bayesian compressive sensing. *IEEE Journal of Selected Topics in Signal Processing.* 2015;9(8):1484–1497.

[15] Hack DE, Patton LK, Himed B, *et al.* Detection in passive MIMO radar networks. *IEEE Transactions on Signal Processing.* 2014;62(11):2999–3012.

[16] Krysik P and Kulpa K. The use of a GSM-based passive radar for sea target detection. In: *2012 9th European Radar Conference (EuRAD)*, IEEE; 2012. p. 142–145.

[17] Yarman CE and Yazici B. Synthetic aperture hitchhiker imaging. *IEEE Transactions on Image Processing.* 2008;17(11):2156–2173.

[18] Wang L, Yarman CE, and Yazici B. Doppler-hitchhiker: a novel passive synthetic aperture radar using ultranarrowband sources of opportunity. *IEEE Transactions on Geoscience and Remote Sensing.* 2011;49(10): 3521–3537.

[19] Wang L and Yazici B. Passive imaging of moving targets exploiting multiple scattering using sparse distributed apertures. *Inverse Problems.* 2012;28(12): 125009.

[20] Wang L and Yazici B. Ground moving target imaging using ultranarrowband continuous wave synthetic aperture radar. *IEEE Transactions on Geoscience and Remote Sensing.* 2013;51(9):4893–4910.

[21] Wacks S and Yazici B. Bistatic Doppler-SAR DPCA imaging of ground moving targets. In: *2014 IEEE Radar Conference*, IEEE; 2014. p. 1071–1074.

[22] Wacks S and Yazici B. Passive synthetic aperture hitchhiker imaging of ground moving targets – part 1: Image formation and velocity estimation. *IEEE Transactions on Image Processing.* 2014;23(6):2487–2500.
[23] Wacks S and Yazici B. Passive synthetic aperture hitchhiker imaging of ground moving targets part 2: Performance analysis. *IEEE Transactions on Image Processing.* 2014;23(9):4126–4138.
[24] Wacks S and Yazici B. Doppler-DPCA and Doppler-ATI: novel SAR modalities for imaging of moving targets using ultra-narrowband waveforms. *IEEE Transactions on Computational Imaging.* 2018;4(1):125–136.
[25] Yazici B, Son IY, and Yanik HC. Doppler synthetic aperture radar interferometry: a novel SAR interferometry for height mapping using ultra-narrowband waveforms. *Inverse Problems.* 2018;34(5):055003.
[26] Mason E, Son IY, and Yazici B. Passive synthetic aperture radar imaging using low-rank matrix recovery methods. *IEEE Journal of Selected Topics in Signal Processing.* 2015;9(8):1570–1582.
[27] Mason E and Yazici B. Performance analysis of convex LRMR based passive SAR imaging. CoRR. 2017;abs/1711.03232.
[28] Mason E and Yazici B. Robustness of LRMR based passive radar imaging to phase errors. In: *Proceedings of EUSAR 2016: 11th European Conference on Synthetic Aperture Radar*, VDE; 2016. p. 1–4.
[29] Yonel B, Son IY, and Yazici B. Generalized Wirtinger flow for passive polarimetrie reconstruction of extended dipole targets. In: *2018 IEEE Radar Conference*; 2018. p. 1353–1358.
[30] Yonel B and Yazici B. A generalization of Wirtinger flow for exact interferometric inversion. *SIAM Journal on Imaging Sciences.* 2019;12(4): 2119–2164.
[31] Yonel B, Son IY, and Yazici B. Exact multistatic interferometric imaging via generalized Wirtinger flow. *IEEE Transactions on Computational Imaging.* 2020:1–16.
[32] Son I, Yonel B, and Yazici B. Multistatic passive radar imaging using generalized Wirtinger flow. In: *2019 20th International Radar Symposium (IRS)*, IEEE; 2019. p. 1–7.
[33] Yonel B and Yazici B. A Deterministic Convergence Framework for Exact Non-Convex Phase Retrieval. *IEEE Transactions on Signal Processing*, IEEE. 2020;68:4612–4626.
[34] Glorot X and Bengio Y. Understanding the difficulty of training deep feedforward neural networks. In: *Proceedings of the 13th International Conference on Artificial Intelligence and Statistics*, PMLR; 2010. p. 249–256.
[35] Krizhevsky A, Sutskever I, and Hinton GE. ImageNet classification with deep convolutional neural networks. In: *Advances in Neural Information Processing Systems 25*, Curran Associates, Inc.; 2012. p. 1097–1105.
[36] Graves A, Mohamed A, and Hinton G. Speech recognition with deep recurrent neural networks. In: *2013 IEEE International Conference on Acoustics, Speech and Signal Processing*, IEEE; 2013. p. 6645–6649.

[37] Zaremba W, Sutskever I, and Vinyals O. Recurrent Neural Network Regularization; 2014. p. 1–8. *arXiv preprint arXiv:14092329.*

[38] Chollet F. Xception: deep learning with depthwise separable convolutions. In: *Proceedings of the IEEE Conference on Computer Vision and Pattern Recognition (CVPR)*, IEEE; 2017. p. 1251–1258.

[39] Nguyen LD, Lin D, Lin Z, *et al.* Deep CNNs for microscopic image classification by exploiting transfer learning and feature concatenation. In: *2018 IEEE International Symposium on Circuits and Systems (ISCAS)*, IEEE; 2018. p. 1–5.

[40] Wang X, Gao L, Song J, *et al.* Beyond frame-level CNN: saliency-aware 3-D CNN with LSTM for video action recognition. *IEEE Signal Processing Letters.* 2016;24(4):510–514.

[41] Jain SD, Xiong B, and Grauman K. Fusionseg: learning to combine motion and appearance for fully automatic segmentation of generic objects in videos. In: *2017 IEEE Conference on Computer Vision and Pattern Recognition (CVPR)*, IEEE; 2017. p. 2117–2126.

[42] Dong D, Wu H, He W, *et al.* Multi-task learning for multiple language translation. In: *Proceedings of the 53rd Annual Meeting of the Association for Computational Linguistics and the 7th International Joint Conference on Natural Language Processing (Volume 1: Long Papers).* Association for Computational Linguistics. vol. 1; 2015. p. 1723–1732.

[43] Zoph B, Yuret D, May J, *et al.* Transfer Learning for Low-Resource Neural Machine Translation. In: *EMNLP*; 2016. Preprint.

[44] Hornik K, Stinchcombe M, and White H. Multilayer feedforward networks are universal approximators. *Neural Networks.* 1989;2(5):359–366.

[45] Hornik K. Approximation capabilities of multilayer feedforward networks. *Neural Networks.* 1991;4(2):251–257.

[46] Chen S and Wang H. SAR target recognition based on deep learning. In: *2014 International Conference on Data Science and Advanced Analytics (DSAA)*, IEEE; 2014. p. 541–547.

[47] Wang H, Chen S, Xu F, *et al.* Application of deep-learning algorithms to MSTAR data. In: 2015 *IEEE International Geoscience and Remote Sensing Symposium (IGARSS)*, IEEE; 2015. p. 3743–3745.

[48] Morgan DAE. Deep convolutional neural networks for ATR from SAR imagery. In: Zelnio E, Garber FD, editors. Algorithms for Synthetic Aperture Radar Imagery XXII. *International Society for Optics and Photonics.* SPIE; vol. 9475; 2015. p. 94750F.

[49] Chen S, Wang H, Xu F, *et al.* Target classification using the deep convolutional networks for SAR images. *IEEE Transactions on Geoscience and Remote Sensing.* 2016;54(8):4806–4817.

[50] Ding J, Chen B, Liu H, *et al.* Convolutional neural network with data augmentation for SAR target recognition. *IEEE Geoscience and Remote Sensing Letters.* 2016;13(3):364–368.

[51] Wagner S, Barth K, and Brüggenwirth S. A deep learning SAR ATR system using regularization and prioritized classes. In: *2017 IEEE Radar Conference (RadarConf)*. IEEE; 2017. p. 0772–0777.

[52] Kazemi S, Yonel B, and Yazici B. Deep learning for direct automatic target recognition from SAR data. In: *2019 IEEE Radar Conference (RadarConf)*. IEEE; 2019. p. 1–6.

[53] Huang Z, Pan Z, and Lei B. Transfer learning with deep convolutional neural network for SAR target classification with limited labeled data. *Remote Sensing*. 2017;9(9):907.

[54] Pei J, Huang Y, Huo W, *et al.* SAR automatic target recognition based on multiview deep learning framework. *IEEE Transactions on Geoscience and Remote Sensing*. 2018;56(4):2196–2210.

[55] Gregor K and LeCun Y. Learning fast approximations of sparse coding. In: *Proceedings of the 27th International Conference on International Conference on Machine Learning Omnipress*; 2010. p. 399–406.

[56] Borgerding M and Schniter P. Onsager-corrected deep learning for sparse linear inverse problems. In: *2016 IEEE Global Conference on Signal and Information Processing (GlobalSIP)*, IEEE; 2016. p. 227–231.

[57] Mason E, Yonel B, and Yazici B. Deep learning for radar. In: *2017 IEEE Radar Conference (RadarConf)*. IEEE; 2017. p. 1703–1708.

[58] Mason E, Yonel B, and Yazici B. Deep learning for SAR image formation. In: Algorithms for Synthetic Aperture Radar Imagery XXIV. vol. 10201. *International Society for Optics and Photonics*. SPIE; 2017. p. 11–21.

[59] Yonel B, Mason E, and Yazici B. Deep learning for passive synthetic aperture radar. *IEEE Journal of Selected Topics in Signal Processing*. 2018;12(1): 90–103.

[60] Yonel B, Mason E, and Yazici B. Deep learning for waveform estimation in passive synthetic aperture radar. In: *2018 IEEE Radar Conference (RadarConf)*. IEEE; 2018. p. 1395–1400.

[61] Yonel B, Mason E, and Yazici B. Deep learning for waveform estimation and imaging in passive radar. *IET Radar, Sonar & Navigation*. 2019;13(6): 915–926.

[62] LeCun Y, Bengio Y, and Hinton G. *Deep learning*. Nature. 2015;521(7553): 436–444.

[63] Cheney M and Borden B. Fundamentals of Radar Imaging. *CBMS-NSF Regional Conference Series in Applied Mathematics*. Philadelphia, PA: Society for Industrial and Applied Mathematics (SIAM); 2009.

[64] Combettes PL and Wajs VR. Signal recovery by proximal forward-backward splitting. *Multiscale Modeling & Simulation*. 2005;4(4):1168–1200.

[65] Rumelhart DE, Hinton GE, and Williams RJ. Learning Internal Representations by Error Propagation. In: Rumelhart DE, Mcclelland JL, editors. *Parallel Distributed Processing: Explorations in the Microstructure of Cognition*, Volume 1: Foundations. Cambridge, MA: MIT Press; 1986. p. 318–362.

[66] Tan CC and Eswaran C. Reconstruction of handwritten digit images using autoencoder neural networks. In: *2008 Canadian Conference on Electrical and Computer Engineering*, IEEE; 2008. p. 000465–000470.

[67] Yonel B, Mason E, and Yazici B. Deep learning for waveform estimation in passive synthetic aperture radar imaging. In: *Algorithms for Synthetic Aperture Radar Imagery XXV.* vol. 10647; 2018. p. 106470E.

[68] Howland PE, Maksimiuk D, and Reitsma G. FM radio based bistatic radar. *IEEE Proceedings – Radar, Sonar and Navigation.* 2005;152(3):107–115.

[69] Colone F, Bongioanni C, and Lombardo P. Multifrequency integration in FM radio-based passive bistatic radar. Part I: Target detection. *IEEE Aerospace and Electronic Systems Magazine.* 2013;28(4):28–39.

[70] Colone F, Bongioanni C, and Lombardo P. Multifrequency integration in FM radio-based passive bistatic radar. Part II: Direction of arrival estimation. *IEEE Aerospace and Electronic Systems Magazine.* 2013;28(4):40–47.

[71] Tan DK, Sun H, Lu Y, *et al.* Passive radar using global system for mobile communication signal: theory, implementation and measurements. *IEEE Proceedings – Radar, Sonar and Navigation.* 2005;152(3):116–123.

[72] del Arroyo JRG and Jackson JA. SAR imaging using WiMAX OFDM PHY. In: *2011 IEEE Radar Conference (RadarConf).* IEEE; 2011. p. 129–134.

[73] Harms HA, Davis LM, and Palmer J. Understanding the signal structure in DVBT signals for passive radar detection. In: *2010 IEEE Radar Conference (RadarConf).* IEEE; 2010. p. 532–537.

[74] Wang Q, Hou C, and Lu Y. An experimental study of WiMAX-based passive radar. *IEEE Transactions on Microwave Theory and Techniques.* 2010;58(12): 3502–3510.

Chapter 10
Fusion of deep representations in multistatic radar networks

Jarez Satish Patel,[1] Francesco Fioranelli,[2] Matthew Ritchie[3] and Hugh Griffiths[3]

Multistatic radar is a hybrid area of research, employing a combination of techniques applicable to both monostatic and bistatic radar. Coordinated RF sensing from a networked radar system is a complex challenge, which has limited the real-world commercial drive of fully coherent multistatic radar systems, but nevertheless the potential benefits are driving new innovations in this area. This chapter aims to introduce and address such foundational difficulties associated with the integration of these radars, specifically with regards to the fusion of information between nodes, while simultaneously enabling the achievement of noteworthy classification accuracy, attained through exploiting cutting-edge deep learning approaches. The scope of the research presented in this chapter is based upon numerous experimental campaigns, entirely undertaken by the authors utilising a real multistatic pulsed Doppler radar, named 'NetRAD'. The scenario of principal focus throughout this chapter is human micro-Doppler effects. From this a number of data fusion and classification architectures are proposed and analysed, using classification accuracy as a metric of success. Optimal fusion methods and configurations are identified in this work, while crucially respecting the fundamental constraints of multistatic radar. A distributed network of unique deep-learning models is found to be the ideal approach to this problem. In addition to this study, the integrity of the multistatic system is rigorously tested against an array of jamming measures, later interpreted by a deep network running in tandem, which is then able to counteract the effects of the jamming, in order to maintain classification performance throughout the radar network.

10.1 Introduction

There are many advantages that multistatic radar systems have over traditional monostatic systems, the primary one being is that it significantly increases physical coverage

[1]James Watt School of Engineering, University of Glasgow, Glasgow, United Kingdom
[2]Microwave Sensing Signals and Systems Group, Department of Microelectronics, Delft University of Technology, Delft, The Netherlands
[3]Department of Electronic and Electrical Engineering, University College London, London, United Kingdom

and provides a multi-perspective view of the target, increasing the probability of detection against stealthy targets, as they are illuminated from a range of different angles and the clutter may be less detrimental. Such systems are also robust, as if any radar node becomes physically damaged or experiences harmful interference, there is no abrupt loss of performance throughout the entire system; this is known as graceful degradation [1]. In the face of targeted jamming against a multistatic system, the nature of the distributed nodes permits intercommunication, and through this, corrupted sensor information can be effectively managed. Nonetheless, necessary algorithms first need to be in place, in order to actively detect the presence of jamming and secondly to handle and respond to this disturbance in an appropriate manner. This need for making a centralised decision, based on multiple diverse data streams is known as data fusion and is a well-established branch of computer science concerned with the management of multiple vital data sources. This is introduced and discussed in detail in Section 10.3, applying the relevant ideas to spatially diverse radar systems.

One of the biggest challenges in multistatic radar is the issue of coherency between radar nodes; that is to say the local oscillator distributed to each node must be disciplined to within sub-nanosecond accuracies [2]. This, at the simplest of levels has a direct influence on the ability of the system to range a target precisely. There is also the issue of practical communication between radar nodes and breaking radio silence, specifically if the node is operating covertly (passively). In general, research into coherent multistatic radar systems is relatively immature as military interests have not really taken hold, due to the array of stringent requirements regarding their deployment and the low technology readiness level associated with this; but as these issues are gradually addressed, there are strategic advantages to be gained, which are eventually too great to ignore.

In the context of classification of micro-Doppler signatures, the spectrogram can vary drastically depending on the aspect angle [3], radar parameters, and a multitude of other conditions [4]. This has the effect of attenuating, distorting or skewing the target signal within the resultant spectrogram. Therefore, the design of a robust feature extraction method which can cope under this challenging environment is incredibly difficult. There is also the question of what the most valuable features are, which is strongly related to the type of target that the system is anticipating; this requires a sound understanding of the dynamics and the nature of the target. Consequently, the classification abilities of the feature extraction algorithm are highly dependent on the knowledge and extraction of these identified features [5,6]; is this a limitation that deep learning could seek to resolve?

Over recent years, these aforementioned issues have been progressively investigated through a number of research articles, the first tackling the challenge of distinguishing between potentially armed and unarmed personnel using a multistatic radar [7]. Following from this, different geometries, aspect angles, classification systems [3] and methods of data fusion between the nodes are considered [8]; all monitoring the effect on classification accuracy, where the results of which can be found in Section 10.4. The latest publications in this field investigate the possibilities of deep learning as a robust classifier, altogether detailed in Table 10.1 and explore an array of applicable configurations and fusion topologies. The flexibility of deep neural networks (DNNs) is demonstrated in [9], which looks into detecting whether a drone is carrying a payload or not and finally, the resilience of multistatic systems to targeted enemy jamming is examined in [10], this is discussed in detail in Section 10.5.

10.2 Experimental multistatic radar system

The multistatic radar system (NetRAD) which has been developed over the past 15 years at University College London (UCL) was used to collect all of the experimental data presented in this chapter. It is a coherent pulsed Doppler radar with three separate but identical nodes operating at S-Band [12,13]. The transmitted waveform is a linear frequency modulated chirp signal, which enables stretch processing and pulse compression at the receiver [14]. The operating parameters surrounding the radar system are detailed in Table 10.2 [10]. All of the experiments were performed in a large open field at the UCL sports ground at Shenley during July 2015 for Setup A and March 2016 for Setup B.

10.2.1 Setup A

This setup deployed three radar nodes along a linear baseline with 50-m inter-nodal separation, as depicted in Figure 10.1(a), the field was split up into a grid of six zones, where the objective was to distinguish between armed and unarmed personnel. The purpose of this configuration was to mimic the surveillance of an area divided into zones of interest and to evaluate possible fluctuations in the classification performance

Table 10.1 Deep learning techniques applied to experimental multistatic radar

Technique	Application	Reference	Year
Stacked auto-encoders (SAE)	Human micro-Doppler	[11]	2017
Deep neural network (DNN)	Drone payload characterisation	[9]	2018
DNN	Human micro-Doppler	[10]	2019

Table 10.2 NetRAD operational radar parameters

Radar parameters	Value
Centre frequency (f_c)	2.4 GHz
Transmit power (P_t)	+23 dBm
Bandwidth (BW)	45 MHz
Pulse repetition frequency (f_{PRF})	5 kHz
Pulse length (τ)	0.6 μs
Pulse compression gain (G_{pc})	14.3 dB
Range resolution (ΔR)	3.3 m

between different zones. The subjects all took turns walking from the top edge of the zone towards the bottom edge and in each of the different zones located in front of the radar system towards the linear baseline. Data was collected from each person walking in each zone, generating a total of 360 recordings: 2 people, 3 nodes, 6 zones, and 10 repetitions per person, i.e. 5 'unarmed' and 5 'armed', where the target was carrying a metallic pole of 1 m in length, mimicking the constrained movement of holding a rifle. The antennas used for the experiments were four mesh grid antennas, with further parameters surrounding this detailed in Table 10.3.

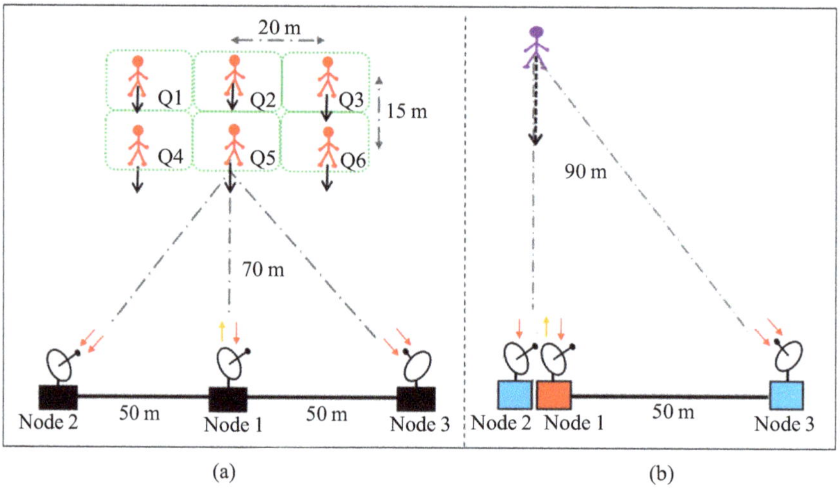

Figure 10.1 Experimental configuration of: (a) setup A and (b) setup B

Table 10.3 Radar parameters used in the experimental setups

Parameters	**Value**	
	Setup A	**Setup B**
Antenna gain (G_{TX}, G_{RX}) (dBi)	24	18
Beamwidth ($\theta_{az} \times \theta_{el}$) (°)	10×8	18×19
Cross range resolution (Φ_{res}) (m)	12[a]	28[b]
Node 1 (Tx/Rx)	VV-Pol	VV-Pol
Node 2 (Rx)	VV-Pol	VH-Pol
Node 3 (Rx)	VV-Pol	VV-Pol
Recording time (t) (s)	5	10

[a]Observed at a distance of 90 m from Node 1 (Tx).
[b]Observed at a distance of 70 m from Node 1 (Tx).

10.2.2 Setup B

This setup deployed the three radar nodes along a linear baseline with 50-m separation, as depicted in Figure 10.1(b); however, as opposed to Setup A, nodes 1 and 2 are colocated and are diverse with regards to polarisation, as to provide additional information. The target was at an average distance of 90 m from the transmitter (Node 1), approximately 103 m from the passive bistatic receiver (PBR) (Node 3) and walked from 100 to 80 m. The target carried a metallic pole in relevant tests of 1 m in length, to mimic the constrained movement of holding a rifle and also a heavy rucksack weighing approximately 15 kg, where combinations of these constituted seven different activities [10,15]. The antennas used were four parabolic dish antennas, with further details stated in Table 10.3. A comprehensive data set was gathered consisting of 10 repeats per person (40 samples) of the seven different activities (280) over the three radar nodes (840 total samples). The recording duration of each sample was 10 s, each of these were matched filtered by performing a Hilbert transformation and correlating against a bank of reference chirp signals, to generate the range time plots.

10.3 Data fusion

The fundamental process of combining information from multiple data sources, characteristically generated from some type of sensor, occurs everywhere around us. The notion of data fusion has aided in the progression and development of a wide range of technical areas, to name but a few: tracking systems, wireless sensor networks, control systems, biometrics, geospatial information systems, and processing systems in healthcare [16]. Indeed, it is clear that data fusion plays a crucial role in a wide range of fields and hence, in modern society, through this essential demand, it has become a heavily researched concept and a well-developed set of techniques.

The purpose of this section is to summarise the vast field of data fusion and present a suitable methodology for approaching tasks in data fusion problems, while simultaneously making the ideas directly applicable to general radar engineering challenges but more specifically, multistatic systems. Towards the end of this section, implementations of these data fusion concepts applied to the real experimental multistatic radar system introduced in Section 10.2, will be deliberated. Assigning strong consideration to relevant data fusion methods that are able to preserve the characteristics and hence, integrity of the input signal, which as will be discovered, directly limits the set of applicable data fusion approaches to radar signal processing.

10.3.1 Significance in multistatic radar systems

Data fusion is a vital process in a multistatic radar system and can be tackled through varying degrees of data sharing. If there was no combination of the sensory data from the multiple nodes, the very essence of the spatially distributed system would be defeated. A multistatic radar operating as a target tracker is perhaps one of the most demanding examples of data fusion one could envisage. The tracking of fast manoeuvring targets, such as a jet fighter places stringent requirements on tracking accuracy, timing, resource management, classification, number of targets, and the critical nature of the task at hand, demands a thorough and substantial data fusion hierarchy, taking into consideration nearly every foundational aspect of data fusion theory [16]. This inherent requirement for a powerful and robust data fusion approach, required for multistatic radar, is considered to be a significant disadvantage when compared to a monostatic radar, as it adds complexity regarding intermediate algorithms, with further measures needing to be in place to deal with an array of diverse scenarios. Consequently, this leads to an increase in deployment time, through having to test and develop additional algorithms, and with respect to the interest of a company, this drives up manufacturing and operational maintenance costs throughout the lifespan of the radar system. These factors are certainly undesirable; however, through extended research into these multi-sensor configurations [17,18], the advantages over other radar architectures will be made increasingly apparent, developing a stronger argument for multistatic systems in the process.

10.3.2 Data fusion methods

The theme of data fusion can be divided into an array of categories depending on the stage at which the fusion process takes place; this is depicted in Figure 10.2, which has been adapted from [19], where symbols have been assigned to the elements for future notations throughout this chapter and to make the flow of information intuitively clear. First, the raw signal x is initially processed by some function (P), which typically transforms the input into an image, either range time or Doppler-time (img), from this a feature extraction algorithm (F), operates to extract relevant features (m), which are deemed to be useful to the overall problem, then the features are arranged in a feature space, as to allow the delineation function (K), to distinguish and hence make a decision regarding the class or category of the input data sequence. From this process, it is evident that there is no optimal or standardised process for performing data fusion in a general sense, as it is highly application specific, with the most tuned element being the feature extraction algorithm. This is the case as it has the highest information compression ratio, as the input image (consisting of a matrix of pixels) must be converted into a vector of useful values (m). The imaging and delineation stages are not as application dependent, as techniques and approaches can be borrowed from parallel implementations amongst other fields, though they still have a noticeable effect on the success of the overall system.

In the following section, the four types of data fusion methods occurring at different levels throughout a complete processing system will be discussed. This is demonstrated in a generalised fashion; however, it should be noted that statements made regarding data fusion methodologies will not be applicable for every possible area, as it is simply too far reaching. The intention is to give a general impression of the impact of the methodology in the context of a complete processing chain utilising a data fusion approach. Following this, potential implementation in a multistatic radar system from an engineer's perspective will be discussed, along with the challenges and considerations that may arise while doing so.

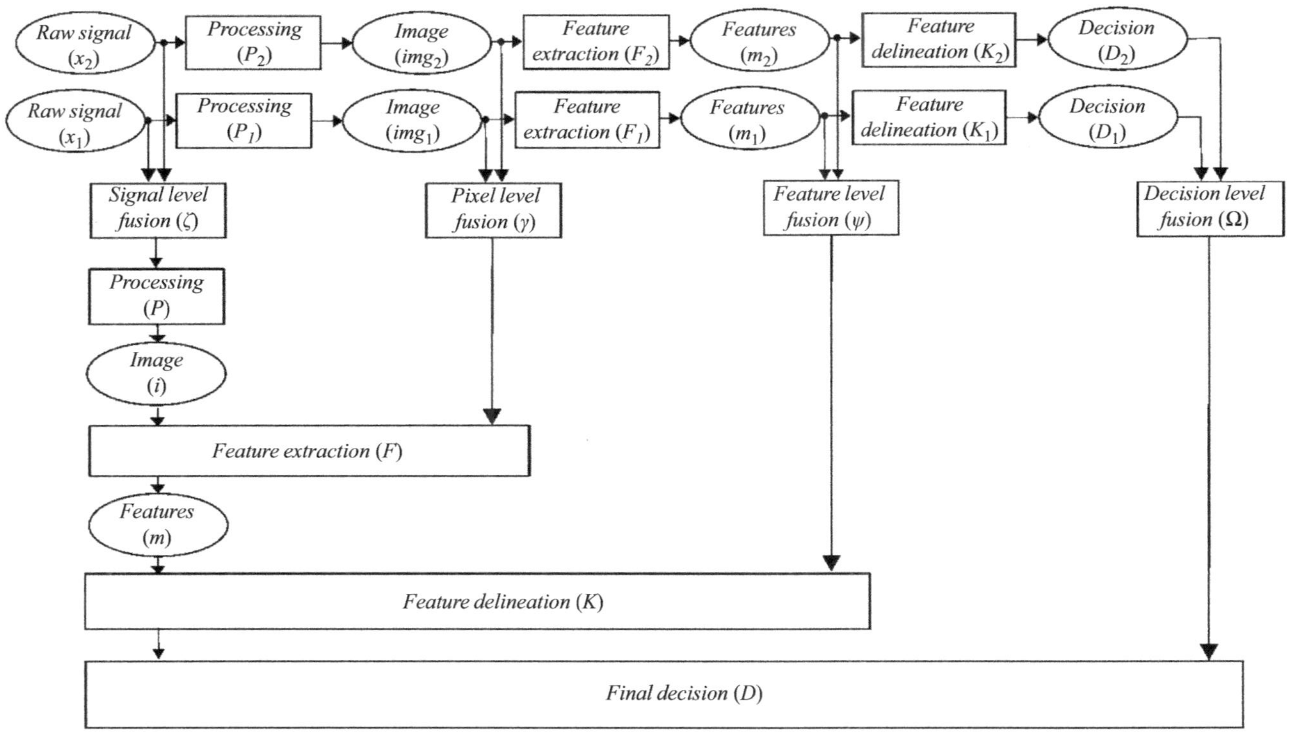

Figure 10.2 Block diagram of methods of performing data fusion throughout a generalised signal processing chain

In a multistatic system, the very opportunity of being able to fuse multiple streams of diverse input data is extremely powerful if implemented in the right way. To first approach this problem, the radar algorithm/system development engineer must identify suitable locations in which data streams can be fused, as to benefit the final decision. Initial identification will primarily be dictated by how much information the multistatic system can share through the projected or established data link. Once this is known, secondary determination can occur through evaluating a structure as in Figure 10.2, it is very important to recognise the notion that data/information rate decreases as the signal progresses through the fusion chain. Lastly, after the engineer has decided what signals to fuse, it must also be decided where this fusion process will take place; this is explored further in Section 10.3.3.

10.3.2.1 Signal level fusion

This entails combining raw data at the lowest possible signal level, occurring immediately after the data is first acquired by the sensing system. This method, for reasons that will become apparent, is the approach that often receives the least attention, as the fusion algorithms involved at this preliminary stage are typically well defined and quite rigid. In the simplest of forms, signal level fusion is the superposition of the input signals; at higher levels, it can be the combination of in phase and quadrature (IQ) components, up to the process of match filtering (widely utilised in radar) which correlates an unknown signal with a known signal, as to detect the presence of one signal in the other [20]. It is possible to perform these fusion techniques locally at the radar node, as the raw data need not be distributed.

A significant drawback to this low-level fusion scheme is that it demands maximum data throughput, as the raw data must be sent to a centralised node for data fusion, so that a final decision can be made. In the context of a multistatic radar, this presents a number of issues, first and foremost, the aforementioned problem of data sharing between nodes, which in a pulsed radar would at least be proportional to twice the chirp bandwidth; this is likely to be in the hundreds of megabits per second (MBPS) range. Related to this limitation; depending on the distance between the radar nodes, time delay and latency can quickly become an issue, for instance, if a target was detected in a multistatic system and the raw data was to be shared between radar sites, the time taken to transfer data would at minimum be in the order of tens of milliseconds (ms). Adding on processing overhead and integration into the central data pool, there may be approximately another tens of ms or more delay in the track update. If the multistatic radar was tracking an aircraft, this update delay would directly translate to an error in known position, and it is easy to see how this is a critical measure. Needless to say, it is not impossible to minimise this update delay to meet the requirement of the desired application; however, the message is evident, in that limitations in other technologies can be approached rather quickly, i.e. in networking and communication systems.

10.3.2.2 Pixel level fusion

After the raw data has been processed by the function P and it has been generated into an image of a suitable quality for later stages; pixel level fusion can now take place.

This is done on a pixel-by-pixel basis, where the pixel value is determined from a set of pixels of other images from concurrent data streams. Weighted combination is the simplest form of image fusion, whereby the weighted average is taken over a pool of images, hence producing the final image; this is expressed in (Equation 10.1) [21]; where n symbolises the number of input channels, w_n, depicts the weight and i and j are the pixel numbers in the two-dimensional (2D) coordinate plane. This weighted combination approach serves as a foundation for more complicated methods of pixel level fusion, as the selection of weights (w_n) in the summation can be based upon a range of measures, empirical constants can also be injected and tuned, all with the objective of improving the worth of the final image.

$$IMG_{out}(i,j) = \sum_{n=1}^{N} w_n \cdot IMG_n(i,j), \quad \sum_{n=1}^{N} w_n = 1 \tag{10.1}$$

Another approach to pixel level fusion is the Brovey transform (BT) which is an arithmetic combination, involving the construction of an image based on three RGB constituents [22]. The fused image is formed by multiplying the panchromatic (PAN) band by each RGB element [23], signified by c and dividing by the sum of all three RGB components, signified as b in the summations.

$$IMG_{fused}(i,j) = \sum_{c=1}^{C} \frac{IMG_{PAN}(i,j) \cdot IMG_c(i,j)}{\sum_{b=1}^{B} IMG_b(i,j)} \tag{10.2}$$

Pixel level fusion in multistatic radar may seem obvious, in that one could simply collate the images from each node to form the fused image; however it must not be forgotten that the characteristics of these images are as a function of aspect/bistatic angle and hence position. Therefore, simply combining these images would only destroy the spatial dependencies contained within each image, defeating a principal advantage of multistatic radar. The engineer must also place careful consideration in the fact that the image could be any combination of a function of time, frequency or any other variable relating to a real-world dimension. These circumstances originate from the nature of the radar acquisition that is processed in real time, as opposed to typical imaging problems which are usually a 'snapshot' of some scene. This detail inhibits the number of potential pixel fusion techniques that can be applied to the radar image. Of course, this does not entirely exclude multistatic radar from exploiting these methods, as these techniques may be applied to multistatic synthetic aperture radar (SAR) and inverse synthetic aperture (ISAR) systems, typically space-borne, which produce a high resolution 2D image of the measurement scene [24,25].

The engineer must be vigilant and creative in deriving solutions, bearing in mind this information. For example, in designing a multistatic fusion system, the designer may seek to adapt angular compensation and data warping techniques [26] applied to airborne bistatic radar and ISAR systems [27,28]. The objective of this compensation would be to decorrelate the spatial information of the multistatic configuration such that the data streams from each of the nodes (Figure 10.8) can be combined uniformly, enabling fusion at the pixel level. The employed technique would perform a real-time adaptive compensation, critically, being a function of the observed bistatic angle [29], which would vary as a function of the position of the target and would have to be estimated, potentially using a maximum likelihood estimator to determine the approximate angles of arrival [30] and hence the corresponding warping matrix.

Pixel fusion is straightforward in a monostatic radar, as physical geometries are no longer a factor, for example if the radar is polarimetrically diverse, the different polarisation combinations, horizontal (H-Pol) and vertical (V-Pol), can be fused together; this is a commonly employed technique in SAR imaging [31,32].

10.3.2.3 Feature level fusion

Fusion at this stage involves the combination of features or groups of features gathered from separate input data streams (x_1, x_2, $\ldots x_n$); the process of identifying and combining the relevant features is a critical undertaking, requiring careful planning. The feature extraction algorithm (F) is usually designed to extract a set of features from the input image, which are then grouped together and treated collectively as a class, then they are progressively fed into the classifier. The simplest method of feature level fusion is concatenating the input feature vector across the multiple data streams, producing a single vector (m). This may be perfectly adequate for systems where the feature distributions from the data streams are easily separable in the feature space; this is expressed through the Mahalanobis distance [33] and is a basis through which a classification algorithm such as principle component analysis would operate on. However, if the features are fundamentally derived from a data stream of the same nature, and the purpose of the multiple streams is to provide diversity to the system, then this basic approach will not suffice. If the features were treated as separate classes but still fed into the same classifier, this could actually harm classification performance as the Mahalanobis distance may be too small and cross-contamination between the feature distributions will occur, making distinguishing between classes very difficult.

Feature selection is a technique which is heavily researched in the field of data science, as it has substantial repercussions in classification throughout later stages in the processing system. It also requires a thorough understanding into the mechanics and the expected variability within data sets. However, it is clear through the methods employed that it is ideal to allow the algorithm to intelligently determine the optimum way to process the features to maximise the classification accuracy. The process of feature selection can be broken down into three dissimilar procedures for obtaining the optimal set of features; these procedures are determined by the way in which the metric of success in the system is measured.

Filter methods

In this procedure, the optimal set of features are chosen independent of the learning paradigm, that is to say the metric for success is the degree of class separability in the feature space, which may be observed through correlation, Euclidean distance or the Bhattacharyya distance [34]. A popular approach to performing this selection is the feature ranking method, where the features are expressed in the feature space and the distance between features are evaluated and hence ranked; from this they are prioritised accordingly. If a large feature set is used, where identification of harmful features is difficult, this method is quick to locate them, as the computational cost of this analysis is relatively inexpensive. The disadvantage to this is that, seemingly unusable features are quick to be rejected, reflected through the associated rank. Therefore, these may not have the full opportunity to explore correlation with the most valuable features already identified in the data set. It is important to note that as the selection of features is independent of the classifier, this method is more generalised as it would not be unusual to expect the determined feature set to be approximately optimal across different classifiers (K), as feature space separability is the means in which they operate upon in practice, regardless.

Wrapper methods

This approach to determining the feature set is particularly popular, if computational cost is not of significant concern and the classifier type is not expected to be changed during operation, or throughout the system lifetime. It is also more thorough compared to filter methods, taking a larger number of factors into consideration during the pursuit of the optimal feature set. The process is depicted in Figure 10.3, which is adapted from [35] and consists of a training set being injected into the feature selection search method (S) progressively; the algorithm may also have the ability to make intelligent decisions at its core; the revised feature set is then passed into the classifier (K), where the feature boundaries are delineated. Finally, an unseen test set is called, expressed by the determined feature set (m_s) and passed into the classifier, recording the classification decision (d^K) and over multiple samples, the overall classification accuracy. This may be recorded and passed into the learning algorithm (ι) to determine another suitable feature set for testing. The general idea of analysing the input data, to make an informed modification to the way the input data is handled by the processing system, is called ensemble learning [36].

If the feature set is sufficiently small, a brute force approach such as sequential forward search may be employed to determine the best feature set. The number of unique feature combinations is described by (Equation 10.3), where n represents the number of features and r signifies the points at which they are taken over (increasing to meet the total number of features n), this simplifies conveniently to a power of two,

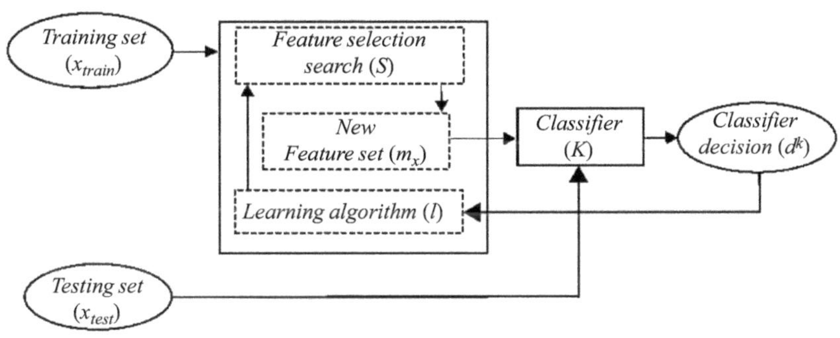

Figure 10.3 Feature selection and learning topology for wrapper methods

allowing practicability to be quickly evaluated. It is termed the curse of dimensionality, when the organisation of the high dimensional feature space becomes unfeasible [37].

$$C_n^r = \sum_{r=1}^{n} \frac{n!}{(n-r)!r!} = 2^n - 1 \tag{10.3}$$

Weighted feature fusion is an approach to combining different feature vectors, knowledge of the input data is utilised to determine the appropriate weights, and these are then applied on the unseen data stream in operation. It should be noted that for this to be effective, the number of samples available must be large, in order to give the determined weights statistical strength.

Embedded methods

In this approach, feature selection is performed during the training stages of the system and is specific to the classifier; the most popular example of this would be the decision tree or regression tree, if the responses are numeric. This learning model is intuitive to follow and mimic the human decision process quite well – it begins from the root node and operates in a top-down fashion. At each node a feature is chosen that best splits the set of test items; there are a number of bases for determining this metric, for example the Gini impurity, information entropy and in continuous cases variance reduction [38]. Again, these methods require large data sets to bring significance to the environment and generally struggle coping with diverse data and hence over fit the resultant model.

10.3.2.4 Decision level fusion

The final stage in the data fusion processing chain is at the decision level, and the decisions from multiple classifiers or diverse data streams are utilised to make an informed final decision (D). This stage is pivotal as the information entropy of the data is at its maximum and manipulation occurs at the highest level, in terms of data abstraction. It is possible for the decision at this stage to be broken down further, as depending on the classification process, a probabilistic confidence vector may be generated across the feature set; this is processed by (C_n) yielding the confidence in the class value (p^C); fusing this information is known as rank level fusion. When this is employed, the classifier decision (d^K) can also be fused, which is termed match level fusion. This is depicted in Figure 10.4, and these methods are discussed in detail later. It is important to note the circumstances in which these fusion methods can arise, and it is of course application dependant; in addition to this, fusion can extend in two dimensions and across common or diverse data streams, this is by no means constrained by that depicted in Figure 10.4.

Class compression is the notion of combining the decisions from a single classifier (K) to enhance and form the final decision (D). The premise of this undertaking is inherent to the nature of the class targets, what they represent and what the end-user would like to infer. This method of decision fusion can be used interchangeably and could vary during operation, as the user may wish to change what they would like to classify or what to determine from the environment. This functionality is most commonly decided when the processing system is in the initial design phase, as it is

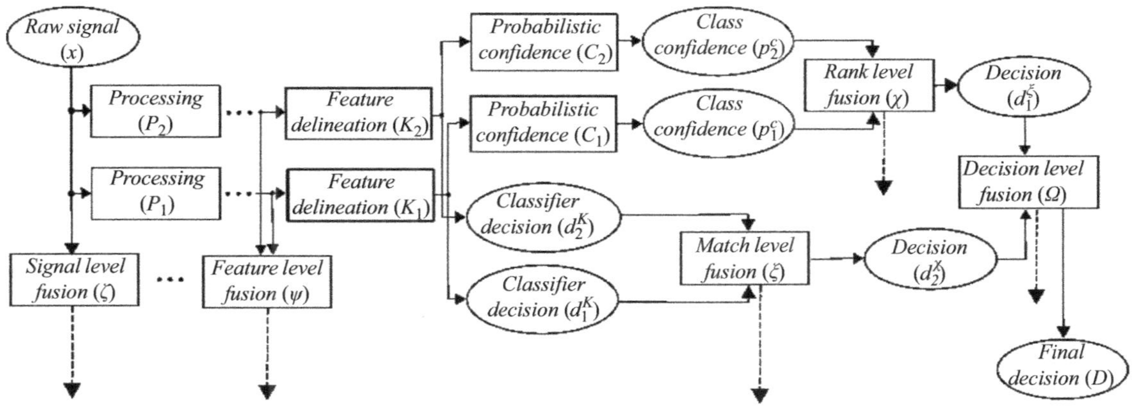

Figure 10.4 Decision level topology involving match and rank level fusion

difficult to integrate after the system has been deployed. Typical examples of decision level fusion may be majority voting, weighted voting or based upon statistical decision theory, where models are derived based on prior knowledge, which again is a form of information ensemble. This may be Bayesian estimation, Dempster–Shafer reasoning or even fuzzy logic [39].

Match level fusion

This form of fusion exists when multiple data streams, derived from the same raw signal input, or the same entity are exploited and hence combined. It should be noted that match level fusion can inherit any technique from decision level fusion, except anything which requires the probabilistic confidence to be known. A widely used implementation of match fusion is in biometrics, where an image of a face and a fingerprint, belonging to the same individual (class), may be independently classified and then having the classifier decisions (match scores) later combined, expressing the final class [40].

The simplest example of match level fusion would be majority voting; decisions made in each stream would be compared to the parallel decisions and the decision with the most votes would become the final result. Another example could leverage processing diversity, whereby the signal (x) is processed (P_n) in a multitude of ways, creating a new classification stream from a common input; in this case, it would be mandatory to employ match level fusion; otherwise, the parallel processing chains would be futile. This approach can significantly enhance the amount of information available and hence performance in the processing system. However, the complexity and the decision time of the system would be adversely affected.

Rank level fusion

This process considers the decision (d^K) and decomposes it using (C), as to produce the confidence in the class (p^C). It is possible for C to exist inside the classifier or to operate in tandem, either way the distinction of the class decision (d^K) and the confidence in the class is noteworthy. Rank level fusion typically involves statistical processing methods borrowed from decision fusion [16], as operation is performed on the probabilistic confidence vector which is composed of a set of real numbers ($p^C \in \mathrm{R}$). The simplest example would be a ranked voting system, as the decision with the most confidence would become the final outcome, this does however theoretically waste the information from the other data streams and is, therefore, not ideal; but there is surely a multitude of other approaches to this task.

10.3.3 Data fusion architectures

If the sensors or data fusion cores are spatially diverse, as they are in a multistatic radar, then the locations in which data fusion occurs is of great importance to the design engineer, as voiced in Section 10.3.1. Furthermore, the same logic can be applied to the entirety of the data fusion methods presented in Section 10.3.2, where the locations of fusion can vary throughout the progression of the information in the system and can simultaneously be shared across the deployed sites. There are three fundamental architectures that will be discussed, each of which will be reflected in detail, exploring their associated advantages and disadvantages, alongside how they can be applied to a prospective multistatic radar system. This is visually depicted in Figure 10.5, which has been adapted from [16], where circular blocks represent data acquisition nodes or sensors, arrows are the flow of data necessary for fusion, square blocks are fusion cores and hexagons are fully reconfigurable elements governed by some entity.

10.3.3.1 Centralised fusion

This architecture consists of multiple distributed sensors where data is sent to a single processing core and fusion solely takes place in this location, which is shown in Figure 10.5(a). This topology is particularly suited to sensors that operate on the same basis or in a similar fashion, with the desired outcome requiring all the data to be present before analysis or a decision can be made. This architecture demands high data throughput in the communication links and low latency is a pressing issue.

10.3.3.2 Distributed fusion

This setup is similar to the centralised fusion approach; however, data fusion takes place in two stages, first locally at each sensor node and then subsequently within the central processing core, yielding a collective final decision. This is useful for systems consisting of multiple diverse sensors, such as a smart structural monitoring,

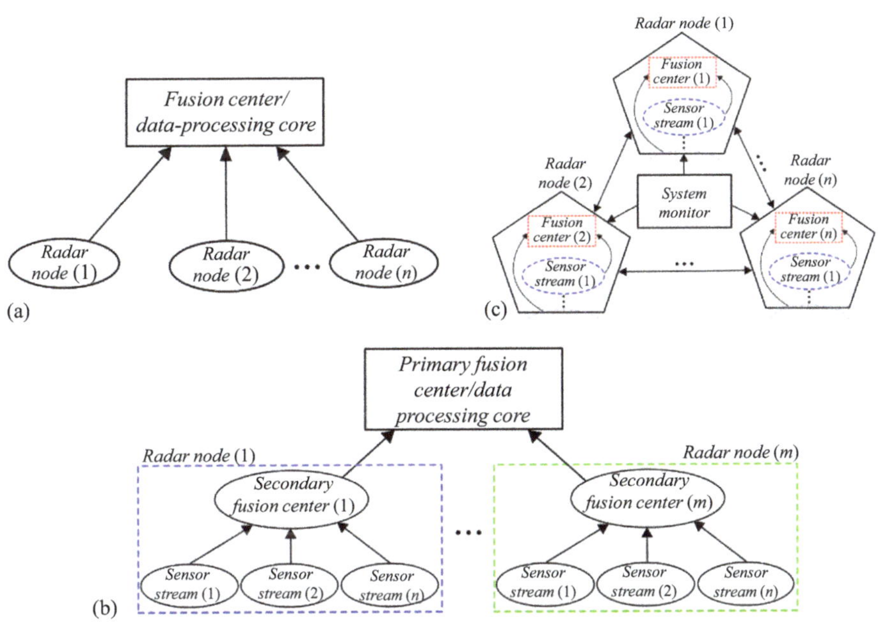

Figure 10.5 Data fusion architectures: (a) centralised fusion, (b) distributed fusion and (c) hybrid fusion

aircrafts and large ships, where it would be irrational to send the information from all the sensors individually to the processing core. This topology relaxes demand on the communication link between the secondary and primary fusion centre, as at this stage it could already perhaps be sharing a decision that would require minimal bandwidth and radio time.

10.3.3.3 Hybrid fusion

This configuration employs a mixture of elements from both of the schemes discussed previously and is primarily utilised in cases where the data throughput and computational workload which is feasible at each location throughout the system is carefully provisioned and of principle concern. In addition to this, during physical operation, the locations at which data fusion occurs are intelligently managed by a monitoring system that has knowledge of the current data rates and capacity of the established data links. Through this, it is able to intelligently reconfigure the system to balance real life sensor demands and data rates within the communication links, which also reflects on the power consumption of each node in the system. It is indeed possible for a multistatic system to require this incredibly demanding and complex topology, as the radar system may be exploiting mobile illuminators of opportunity or consisting of a network of mobile radar nodes.

10.3.4 Voting system implementations

Preliminary research into classifying human micro-Doppler signatures, using the multistatic radar system as described in Section 10.2.2 which employs a voting scheme, was first witnessed in [8]. In this work, involving unarmed and potentially armed targets operating in the field over three diverse aspect angles, the features from the resultant spectrograms were extracted using singular value decomposition (SVD), which is a method from linear algebra that performs matrix factorisation. It is a powerful extraction method as the characteristic shape of the spectrogram signature is respected, allowing parameters with physical meaning, such as target velocity, spectrum periodicity and spectrum bandwidth to be retained and inferred from the resultant SVD vectors. The SVD of a matrix (**M**) can be expressed as $\mathbf{M}=\mathbf{U}\mathbf{S}\mathbf{V}^T$, where **U** is the matrix containing the left singular vectors, **S** is a diagonal matrix with the singular values of **M** and $\mathbf{V}^T$ is a transposition of the right singular vectors (note that this expression is different for complex matrices of **M**). Features such as standard deviation, periodicity and amplitude, extracted from the valuable **U** and **V** singular vectors, were then fed into a single Naïve Bayes (NB) classifier, employing a virtual centralised fusion architecture requiring raw data to be shared with the monostatic node, serving as the localised fusion core. This is also a simple example of feature level fusion, as the feature sets from the three input streams are pooled and fed into an individual classifier. An alternate distributed fusion scheme was also implemented, where unique classifiers were disseminated and trained at each of the three radar nodes, with only the decision for each node being shared, allowing for decision level fusion to occur at the core processor. As this was a binary problem, a voting scheme was straightforward to implement on three radar nodes, with the majority decision being at least two out of three. The benefits of the two architectures are compared, with the metric for success being the classification accuracy, with the confusion matrices presented in Table 10.4.

The work undertaken in [3] employs a similar experimental setup to [8]; however, there are now eight aspect angles and a pair of empirical features (mean period and radar cross section (RCS) ratio), which are now used as inputs to the classifier, instead of the SVD-based features. Further to this, a voting with threshold system is employed as an incremental improvement to the binary voting scheme. The difference is that the level of confidence in the decision is utilised through function (C) and hence producing the probabilistic confidence value (p^C), as depicted in Figure 10.4; this is an initial demonstration of rank level fusion (χ) in multistatic radar. In case the two nodes agree on the same decision, the confidence is compared to the threshold; if it is greater, this becomes the final decision. If it is below the threshold and the confidence of the third node is above the threshold, this becomes the final decision. The threshold was experimentally determined to be 0.65, with this approach addressing the issue of two nodes having reached an incorrect decision due to the aspect angles being unfavourable relative to the other node. It should also be noted that the virtual configuration of the classifiers (hence fusion architectures) is the same as in [8]. The classification accuracy obtained using this additional method is detailed in Table 10.5.

In [41], a range of different classifiers were tested along with the fusion methods proposed in [3], these were diagonal linear discriminant analysis (DLDA), diagonal quadratic discriminant analysis (DQDA), NB, nearest neighbours of three samples (NN3) and five samples (NN5) and finally classification tree (CT). The features used were based on the centroid of the target, namely centroid Doppler and bandwidth; these were trained on 20% of the total available samples, with the results of this analysis presented in Table 10.6.

Table 10.4 Confusion matrices for the classification results obtained using the extracted SVD-derived features

Confusion matrix (%)		Monostatic data only		All multistatic data		Binary voting	
		Unarmed	**Armed**	**Unarmed**	**Armed**	**Unarmed**	**Armed**
Angle 1	Unarmed	95.1	4.9	82.4	17.6	94.4	5.6
	Armed	3.6	96.4	4.9	95.1	0.0	100
Angle 2	Unarmed	74.1	25.9	79.7	20.3	85.8	14.2
	Armed	9.1	90.9	7.7	92.3	4.8	95.2
Angle 3	Unarmed	80.8	19.2	77.3	22.7	89.6	10.4
	Armed	18.7	81.3	13.5	86.5	6.4	93.6

Source: Classification of unarmed/armed personnel using the NetRAD multistatic radar for micro-Doppler and singular value decomposition features [8]

Table 10.5 Classification accuracy as function of aspect angle for four different configurations of classifiers

Aspect angle to mono node α	Monostatic data only	All multistatic data	Binary voting	Voting with threshold
0°	85.39	84.52	88.00	88.22
30°	90.50	89.26	90.72	90.94
60°	79.44	84.22	90.22	90.22
120°	73.83	80.06	82.28	85.44
180°	81.83	89.61	89.17	90.61
210°	79.83	87.72	88.94	90.11
240°	90.67	90.89	97.50	97.44
300°	87.00	86.30	89.61	91.39
AVG	83.56	86.57	89.56	90.55

Source: Aspect angle dependence and multistatic data fusion for micro-Doppler classification of armed/unarmed personnel [3]

Table 10.6 Classification accuracy for different classifier types and data fusion topologies

Classifier type	Monostatic data only	All multistatic data	Binary voting	Voting with threshold
DLDA	91.6	76.7	91.1	92.2
DQDA	88.6	75.8	89.7	90.3
NB	84.1	75.3	84.9	85.3
NN3	89.8	76.9	87.8	88.0
NN5	89.0	75.6	86.2	88.3
CT	85.5	77.7	86.7	86.7

Source: Performance analysis of centroid and SVD features for personnel recognition using multistatic micro-Doppler [41]

10.3.5 Weighted ensemble

The work in [10] employs the experimental setup introduced in Section 10.2.1, utilising a weighted ensemble approach on the decisions from each of the classifiers. The preliminary stage of this work utilises a feed forward network (FFN), with seven empirical features obtained using a bespoke feature extraction algorithm; these were used as inputs to the network that was trained on 60% of the available data. The difference in this particular classification framework is that FFNs were trained at each node and for each square area in the field (zone). This was done to overcome the spatial dependency the spectrogram has on the recorded position. Each NN was allocated a unique weight that is related to the performance of the individual network in response to the pre-allocated validation data (20%). The weighting matrix

($\mathbf{W}_{N,Z}$) is derived through directly mapping the result of this analysis, ($\mathbf{Y}_{N,Z}$) onto an exponential function; this is given by the following equation:

$$\mathbf{W_{N,Z}} = \frac{\alpha \cdot \mathbf{Y_{N,Z}}}{100.1 - \mathbf{Y_{N,Z}}}, \quad \alpha = 3 \tag{10.4}$$

Mapping the accuracy onto a non-linear function allows NNs which achieved high-classification accuracies in the initial testing stages to be favoured over ones that did not. The value of alpha allows the steepness of the curve to be tuned and for the value chosen, accuracies above the 95% region are favoured, where the radar nodes were often in contention. This value was determined coarsely by sweeping the parameter and observing the highest achieved accuracy; this should ideally be done over a much larger data set to give it statistical significance and relevance in a real environment. As this work is based around a binary problem, the predictions ($\mathbf{P}_{N,Z}$) were given positive and negative notations based on the armed or unarmed case. This allows decision of a particular node to constructively or destructively interfere with the decisions; the final decision (***D***) can be found by simply taking the dot product of $\mathbf{P}_{N,Z}$ with the weighting matrix $\mathbf{W}_{N,Z}$ over the three radar nodes.

10.3.6 Match and rank level fusion implementations

The work in [9], uses the experimental setup as introduced in Section 10.2.1, however the target is a drone instead of a human and is operating 60 m away from the monostatic node. Three different time-frequency representations of the target are generated: spectrogram, a result of the short-time Fourier transform (STFT); the cepstrogram [42] and the cadence velocity diagram (CVD). Their relationship to each other is stated in discrete time in (10.5), (10.6) and (10.7), respectively. In these equations, w is the window function, n is the discrete time sample, f is the frequency bin, m is the short-time bin, f_s is the sampling frequency, N is the total number of samples, q is the unique 'quefrency' variable and f_{cad} is the cadence frequency. The intensity plots of the transformations are then passed into a single DNN based on the AlexNet architecture [43], which is trained to treat each transformation as a separate class, as shown in Figure 10.6. Finally, the match score is generated for each representation stream, and match fusion occurs combining the information from these three domains. It is also possible for class compression to take place, as in this case, there are six classes with increasing masses of payloads; if the end-user is only interested as to whether the drone is carrying any payload, five of the classes can be compressed into a single payload class. This can potentially raise the classification accuracy, as the DNN has been trained to detect progressive variations (in the form of individual classes) rather than a generalised signature of whether the drone is carrying a payload or not (two distinct classes).

$$STFT[f, m] = \sum_{n=0}^{N-1} x[n] \cdot w[n-m] \cdot e^{-j(2\pi nk/N)}, \quad \text{where } k = \frac{Nf}{f_s} \tag{10.5}$$

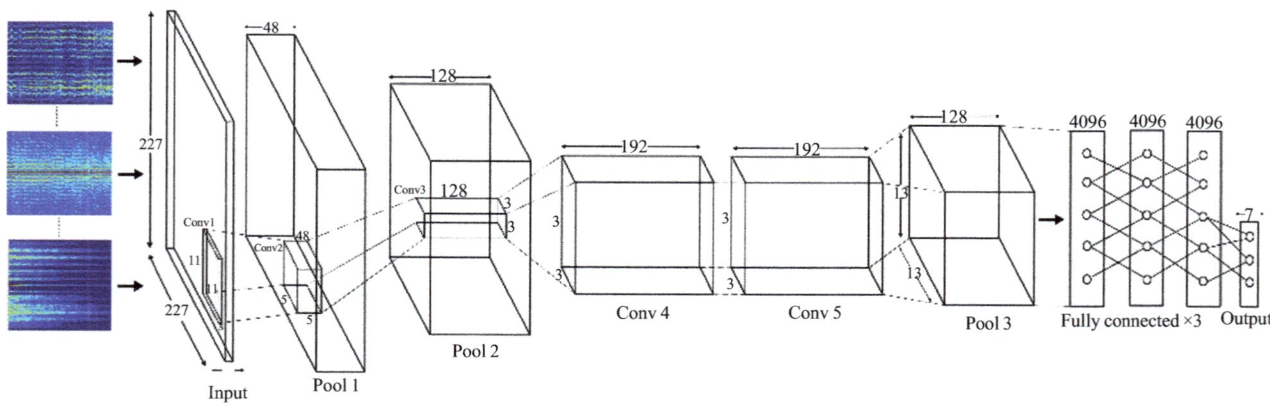

Figure 10.6 AlexNet architecture with multi-transformation inputs

$$Ceps[q,n] = \left| \frac{1}{N} \sum_{n=0}^{N-1} \left\{ \log \left(|STFT[f,n]|^2 + 1 \right) \right\} \cdot e^{j(2\pi nk/N)} \right|^2 \tag{10.6}$$

$$CVD[f, f_{cad}] = \left| \sum_{n=0}^{N-1} STFT[f,n] \cdot e^{-j(2\pi n k_{cad}/N)} \right| \tag{10.7}$$

Research efforts have recently been directed into how to best fuse information and decisions from multiple radar nodes in an efficient manner from a data throughput conscious standpoint [10]. This work utilises the experimental setup introduced in Section 10.2.2 and explores seven combinations of walking with a metallic pole, simulating a rifle, while also carrying a rucksack. The spectrogram samples were processed in much the same way as [9] (Equation (10.5)) and were split into 10 slices (1.0 s each) yielding a total of 8,400 spectrogram samples across the whole system. To provide the DNN with a further representation of the input sample, the CVD was produced by taking the FFT across each successive Doppler bin (Equation (10.7)). The cepstrogram transformation that reveals log-frequency change over time was investigated, but the plots obtained were found to be unfavourable for classification; therefore, it was held out (Equation (10.6)).

Multiple fusion configurations were tested with an array of additional fusion methods taking place in the process. The high level fusion structure is depicted in Figure 10.7, where the first configuration is a voting system positioned at the rank

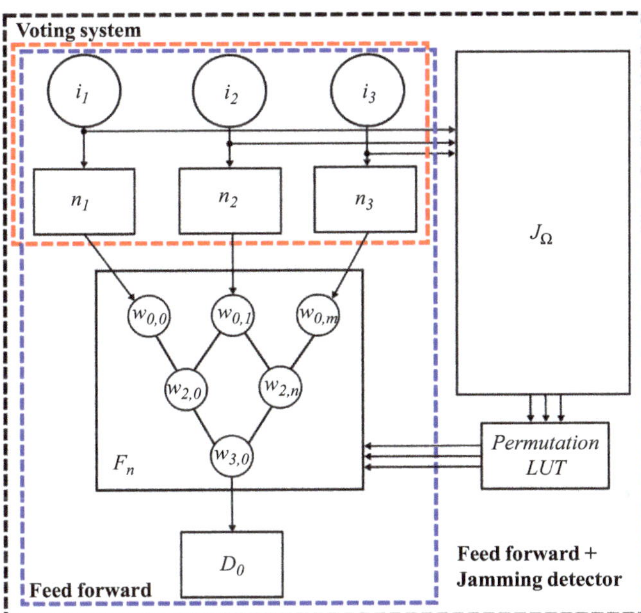

Figure 10.7 Diagram of progressive network configurations implemented in [10]

level fusion stage and the decisions with the highest confidence made at the DNN of each radar node are of precedence. The second utilises an FFN trained on the validation data reserved for the training phases of the DNN. The third utilises a further DNN serving as a jamming detector (J_Ω) for the input images; it is mentioned in this section as it leverages match level fusion and class compression, specifically in (J_Ω) and produces a noteworthy result. These three approaches to fusing the decisions from multiple radar nodes are expanded further and discussed next, with more details given to the auxiliary fusion processes involved.

10.3.6.1 Rank level voting system

The simplest network architecture from [10] is depicted in the red dashed box as seen Figure 10.7. It consists of three independently trained DNNs (n_x), where input image pools are represented by (i_x) and are delivered concurrently to the respective networks. As each network has an output of 14 classes and the CVD is a direct transformation of the spectrogram, this means that seven classes have a corresponding pair. Therefore, at each network an internal vote is made between the two transformations as to which one is the most confident in their decision; this is an example of rank level fusion. The intermediate prediction (aggregate) is composed of the most confident prediction of each network for each input test image. Finally, the predictions from the three nodes are then compared and a similar ranked fusion procedure takes place, whereby the final decision is dominated by the most confident radar node.

10.3.6.2 Rank level FFN fusion system

A method of combining the decisions from each of the three radar nodes was determined, whereby a FFN (F_n) consisting of 21 hidden units was placed between the probabilistic output of each of the three DNNs and funnelled back to the original classes and hence the final decision (D_0). Each of the DNNs have 14 outputs, using the same ranked voting process as before; this is compressed down to 7 and then aggregated to form a 21-class output. This is then input to the FFN which is trained on the unused 10% validation data pool; from this the appropriate internal weights ($w_{m,n}$) are systematically determined. The FFN configuration is portrayed by the blue dashed box in Figure 10.7.

10.3.6.3 Match fusion and class compression in J_Ω

The final network configuration consists of an amalgamation of elements from the two previous designs, but in addition to this, there is a further network (J_Ω) running in parallel, which can detect the presence of jamming in any of the three radar nodes (introduced in detail in Section 10.5). Six types of synthetic jamming techniques were injected directly into the range time data for each of the three discrete jamming levels, leading to 18 jamming classes. This included the spectrogram and CVD versions which lead to total of 38 classes and includes the baseline transformations which have no jamming applied (2 classes). The jamming detector achieved an accuracy of 86% over the testing set, which consisted of 127,680 images ($19\times2\times8{,}400\times40\%$), and after match level fusion was performed, an accuracy of 95.5% was attained. This achieved quite a substantial performance gain of approximately 10%, considering the large number of samples in the testing pool.

Due to the nature of the scenario it was necessary to leverage class compression in order to distinguish between the three discrete jamming levels. This was possible as the knowledge of the jamming type was of irrelevance, and it was more important to determine the degree in which the system is being jammed, i.e. the severity. Therefore, within the permutation look up table (LUT), the pseudo-classes are converted to the discrete jamming level classes, offering a performance gain that can be seen unintentionally through Figure 10.12(b), where the permutations from 1 to 16 are hampered by cases of nil jamming, which J_Ω was only 89% successful in classifying. Further details regarding what this graph represents in this scenario can be found in Section 10.5.2. In a practical setup, such a case of class imbalance should ideally be avoided; however, the example given before demonstrates the performance gains which can be acquired from the class compression technique.

10.3.7 Summary

Fusion of the information from radar nodes in a multistatic radar network is a chief undertaking for the system design engineer, as there is a plethora of techniques, methods and configurations, which must be considered before the processing system is deployed. This section has presented a detailed exploration into data fusion methods and architectures, while also discussing the relevance of this information in multistatic radar. Following from this, implementations of these techniques applied to a real multistatic radar system have been introduced, presenting the findings from various experimental campaigns.

The voting system was first applied to multistatic radar data at the decision level in [8] and employed a virtual centralised and distributed fusion architecture. It revealed that the centralised scheme was unideal for multistatic radar as reflected in the results; however, the distributed fusion scheme with binary voting occurring at the fusion centre proved that superior results could be obtained. Further to this in [3], a threshold was introduced into the voting system process, which was able to slightly enhance the results through utilising match level fusion. In [41], a range of different classifiers were tested and confirmed the operating topologies stated before. In [10], a weighted ensemble was utilised in order to generate a predefined weighting matrix, related to the success of a classifier in each zone; this work made the issue of spatial dependency in classification systems especially clear. Finally, implementations of decision level fusion, specifically match and rank, were presented [9,10]. These works highlighted the impact that careful data fusion design can have on a complete processing system, where a variety of methods were applied from the fusion chain, as depicted in Figures 10.2 and 10.4. In particular decision level methods, whereas prior works [3,8,41] employed signal level fusion (matched filter), feature level fusion and elementary decision methods.

The optimal data fusion architecture in a multistatic system is clearly a distributed fusion scheme, as a wide variety of decision fusion methods become available to the user; but most importantly, it can fully support low bandwidth data communication between nodes, as only probabilistic vectors or final decisions need to be shared. This consideration also benefits covert PBRs, where they may only wish to communicate in short bursts of packets, rather than continuous streams, which could compromise the integrity of the receivers' location.

Future work would involve thorough investigations regarding the implementation of hybrid fusion configurations in multistatic systems; where the radar might contain a node which is mobile, in which case it would benefit from being fully reconfigurable during operation, as the requirements of the system change with time amongst many other environmental variables. In addition to this, further research must be undertaken regarding fusion techniques, perhaps at the pixel level and at the feature level, as this is currently underutilised in the data fusion processing chain and could potentially benefit from processing diversity.

10.4 Deep neural network implementations

Deep learning applied to multistatic radar is an emerging area of research and brings a number of additional challenges over implementing a DNN in a traditional monostatic radar. Specifically, with regards to data sharing, this means that the DNN must be distributed and operate independently at each radar node, unless there is an established high bandwidth data communication link and usage of the data link is sanctioned. In the following section, a number of implementations of DNNs applied to the experimental setups in Section 10.2 will be analysed and discussed. The first presents the application of stacked auto-encoders (AEs) (SAEs) in a data set where the target operates within a number of different spatial locations (zones). The second explores details surrounding the employment of AlexNet [43], in classifying a micro-drone carrying increasing payloads. The last, discusses the coexistence of three DNNs operating together to achieve classification, along with a parallel over watch jamming detector, detailed further in Section 10.5.2.3.

10.4.1 Spatially diverse human micro-Doppler

As introduced in Section 10.2, the work in [10] utilises a SAE deployed at each node and each zone. An appropriate compromise was made between processing feasibility and quality of results; a good candidate to meet these needs was the AE, these have been utilised extensively in many image analysis and classification problems [44,45] but have eventually been replaced by more complex algorithms in recent years. Each individual SAE was composed of four auto-encoding layers, employing a sigmoid activation function, followed by a soft-max regression layer which together forms the DNN. The number of neurons contained within each of the four hidden layers consisted of 1,000, 500, 100 and 20, respectively; finally, the soft-max layer is reached consisting of two elements. The AEs were trained using a scaled conjugate weighted decent algorithm, with the cost function derived from a combination of the mean square error and the sparsity. The first layer ran for 30 epochs and was then frozen, with the subsequent layers running for a further 100 epochs. The input is a 1950 element matrix which is a spectrogram of duration 5 s split into four slices. As it propagates through the network the dimensionality of the spectrogram is reduced until it ultimately becomes a two-element matrix representing the case of either armed or unarmed.

The results show high classification accuracies obtained for this simple binary problem and slight performance gains from the weighted decision fusion process. However, the experiment would have benefited further from a larger data set, giving statistical power to the problem, allowing sturdier conclusions to be drawn. Nonetheless, an interesting scenario was tested whereby all the data, from all the zones and nodes, was pooled together and a large DNN was trained on 20% of the available samples; this achieved an 83.3% classification accuracy. In addition to this, the work details the results of the feature set applied to an FFN. An example of the obtained log-normalised range time, Doppler time and CVD plots for a human performing the same activity is shown in Figure 10.8, across each of the radar nodes.

10.4.2 Micro-drone payload classification

The work in [9], as introduced in Sections 10.1 and 10.2, dealt with two scenarios involving a micro-drone which was either flying or hovering, while simultaneously carrying payloads of increasing mass from 0 to 600 g. The hovering data set consisted of 6, 30 s long recordings for each node (6×3), cut into 5-s windows and split at 0 Hz (216 in total). The class imbalance had to be rectified, so this resulted in the no payload class increasing in size to 180 samples through the additive white Gaussian noise augmentation process, to equal the payload class (total 360). This was then split into the data groups, so 432 samples for training (after the Y translation), 36 samples for validation and 108 for testing. It should be noted that this is the number of samples for each transformation, and all three are being applied to the CNN. This was repeated five times with the mean accuracy given for all. An average accuracy of 95.1% was achieved, 97.2% for the spectrograms, 93.5% for the cepstrograms and 94.4% for the CVDs.

The flying data set consisted of 5 × 30 s recordings cut and split in the same way resulting in 180 samples. The class imbalance was (1:4) so the no payload class increased to 144 samples (total 288). This was then split into data groups, 344 samples for training (after translation), 28 for validation and 88 for testing. The testing procedure is repeated as detailed before. An accuracy of 96.6% was achieved, 97.7% for the spectrograms, 100% for the cepstrograms and 92.0% for the CVDs. A crucial transfer learning test was also performed by applying the flying test data to the hovering CNN and vice versa, an average accuracy of 78% was achieved for both situations, indicating that neither network grossly over fitted to their associated training set. It should be noted that the multiple domain representations of the micro-drone for the six different classes were visually challenging and could barely be distinguished by the human eye. The DNN was able to identify extremely subtle variations in the rotor blade modulation from the drone and hence classify between the differing payload classes. An example of log-normalised Doppler time, quefrency time and CVD plots for a drone hovering and flying with no payload is shown in Figure 10.9.

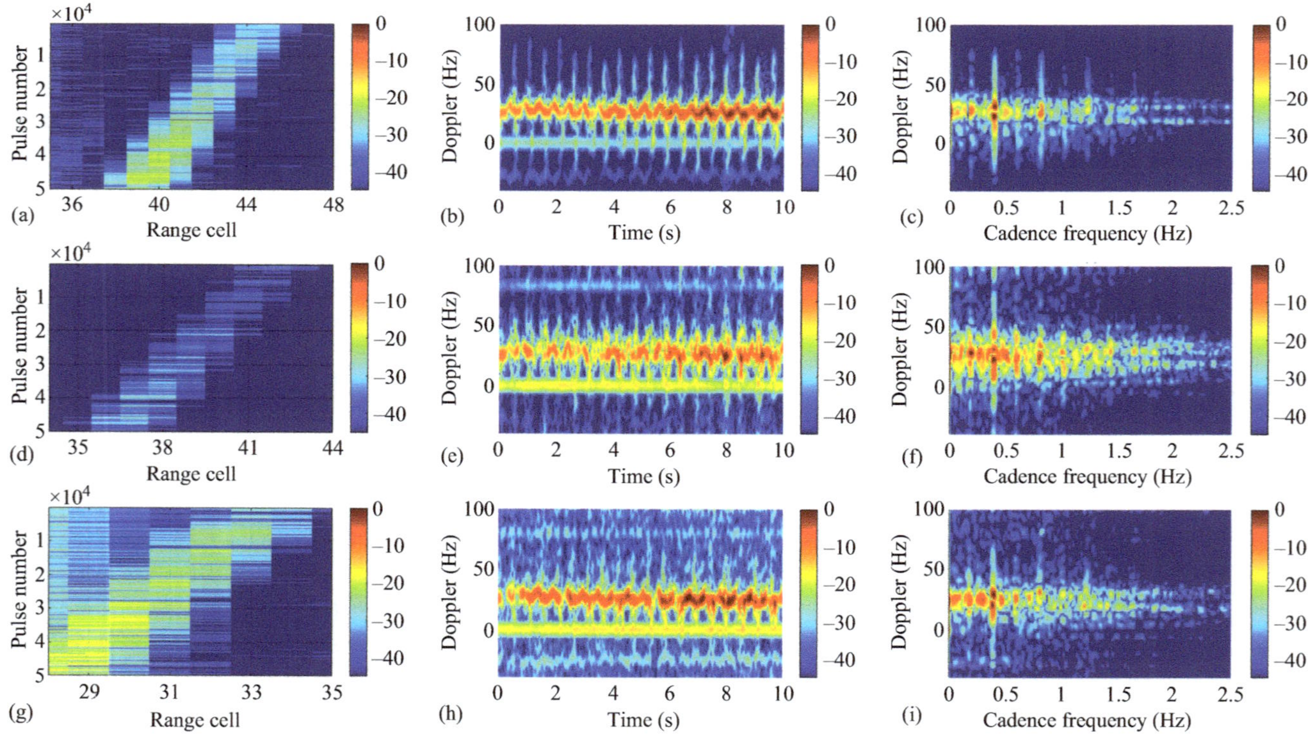

Figure 10.8 Radar imagery of a human target as perceived by each radar node (1,2,3) presented in each row, respectively, ordered as range time (a), (d) and (g); Doppler time (b), (e) and (h); and CVD (c), (f) and (i)

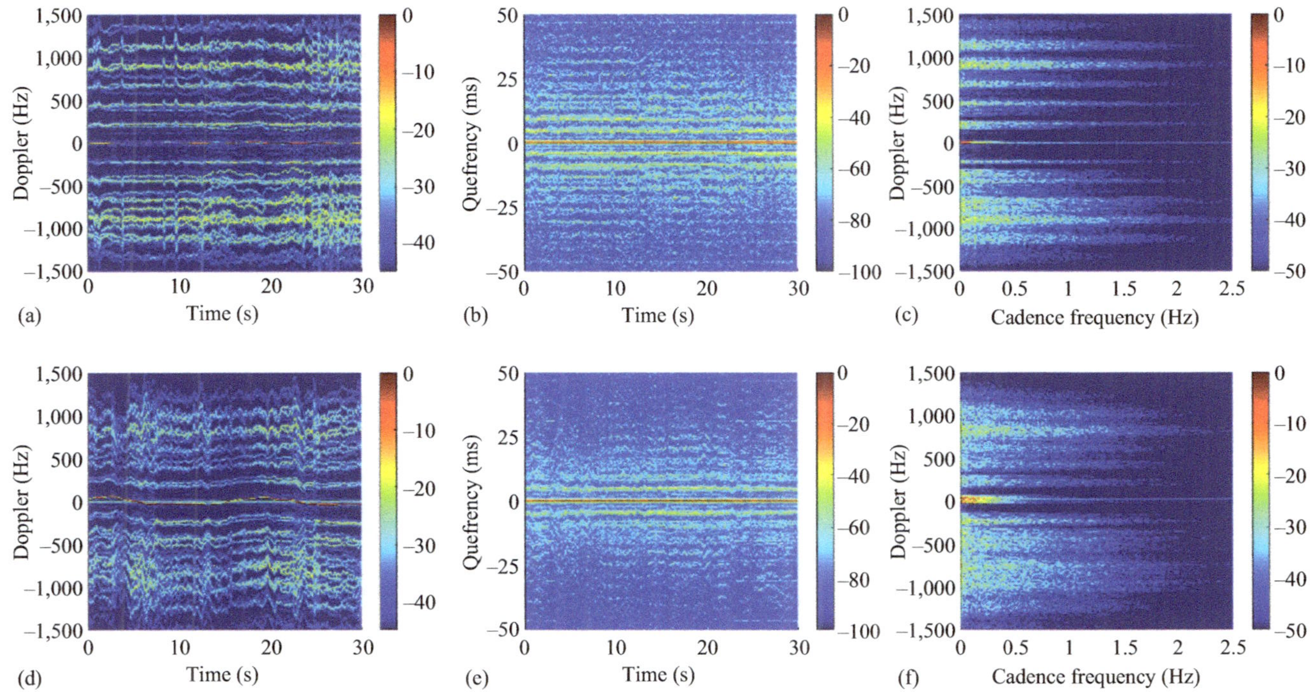

Figure 10.9 Radar imagery of a drone target, hovering in the first row and flying 30 m towards Node 1 in the second row; ordered as Doppler time (a) and (d), quefrency time (cepstrogram) (b) and (e) and CVD (c) and (f)

10.4.3 Presence of jamming in human micro-Doppler

In [10], the deep learning architecture chosen for implementation was again based upon the pre-trained model AlexNet. This was chosen as it significantly outperforms other models, such as GoogleNet, Resnet50, Resnet101 and InceptionResnetV2, in a multitude of aspects such as convergence time, network simplicity and training time (for this scenario). DNNs were trained at each radar node meaning that a total of 2,800 samples were available and from this, the samples were split into 60% training, 10% validation and 30% test pools. The training data was augmented in real time and consisted of mirroring the spectrograms and CVDs in the X and Y planes; any other combination of rotations, shearing, scaling and translations only hindered the obtained classification accuracy. In the context of this work, tuning the specific architecture of the deep learning model was not a primary focus as it would have led to significantly longer development times; accepting this enabled creativity regarding combinations of these networks, applicable to implementation in such multistatic systems.

The learning rate for each DNN was set to 1×10^{-4}, the mini-batch size to 16 and the maximum epochs to 45, this led to a training time of approximately 3 h each. As mentioned in Section 10.3.6, there are three experimental scenarios that are investigated in this work; the results of two of which will be presented, with the results from the jamming mitigation system detailed in Section 10.5.2.3.

10.4.3.1 Rank level voting system results

The accuracy achieved using the unjammed test data and the voting topology is detailed in Table 10.7. The average accuracy gained from the intermediate voting stage over the simple aggregate average was 6.2%. Finally, an average of 14.3% is gained utilising the final voting stage. The confusion matrix for the final vote is shown in Figure 10.10(a), with the class accuracy having a standard deviation of 8.42.

10.4.3.2 Rank level FFN fusion system results

The FFN obtains the same intermediate classification accuracies as in Table 10.7, except for the final decision, where 90.7% is obtained. The confusion matrix for this test is shown in Figure 10.10(b) and an improvement of 3.9% is demonstrated over the voting system methodology. The accuracy across the classes appears to be more stable

Table 10.7 Voting system classification accuracy

Scenario (%)	Node 1	Node 2	Node 3
Average	73.3	75.2	68.9
Spectrogram	78.1	79.1	75.6
CVD	68.5	71.3	61.8
Internal vote	79.2	80.7	76.1
Final vote		86.8	

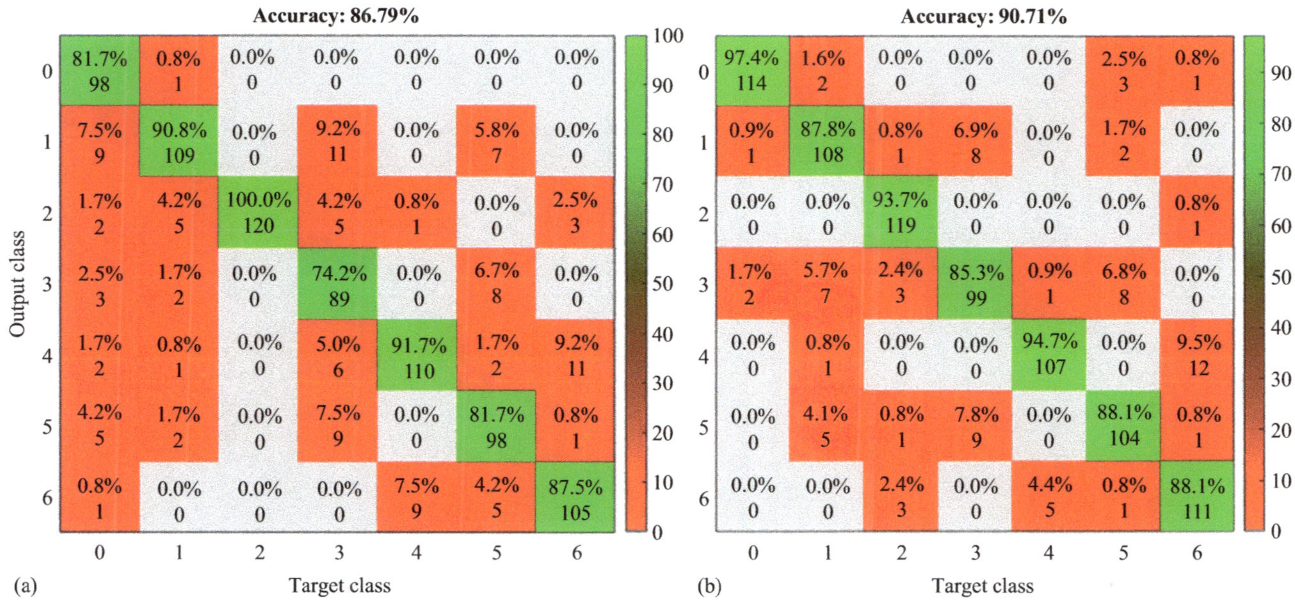

Figure 10.10 Confusion matrix for the: (a) ranked system and (b) FFN system

consistently staying over 85% and this is confirmed through the standard deviation of 4.49.

10.4.4 Summary

This section has presented and discussed implementations of DNNs, applied to the emerging area of classification in multistatic radar networks through three publications, all based on classifying data gathered from the pulsed radar system NetRAD. The first [10] utilised AE's compiled and stacked together to form an SAE, which proved to be quite a robust classifier for this simple binary problem. The second [9] exploited more advanced pre-trained models based on AlexNet, which achieved impressive classification accuracies, considering the visual complexity of the multi-domain representations. The third again used AlexNet to classify between seven cases of a human target holding a simulated rifle and a rucksack, a similar network configuration was employed in [10]; however, more consideration was given to the fusion of the decisions from internal and external stages. From this it was discovered that it is better not to pool together samples across diverse sensors, which would have the effect of blurring the classes within the DNN rather than combining and compressing the decisions down throughout later stages. The disadvantage to this, however, is the increased complexity of the DNN and the surrounding decision-making mechanisms specifically, the generalisation capabilities of the DNN to particular cases, for example in [9] payload or no payload. The end-user will also be prevented from analysing neural activations, which reveal the difference between these two cases, as the DNN will have been taught to distinguish features at each step, rather than the individual two cases.

A wide range of options and parameters was explored in order to improve the classification accuracy of the collective DNNs. The pre-trained DNN AlexNet was found to be superior in terms of classification accuracy, complexity and training time; in addition to this, only X–Y reflections were found to be beneficial when augmenting the training data, which is logical, as the transformations have strong kinematic meaning, varying as a function of time. Differing dwell times of (0.5, 1 and 2.5 s) were also briefly investigated, and a dwell time of 1 s was found to be the most suitable for this specific classification problem. This represents a challenging trade-off between generating a sufficiently sized data set for the DNN and also that each image contains enough visual cues for classification to occur. A further frequency transformation called the cepstrogram [42], which emphasises periodic details in the spectrogram [9], was also experimented with. This was found to not contain enough information as to warrant its implementation as a useful transformation in the scope of this classification problem.

As these classification works are in the initial phases, there is great scope for future work to be undertaken. The most distinguished outcome in any classification driven problem is to obtain higher accuracies. This could potentially be achieved in a multitude of ways: to design and train a custom DNN architecture, enabling the weights for each layer to be determined from scratch, rather than predefined matrices as it is using AlexNet. The training time will be significantly longer, but the advantages to be gained may be favourable. Impressive results were observed in the case of the jamming detector (Section 10.3.6.3), where the training sets were significantly larger and it was trained over a long period. An inquiry into the generation of synthetic data sets may also be of interest, such as the use of generative adversarial networks (GANs), where their use has already been proven in the context of human micro-Doppler signatures [46].

10.5 Jamming effects in multistatic radar

A principal advantage of multistatic radar over typical monostatic systems is the ability to continue to operate under the effects of enemy jamming targeting a specific radar node, as there are multiple sources of information. Though it is not impossible for a monostatic system to mitigate or tolerate the effects of jamming, as the receiver bandwidth, operating frequency or polarisation [1,47,48] of the system could perhaps be interchanged, upon recognition of the tell-tale effects. However, there is indeed a strong argument that a multistatic system could handle this disruption with improved elegance and better efficiency, as opposed to its counterpart.

This section presents research efforts focused into addressing the presence of these effects and showcasing the aforementioned idea that a multistatic system is better equipped to deal with these effects. The first inspects the immediate effects that spectrogram degradation can have on classification performance [15]. The second synthetically injects a diverse array of jamming measures into the range time data, observing the effects and presenting a DNN topology to actively detect and compensate for the presence of this targeted effect, with the principle aim to maintain classification performance.

10.5.1 Spectrogram degradation

The work in [15] explored classification performance with degraded radar data as a function of the manipulated number of pulses in the range time domain. This utilised a feature extraction algorithm that exploited eight empirical and statistical features to perform classification, and as expected accuracy is hindered as the severity of degradation is increased. This work also explored the variation in accuracy using different classifiers, namely NB, k-nearest neighbours and SVM. The SVM classifier leveraging multistatic data was found to perform approximately 8% better than the equivalent classifier using only monostatic data.

10.5.2 Jamming resilient systems

The ability for a radar system to be resilient to jamming effects is incredibly crucial, as very often lives will be dependent on the continued uninterrupted operation of the radar system. This section will be broken down into three parts: potential jamming techniques that are sensible to be applied to the multistatic radar system previously described throughout this chapter; the design and integration of a jamming detection system (J_Ω) based on a DNN architecture; and the results and analysis of this implementation, tested with the same data set as discussed in Section 10.3.6.3 [10].

10.5.2.1 Synthetic jamming techniques

To simulate potential scenarios of jamming, a methodology was devised whereby commonly utilised and well understood electronic attack (EA) methods were borrowed from the field of electronic warfare (EW) [46,48] and adapted to suit the experimental setup, though it should be noted that these techniques are by no means restricted or exclusive to the experimental radar system. A summary of the implemented techniques and their corresponding class identifiers are detailed in Table 10.8. In each of the listed techniques, three levels of jamming severity were employed; this is justified in Section 10.5.2.3. There are 6 jamming implementations, 3 levels (18) and a further class (ς_0) to represent an uncorrupted case (19) and two transformations types, leading to a total of 38 jamming classes. The jamming to signal ratio (JSR) was determined for each step and was evaluated based on rearranging (10.8), which ensures that the values are feasible for a realistic jammer; this represents a self-protection jammer (SPJ).

$$\overline{P_J} = \frac{J_{eff} J_{SR} \tau f_{PRF} P^R_{pk} G^2_R B^J_W G_{pc}}{G^2_J B^R_W} \cdot \frac{\sigma}{4\pi R^2} \qquad (10.8)$$

Table 10.8 Combinations of jamming types, techniques and a description of the method of implementation both virtually and physically

Jamming		Description of implementation and practical deployment
Type	**Technique**	
Escort	Targeted pulse compression (ς_1–ς_3)	High level of jamming whereby the transmitted waveform is learnt and it retransmits modifying or removing or the initial frequency modulation – *a dedicated vehicle retrofitted for electronic attack*
Self-protection	Spot (ς_4–ς_6)	Random noise injected throughout the range cells that the target is operating within – *a simple dirty handheld jammer*
Stand off	Cover (ς_7–ς_9)	Noise superimposed onto random frequency modulations – *dedicated kit that another soldier could carry*
	Blinking (ς_{10}–ς_{12})	Similar to spot jamming, however, jamming is alternated at 2 Hz; in this scenario there is no second target in the field – *dedicated kit to provide coordinated jamming*
Stand in	Unintentional /inadvertent (ς_{13}–ς_{15})	Low level of jamming, whereby random bursts of noise are observed extending across the entire processing range of radar – *communications within the area of detection are inadvertently in operation during the time of recording (such as Wi-Fi)*
	Deceptive (ς_{16}–ς_{18})	Very high level of jamming, with similar methodology to targeted pulse compression; however, the position of the target is constantly shuffled between range bin extents on a pulse to pulse basis. Consequently, the CFAR detector output becomes erratic – *an advanced device inconspicuously deployed close to the radar, providing intelligent directed jamming into that node*

where P_J represents the average transmit power, G_J the antenna gain and B_W^J the bandwidth, collectively for the jamming device. B_W^R is the bandwidth, τ is pulse length, p_{pk}^R is the peak transmit power, f_{PRF} is the PRF, G_{pc} is the pulse compression gain, G_R is the antenna gain, R, is range and σ is the RCS, all of these collectively applying the specifications of the pulsed radar system Table 10.2. For the three levels of jamming, the average transmit power calculated using (Equation 10.8) was 63 μW, 630 μW and 6.3 mW, respectively; using a jamming antenna with a reasonable gain of 10 dBi, a target RCS of 0 dBsm [1] and at a range of 90 m. This was obtained under the following assumptions: the noise temperature, receiver bandwidths and combined losses of the radar and jamming device are equal and that the jamming efficiency (J_{eff}) is 100%, meaning the device is jamming over the full radar bandwidth. The calculated average power levels in order to generate the desired JSR's are feasible for any of the jamming devices/systems listed in Table 10.8. This quick analysis delivers an understanding of the roles each radar parameter plays in the context of the experimental setup.

The jamming signals were injected into the range time domain of the radar data, and the constant false alarm rate (CFAR) detector could operate mostly unaffected in all cases, but under the most extreme jamming conditions. In this case the range cells were not able to be isolated; therefore, the STFT was applied over all range cells to obtain the micro-Doppler signature (60 cells). This was necessary as otherwise completely blank spectrograms would be produced, meaning the targeted jamming had done its job and leaving little for the system to respond to. In the context of a real system employing a robust CFAR detector, it may have some further algorithms to be able to cope with various forms of jamming. However, successful classification in later stages would still not be guaranteed due to the distortion present in the resultant spectrograms, as depicted in Figure 10.11. Therefore, the rationale behind this implemented penalty is a reasonable compromise between having an operational CFAR and degraded spectrograms.

It should be noted that in the case of targeted pulse compression, the idea of direct jamming injection does not apply and the method of degradation is not trivial. This is because it is not possible to directly change the modulation of the radar data after the experiments had taken place, the closest equivalent was to intervene in the match filtering process. This was achieved by altering the frequency response of the reference chirp and defining three worsening levels akin to the potential learning stages taking place in the EA escort vehicle described in Table 10.8. The first level was defined as a frequency chirp extending over half the true bandwidth of the real chirp signal (22.5 MHz), the second being a CW signal at 2.4 GHz and the final level being no match filtering process taking place at all; as the chirp modulation has been completely characterised by the jammer and effectively reverses the pulse compression gain of 14.3 dB, as in Table 10.2.

10.5.2.2 Jamming detection system

As J_Ω is based on a DNN architecture, this enables it to detect and distinguish fine details within a wide variety of the jamming measures; it also has the ability to discriminate between the three levels of jamming severity. To test the system, a LUT of permutations was devised consisting of 64 unique combinations (4^3), which is

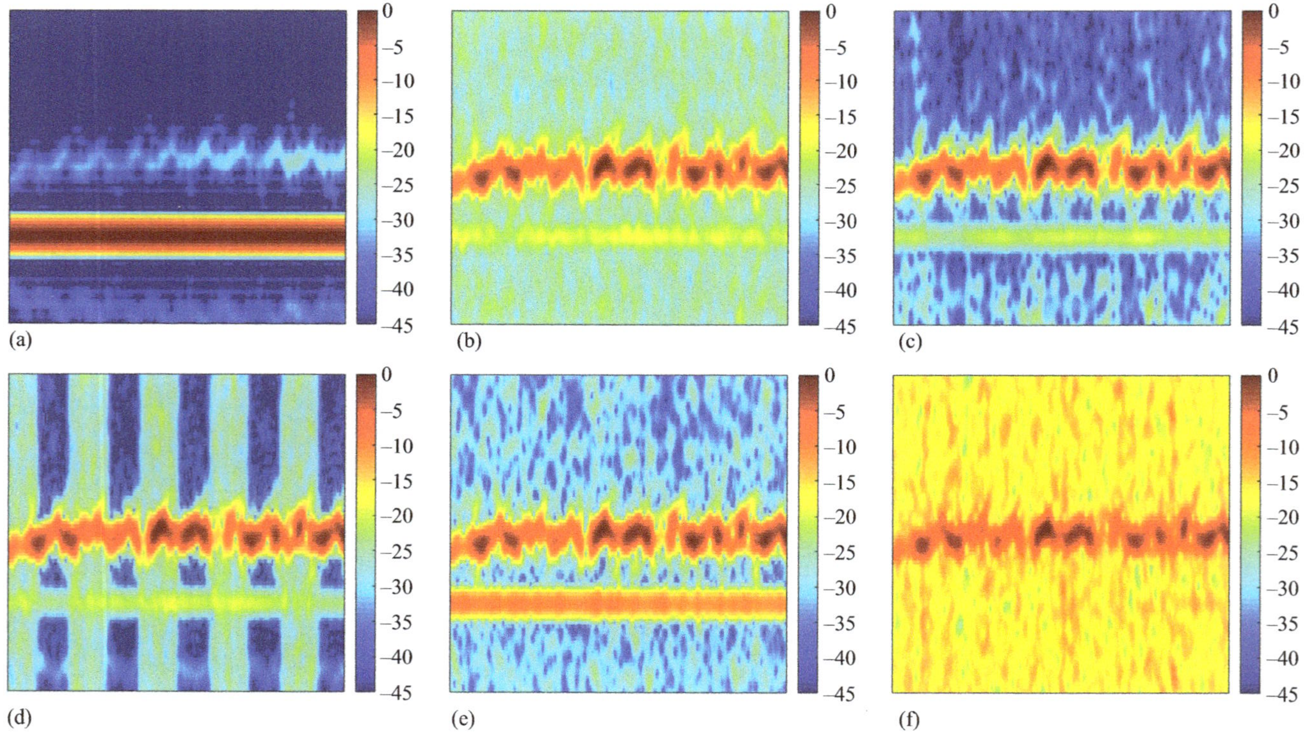

Figure 10.11 *Collage of spectrograms after synthetic jamming injection into the range time domain (time on the x-axis); the correlation with the JSR is indicated in brackets:(a) targeted pulse compression (no matched filter), (b) spot (20 dB), (c) cover (random modulations of amplitude proportional to the noise level, 20 dB), (d) blinking (20 dB), (e) unintentional (number of corrupted pulses proportional to noise level, 20 dB), and (f) deceptive (number and degree of shuffled pulses across range cells proportional to the noise level 20 dB) [10]*

based on the three jamming levels plus one uncorrupted (4), to the power of the number of radar nodes (3), this can be seen in the x axis of Figure 10.12(a). For each of the permutations, a unique FFN was trained on the equivalent jammed images based on the reserved 10% validation data. This was repeated over 5,000 iterations taking the network achieving the highest classification accuracy using the 30% testing data. With this bank of 64 FFNs, the jamming detector is able to analyse the input image and determine the class and hence the level of jamming. From this, it can then switch in the appropriate FFN such that the particular jammed node is weighted appropriately as to retain confidence in the final classification decision (D_0) from Figure 10.7.

10.5.2.3 Jamming mitigation

Figure 10.12(a) shows the classification accuracy of the trained FFN bank when the equivalent jammed images from the testing image set are injected into the network. It is important to note that no jamming detection occurs as they have been injected systematically into the assigned jamming permutations. The classification accuracy of the jamming detector is shown in Figure 10.12(b); this indicates how well the detector is able to match the predicted jamming state to the correct jamming permutation for all three radar nodes. The first 16 permutations exhibit a significant dip of approximately 20%, as the jamming detector only achieves an accuracy of 89% for cases of nil jamming. This is made worse by the fact that there is only one class for this case as opposed to six for each other forms of jamming, as highlighted in Section 10.5.2.1. It should also be noted that even if the incorrect FFN is called from the reference bank, outright classification failure is not guaranteed; therefore, Figure 10.12(a) and (b) do not simply multiply to produce Figure 10.13(a).

Finally, the jamming detection network (J_Ω) was trained over a period of 5 days and was manually stopped when the validation accuracy saturated at 86%. This was more than acceptable considering the large volume of training images used (191,520). Following this, the 38 classes were compressed back down to the original 19 classes, with the results of this stated in Section 10.3.6.3. As mentioned before, three levels of jamming severity were chosen; the SNR on average of each spectrogram was 37 dB and the third jamming level was chosen to be 30 dB as this caused the classification accuracy to consistently fall below 20% in every case (approaching the random guess accuracy of 14.3%). The other two jamming levels were chosen to be 10 and 20 dB, respectively, which produced an even spread of jamming levels. This resulted in a residual signal-to-jamming ratio (SJR) of approximately 7, 17 and 27 dB over the entire data set.

Figure 10.13(a) depicts the classification accuracy of the complete system, when the jamming detector is running in parallel with the three base DNNs at each node and automatically determining the jamming severity for each successive input image. Following this, it then matches with the believed permutation index and hence the appropriate network from the 64-FNN reference bank. The red trace shows the accuracy obtained from using the aforementioned complete system; a mean gain of 16.2% (green trace) is observed over the voting system (represented by the blue trace). The benefits of the system become evident when the input jamming combinations are

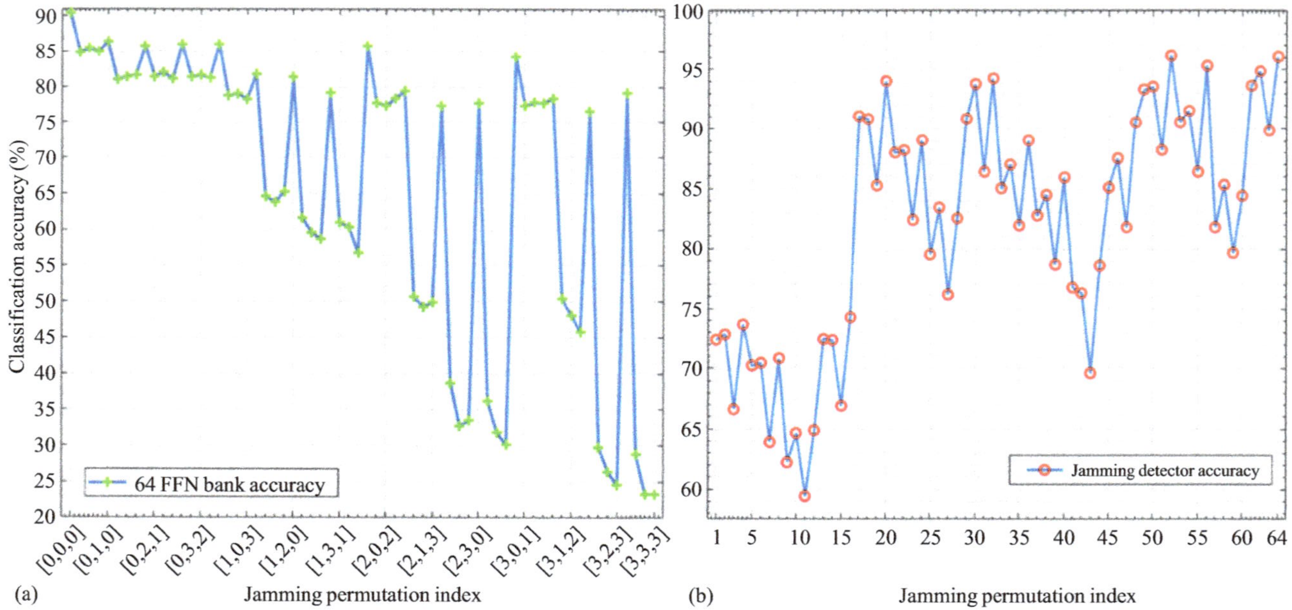

Figure 10.12 Classification accuracy of the (a) 64-FFN bank and (b) jamming detector (J_Ω) over the specified permutation indices

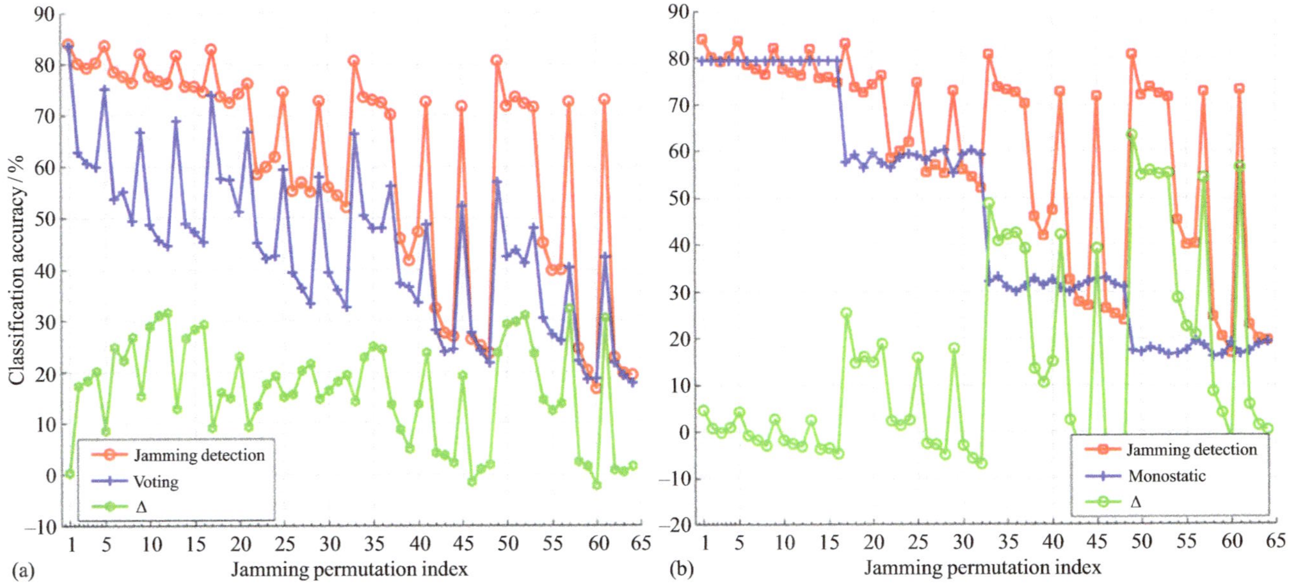

Figure 10.13 Classification accuracy of the jamming detection system compared with (a) the voting system and (b) the monostatic setup

particularly severe (50–64); however, there is only so much that the system can cope with in the presence of serious directed jamming into all three of the radar nodes.

The result of using the predictions from only the monostatic radar node is shown in Figure 10.13(b); the four steps observed are due to the discrete jamming levels and the permutation order used for node one. Though the comparison of every point is strictly not possible as the red trace encompasses permutations across all three radar nodes, it is still clear that the multistatic detection system offers performance gain, as the permutations within each step only mean that the other two nodes are being jammed to a further degree. The benefits of a multistatic system are especially obvious at the permutation boundaries (indices: 1, 16, 32 and 48, where the jamming is maximum with respect to the monostatic node). An improvement of approximately 4%, 25%, 49% and 63% is observed, respectively.

10.5.3 Summary

This section has introduced and demonstrated the application of a DNN to perform active jamming detection in parallel with a classification system, on data gathered from a real multistatic radar, where realistic jamming measures have been identified and verified by calculation and subsequently applied. The results obtained demonstrate that utilisation of FFNs in this sensor system is indeed advantageous even in the simplest of cases; however, this architecture begins to really prove itself in the face of targeted jamming, keeping classification accuracy above 70% for roughly 60% of the permutations, as opposed to the spurious nature of the voting system. However it should be noted that the voting system still has some jamming mitigation capabilities, as the injected jamming may cause the DNN to become unconfident in its final probabilistic decision, therefore handing over responsibility to the other nodes is an attempt to try and salvage a correct decision; this in some cases still functions to a respectable degree.

The notion that multistatic radar provides improved performance over a monostatic system, in the presence of targeted enemy jamming, is thoroughly put through its paces. This is proven through a range of tests and experimentations; but specifically, in Figure 10.13(b), where even just the opportunity of having uncorrupted data is enough to maintain classification accuracy at 80%, as opposed to just 17% for the monostatic node (in the worst cases of jamming). In addition to this, the multistatic setup can also provide small performance gains of a couple per cent due to having three times the amount of information available to the overall system. This would be exaggerated further if the target was in an unfavourable position in the field, but still within the coverage of the multistatic network.

This complete classification system architecture is capable of highly efficient data throughput between sensors, as only probabilistic confidence vectors need to be shared. Although all three input images are input to the single jamming detection classifier, this DNN could easily be duplicated at each node sharing only the degree to which the node thinks it is being jammed between each node. The system could then make a decision about which FFN to switch in at a centralised node, then producing a final decision as to the class of the target in question. The modular nature of this architecture is particularly advantageous and relaxes demands on data throughput between radar nodes, which is a principle concern and can very easily become an issue over long distances.

This section dealt with the effects caused by known jamming techniques and trained the DNN in a supervised learning fashion. In a practical environment it is not entirely feasible to predict every possible effect caused by targeted jamming; therefore, the system may not be able respond in an ideal manner. A thorough examination into this issue is required and will be considered for future work, where the scenario will be treated as an unsupervised learning problem. In addition, it may also be possible to leverage SAEs, to potentially undo or to actively recover the effects of jamming. This is expected to have a similar effect as the topology described in this work, in that it will be effective up until a specific threshold. A direct expansion to this work would be to have a combination of the topologies, whereby the jamming detector would be able to call upon specific SAE recovery cells to restore as much of the original data as possible, prior to classification.

10.6 Conclusion

This chapter has thoroughly explored the necessity and importance of careful planning in the implementation of data fusion methods and architecture within multistatic radar networks. The identification of opportunities for data fusion in a processing system, depicted in this chapter, showcases a multitude of locations for data fusion to occur. These have been progressively integrated into research throughout recent years and is collectively organised, presented and discussed. The works deliberated consisted of the classification of human micro-Doppler signatures, to the characterisation of payloads carried by a micro-drone.

Due to realistic restrictions inherent to the deployment of a multistatic radar, such as data throughput and concealment of sensors, great consideration must be given to the data bandwidth, and the time intervals of communication which would be required by a classification framework. This invalidates centralised fusion schemas employing a single common DNN, as communication requirements would be too great. Instead a distributed fusion scheme was proposed in which unique DNNs are deployed at each radar node, allowing the classification system to overcome spatial dependencies; this is another key issue that must be addressed through further research and would enable a singular DNN model to be trained and located at each node, potentially increasing performance. Hybrid fusion is a topology that would only be reserved for a highly futuristic multistatic radar network, in which every node is intelligently managed in real time and has knowledge of the surrounding environment; through this the fusion methods would change accordingly throughout operation.

Data fusion techniques at the decision level, namely match and rank fusion, were exercised through the experimental setups discussed, and it was shown that classification performance increased for all cases. The aforementioned fusion techniques are well suited for covert multistatic systems as it is desirable to relay information in short bursts of packets. Fusion at lower levels, such as pixel or at the raw signal, should only be considered if there is a physical communication link; this could potentially be achieved through a hybrid topology where radio opportunities of a long window could be identified beforehand and consequently exploited.

Realistic scenarios were devised whereby the experimental multistatic radar is gradually pressured by known targeted jamming techniques. A DNN was enlisted to identify and distinguish between these techniques and converted them to discrete states in terms of the associated JSR. From these decisions, the DNN had the ability to adapt the weights in the data fusion process amongst the radar nodes; this consistently maintained classification for cases of jamming in a single node. But as expected, difficulty is encountered when multiple nodes were being jammed simultaneously, which would practically be an uncommon occurrence.

The applications of DNNs applied to radar classification and processing has been convincingly proven through this research. It has demonstrated impressive abilities to detect fine subtleties within input images, alongside being remarkably generalised and professing the capability to automatically mould itself into a range of scenarios. This allows it to be dropped into any problem where a decision needs to be made in the processing system and provides the designer with the freedom to explore and focus on other aspects of the system design, such as data fusion, which as discovered is a solid requirement in multistatic radar.

References

[1] Griffiths H, Baker C, and Adamy D. *Stimson's Introduction to Airborne Radar.* 3rd ed. Stevenage: Institution of Engineering and Technology; 2014.

[2] Inggs M, Inggs G, Sandenbergh S, Al-Ashwal W, Woodbridge K, and Griffiths H. Multistatic networked radar for sea clutter measurements. *2011 IEEE International Geoscience and Remote Sensing Symposium*, Vancouver, BC, 2011, pp. 4449–4452. doi: 10.1109/IGARSS.2011.6050220.

[3] Ritchie M, Griffiths H, and Fioranelli F. Aspect Angle Dependence and Multistatic Data Fusion for Micro-Doppler Classification of Armed/Unarmed Personnel. *IET Radar, Sonar and Navigation.* 2015.

[4] Du R, Fan Y, and Wang J. Pedestrian and Bicyclist Identification Through Micro Doppler Signature With Different Approaching Aspect Angles. *IEEE Sensors Journal.* 2018.

[5] Li H, Shrestha A, Heidari H, Kernec J, and Fioranelli F. Magnetic and Radar Sensing for Multimodal Remote Health Monitoring. *IEEE Sensors Journal.* 2019;19(20):8979–8989.

[6] Ryu S, Suh J, Baek S, Hong S, and Kim J. Feature-Based Hand Gesture Recognition Using an FMCW Radar and its Temporal Feature Analysis. *IEEE Sensors Journal.* 2018;18(18):7593–7602.

[7] Fioranelli F, Ritchie M, and Griffiths H. Multistatic Human Micro-Doppler Classification of Armed/Unarmed Personnel. *IET Radar, Sonar and Navigation.* 2015;9(7):857–865.

[8] Fioranelli F, Ritchie M, and Griffiths H. Classification of Unarmed/Armed Personnel Using the NetRAD Multistatic Radar for Micro-Doppler and

Singular Value Decomposition Features. *IEEE Geoscience and Remote Sensing Letters*. 2015.

[9] Patel JS, Al-Ameri C, Fioranelli F, and Anderson D, Multi-Time Frequency Analysis and Classification of a Micro Drone Carrying Payloads using Multistatic Radar. *IET International Radar Conference.* 2018.

[10] Patel JS, Fioranelli F, Ritchie M, and Griffiths HD. Fusion of Deep Representations in Multistatic Radar Networks to Counteract the Presence of Synthetic Jamming. *IEEE Sensors Journal.* 2019;19(15):6362–6370.

[11] Patel JS, Fioranelli F, Ritchie M, and Griffiths H. Multistatic Radar Classification of Armed vs Unarmed Personnel using Neural Networks. *Evolving Systems*. 2017;9:135–144.

[12] Derham T, Doughty S, Woodbridge K, and Baker C. Design and Evaluation of a Low-Cost Multistatic Netted Radar System. *IET Radar, Sonar and Navigation.* 2007;1(5):362.

[13] Derham T, Doughty S, Woodbridge K, and Baker CJ. Realisation and evaluation of a low cost netted radar system. *CIE International Conference of Radar Proceedings*, Shanghai. 2006, pp. 1–4.

[14] Jankiraman M. *Design of Multi-Frequency CW Radars*. Raleigh, NC: SciTech Publishing; 2007.

[15] Fioranelli F, Patel JS, Gurbuz SZ, Ritchie M, and Griffiths H. Multistatic human micro-Doppler classification with degraded/jammed radar data. *2019 IEEE Radar Conference (RadarConf)*, Boston, MA, USA. 2019, pp. 1–6.

[16] Raol J. *Multi-Sensor Data Fusion with MATLAB.* Boca Raton, FL: CRC Press; 2010.

[17] Liu L, Popescu M, Skubic M, and Rantz M. An automatic fall detection framework using data fusion of Doppler radar and motion sensor network. *2014 36th Annual International Conference of the IEEE Engineering in Medicine and Biology Society*, Chicago, IL, 2014, pp. 5940–5943.

[18] Erol B, Amin MG, and Boashash B. Range-Doppler radar sensor fusion for fall detection. *2017 IEEE Radar Conference (RadarConf)*, Seattle, WA, 2017, pp. 0819–0824.

[19] Jiang D, Zhuang D, and Huang Y. Survey of Multispectral Image Fusion Techniques in Remote Sensing Applications. Image Fusion and Its Applications, Yufeng Zheng, IntechOpen. doi: 10.5772/10548. Available from: https://www.intechopen.com/books/image-fusion-and-its-applications/survey-of-multispectral-image-fusion-techniques-in-remote-sensing-applications

[20] Levanon N and Mozeson E. *Radar Signals.* Hoboken, NJ: Wiley; 2004.

[21] Yang Y, Han C, Kang X, and Han D. An overview on pixel-level image fusion in remote sensing. *2007 IEEE Intl. Conf. on Automation and Logistics*, Jinan, 2007, pp. 2339–2344.

[22] Tu T. Adjustable Intensity-Hue-Saturation and Brovey Transform Fusion Technique for IKONOS/QuickBird Imagery. *Optical Engineering.* 2005;44(11):116201.

[23] Shadoosti HR. MS and PAN Image fusion by combining brovey and wavelet methods. *Information Technology, Computer and Communication Intl Conf. 2015.* arXiv https://arxiv.org/abs/1701.01996.

[24] Walterscheid I and Brenner AR. Multistatic and multi-aspect SAR data acquisition to improve image interpretation. *2013 IEEE International Geoscience and Remote Sensing Symposium – IGARSS*, Melbourne, VIC, 2013, pp. 4194–4197.

[25] Krishnan V, Swoboda J, Yarman CE, and Yazici B. Multistatic Synthetic Aperture Radar Image Formation. *IEEE Transactions on Image Processing.* 2010;19(5):1290–1306.

[26] Erol B, Gurbuz SZ, and Amin MG. Frequency-warped cepstral heatmaps for deep learning of human motion signatures. 2018 52nd Asilomar Conference on Signals, Systems, and Computers, Pacific Grove, CA, 2018, pp. 1234–1238.

[27] Willis NJ and Griffiths HD. *Advances in Bistatic Radar.* Raleigh, NC: SciTech Publishing; 2007.

[28] Chen VC and Ling H. *Time-Frequency Transforms for Radar Imaging and Signal Analysis.* 1st ed. Artech House, Norwood, MA; 2002.

[29] Himed B, Zhang Y, and Hajjari A. STAP with angle-Doppler compensation for bistatic airborne radars. *Proceedings of the 2002 IEEE Radar Conference (IEEE Cat. No.02CH37322)*, Long Beach, CA, 2002, pp. 311–317.

[30] Tang B, Tang J, Zhang Y, and Zheng Z. Maximum Likelihood Estimation of DOD and DOA for Bistatic MIMO Radar. *Signal Processing.* 2013.

[31] Seng CH, Bouzerdoum A, Phung SL, and Amin MG. A two-stage image fusion method for enhanced through-the-wall radar target detection. *2011 IEEE RadarCon Radar*, Kansas City, MO, 2011, pp. 643–647.

[32] Li F, Zhang J, Zhou L, Du J, and Yan B. A new segmentation method of SAR image based on multi-polarization information fusion. *2009 Fifth Intl. Conf. on Natural Computation*, Tianjin, 2009, pp. 99–103.

[33] Shi H, Liu J, Xue P, *et al.* Improved Relative-transformation Principal Component Analysis Based on Mahalanobis Distance and Its Application for Fault Detection. *Acta Automatica Sinica.* 2014;39(9):1533–1542.

[34] Straka O and Šimandl M. Using the Bhattacharyya Distance in Functional Sampling Density of Particle Filter. *IFAC Proceedings Volumes.* 2005;38(1):1006–1011.

[35] Ved M. Feature Selection in Machine Learning: Variable Ranking and Feature Subset Selection Methods [Internet]. Medium. 2018 [cited 8 February 2019]. Available from: https://medium.com/@mehulved1503/feature-selection-in-machine-learning-variable-ranking-and-feature-subset-selection-methods-89b2896f2220.

[36] Webb GI and Zheng Z. Multistrategy Ensemble Learning: Reducing Error by Combining Ensemble Learning Techniques. *IEEE Transactions on Knowledge and Data Engineering.* 2004;16(8):980–991.

[37] Pudil P, Novovicová J, and Kittler J. Floating Search Methods in Feature Selection. *Pattern Recognition Letters.* 1994;15(11):1119–1125.

[38] Chandrashekar G and Sahin F. A Survey on Feature Selection Methods. *Computers and Electrical Engineering.* 2014;40(1):16–28.

[39] Freedman DD. Overview of decision level fusion techniques for identification and their application. *Proceedings of 1994 American Control Conference – ACC '94*, Baltimore, MD, 1994, vol. 2, pp. 1299–1303.

[40] Gawande U, Zaveri M, and Kapur A. A Novel Algorithm for Feature Level Fusion Using SVM Classifier for Multibiometrics-Based Person Identification. *Applied Computational Intelligence and Soft Computing.* 2013;2013: 1–11.

[41] Fioranelli F, Ritchie M, and Griffiths H. Performance Analysis of Centroid and SVD Features for Personnel Recognition Using Multistatic Micro-Doppler. *IEEE Geoscience and Remote Sensing Letters.* 2016;13(5):725–729.

[42] Bogert BP, Healy MJR, and Tukey JW. The quefrency analysis of time series for echoes: cepstrum, pseudo autocovariance, cross-cepstrum and saphe cracking. *Proc. of Time Series Analysis*, 1963, Wiley, NY, Chapter 15, p. 209–243.

[43] Krizhevsky A, Sutskever I, and Hinton G. ImageNet Classification With Deep Convolutional Neural Networks. *Communications of the ACM.* 2017;60(6): 84–90.

[44] Wu D, Pan W, Xie L, and Huang C. An Adaptive Stacked Denoising Auto-Encoder Architecture for Human Action Recognition. *Applied Mechanics and Materials.* 2014;631–632:403–409.

[45] Ijjina E and Krishna C. Classification of Human Actions Using Pose-Based Features and Stacked Auto Encoder. *Pattern Recognition Letters.* 2016;83:268–277.

[46] Erol B, Gurbuz SZ, and Amin MG. GAN-based synthetic radar micro-Doppler augmentations for improved human activity recognition. *2019 IEEE Radar Conference (RadarConf)*, Boston, MA, 2019, pp. 1–5.

[47] Adamy D. EW 101: *A First Course in Electronic Warfare.* 1st ed. Boston, MA: Artech House; 2001.

[48] Adamy D. EW 102: *A Second Course in Electronic Warfare.* 1st ed. Boston, MA: Artech House; 2004.

Chapter 11
Application of deep learning to radar remote sensing

John Rogers[1], Lucas Cagle[1], John E. Ball[1], Mehmet Kurum[1] and Sevgi Z. Gurbuz[2]

Although the origins of radar can be traced back to the military, since its inception, civilian applications have flourished, especially those related to remote sensing. Applications such as object (e.g., ship or vehicle) detection directly translate from their military counterparts of airborne and ground-based automatic target recognition (ATR). However, most applications involving the remote sensing of the environment fundamentally reverse the way radar backscattering is perceived. In detection and recognition, scattering from any surface other than the object is regarded as "clutter"—undesirable reflections that ought to be removed or suppressed in the data so that the true nature of the object of interest can be ascertained. However, in environmental remote sensing, it is the surface or volume scattering that we seek to understand and exploit. In many cases, remote sensing aims at extracting geophysical properties, which can be related back to the way materials interact with electromagnetic waves. Examples include soil moisture or water concentration, terrain elevation, biomass, mass movement rates, hydrometeor type, plant health, drought tolerance, crop yield, ice layers, and snow thickness. Because deep learning (DL) was originally developed in consideration of real-valued data and optical images, the potential performance, architectures, and optimization of deep neural networks (DNNs) operating on radar remote sensing data must be reassessed. The relationship between geophysical properties, electromagnetic scattering, and the RF data representations used to reveal these relationships creates vast, complex, multidimensional, and time-varying datasets over which DL can be leveraged. Thus, the rich and unique qualities of remote sensing data present new challenges for DL, which has driven much research in this area.

11.1 Open questions in DL for radar remote sensing

11.1.1 What is the best way to exploit multimodal sensing data?

Remote sensing data is often multimodal: in addition to microwave sensors, thermal infrared cameras, and optical sensors, such as ground-based, airborne, and spaceborne

[1]Department of Electrical and Computer Engineering, Mississippi State University, Starkville, MS, USA
[2]Department of Electrical and Computer Engineering, University of Alabama, Tuscaloosa, AL, USA

RGB, panchromatic, multispectral and hyperspectral imagers, and lidar are often utilized. Different sensors interact with the environment in different ways, resulting in datasets that embody characteristics that are different, yet related. Joint domain processing is a critical task for fusion of these complementary data sources, and augments traditional decision fusion techniques. Multimodal remote sensing data creates a unique opportunity for the design of novel architectures that can match and exploit images acquired from different perspectives to generate new data representations that can reveal more about a scene. One example of this could be fusion of electro-optical and synthetic aperture radar (SAR) imagery to construct 3D images [1,2].

Alternatively, multimodal data can provide redundancy for improving the accuracy of algorithms devised for single sensors. While annotation for generating ground truth is often done manually by human experts, DL can potentially be exploited for automatic annotation and cross-referencing between modalities. Alternatively, it would be interesting to investigate the transferability of a network trained on one modality to analyze another modality. A basic example of this is the use of networks pretrained on massive databases of optical images to classify radar micro-Doppler signatures [3]. In computer vision, variants of generative adversarial networks (GANs) [4] have been used for image translation, style transfer, and the generation of photos from paintings. As a community, we have yet to investigate how such innovations in computer vision could be leveraged for remote sensing, or, indeed, how novel architectures could be designed for joint exploitation of multimodal datasets.

11.1.2 How to satisfy the large data needs for training a DL system?

In image-based DL processing, images can be acquired relatively easily using inexpensive cameras or from Internet sources. Most DL networks need a large amount of quality training data. Even with imagery, data augmentation is often used to supplement training data. For images, data augmentation can consist of image rotations, randomly cropping out rectangular portions of the image, and image enhancements such as brightening or darkening, to name a few. These types of operations are standard and easily implemented using TensorFlow or OpenCV. Another approach is to use a high-quality GAN to generate photo-realistic imagery. All of these methods work well for image-based DL systems. The question is how to augment data for training a DL-based radar system? Simulation is a good choice, assuming the simulator has enough fidelity to adequately capture nuances of the real data. Another choice might be adding random IQ noise to currently captured radar signals. Collecting radar data can be expensive, time-consuming, and, in general, clutter and unwanted multipath interference could cause issues. This is currently an open question, and one all DL system designers must wrestle with.

11.1.3 How can prior knowledge on sensor quality be incorporated into joint domain models?

Integral to the question of joint domain processing, however, is the question of how knowledge of the different accuracy, adequacy, and data quality provided by each sensor should be incorporated into novel neural architectures. Different radar systems can have different resolutions in velocity, range, and cross-range. Interferograms acquired using single-pass versus multi-pass SAR systems may have varying degrees of reliability. Observations from different perspectives may affect the accuracy of radar measurements, as might levels of noise and artifacts due to RF electronics. In remote sensing, sensor calibration is key to controlling the quality and radiometric consistency of sensor measurements. Thus, differences between individual sensors must be taken into account when devising joint domain processing techniques. In this regard, the data-driven nature of DNNs provides a unique opportunity for capturing and taking into account potentially dynamic changes in sensor performance.

11.1.4 What is the best way to leverage data, models, and prior knowledge?

Traditional radar remote sensing relies heavily on expert domain knowledge, both for the development of models and for the design of handcrafted features. Examples based on modeling include a plant's potential growth based on measurement of root biomass underground [5], or soil moisture from measurement of the radar backscatter coefficient [6]. Models are powerful tools, which capture behavior based on underlying physical principles; however, they can be very complicated and are often highly nonlinear. Moreover, model outputs can be inaccurate if the models do not adequately capture the intricate interrelationships between input variables.

Since DL networks can learn very complex nonlinear systems, DL provides an opportunity to improve existing models. As an example, consider the problem of classification. In the conventional approach, a human expert (or a set of experts) develops features based on their expert domain knowledge or physical principles. Preprocessing is applied to improve the accuracy of features extracted and suppress sources of noise and interference. However, DNNs can be used in either a supervised or unsupervised learning method to *automatically learn* optimal discriminative features. This often results in the extraction of features that may not ever be considered by experts, or which are too complicated to be directly engineered.

The main benefit of most of the traditional methods is that they are often well founded and other researchers can understand the basic theoretical and physical principles on which the systems are based. On the other hand, a DL system can produce results that compare to or surpass hand-designed systems, but often the true operation internally is somewhat of a mystery, a black-box operator that gets good results. But are these results explainable? Can a researcher really understand what a DL system that has millions of parameters is really doing?

Attempting to solve this problem has created a branch of machine learning called "explainable artificial intelligence" or "explainable AI." An example from traditional machine learning could be the use of feature selection to rank features in terms of their relevance to classification [7–10]. While filter methods devise a metric, such as distance or mutual information, for assessing the degree to which a feature can discriminate between classes, wrapper methods, such as sequential backward elimination [11], systematically compare the classification accuracy offered by feature sets to optimally select a group of features. Examination of which features were selected can thus provide insights into problem at hand.

Visualizations are an important tool for trying to understand why DNNs are effective. Examples include using visualizations [12,13] to gain insight into the function of intermediate feature layers and the operation of the classifier. The activation maps across different layers can be plotted to visually compare what has been learned as the data progresses through the network (Figure 11.1). Alternatively, the neural network can be supplied a randomly generated image, which is then modified by gradient ascent until the softmax score is maximized for a given class [14]. This will result in a visual image of what the network thinks a given class "looks like," as shown in Figure 11.1. More recently, there has been much work in the development of new methods to make models more explainable and interpretable [15,16]. This includes methods, such as layerwise relevance propagation [17], which aims at determining which features contribute most positively to a DNN's output.

DNNs essentially learn hierarchical features at different scales: small-scale features are learned in the initial stages of the network, while increasingly more complex and abstract features are learned as depth increases. This can be leveraged for greater understanding, but doing so requires broadening focus beyond just pixel-level processing, and considering spatiotemporal variations across multiple layers.

The convergence of insights gained by trying to understand DNNs and the benefits of guiding DNNs with physical process models is driving research into the development of hybrid approaches, which attempt to maximize the benefits of data-driven machine learning by incorporating the *a priori* knowledge offered by models. Physics-aware or physics-driven machine learning represents the emergence of a new research thrust within geophysical signal processing [19].

11.1.5 How can DL aide in solving multi-temporal processing challenges?

The problem of processing time-varying data is a challenge that arises in many applications, and for which architectures such as recurrent neural networks (RNNs) have been proposed to address. Remote sensing data differs from other applications, however, in which sequential data typically has low temporal resolution. Data revisit rates are often long, on the order of days or longer. Additionally, ground-truth data is even more expensive as the original remote sensing data to be analyzed, while sequential imagery must also be co-registered in most applications. Moreover, the changes observed between samples of multi-temporal data are often nonlinear and can occur on different timescales depending on the problem at hand: seasons, weeks, or even years. Data analysis of multi-temporal data involves not just change detection, but also detecting and characterizing objects, and tracking changes temporally. Thus, an open question is how DNNs can be advanced to fully address these tasks over short, medium, and long timescales, while also having multifunctional capabilities involving classification, object recognition, and physical modeling.

Many works have investigated the performance of RNNs for the analysis of satellite image time series data for land-use classification with airborne and spaceborne

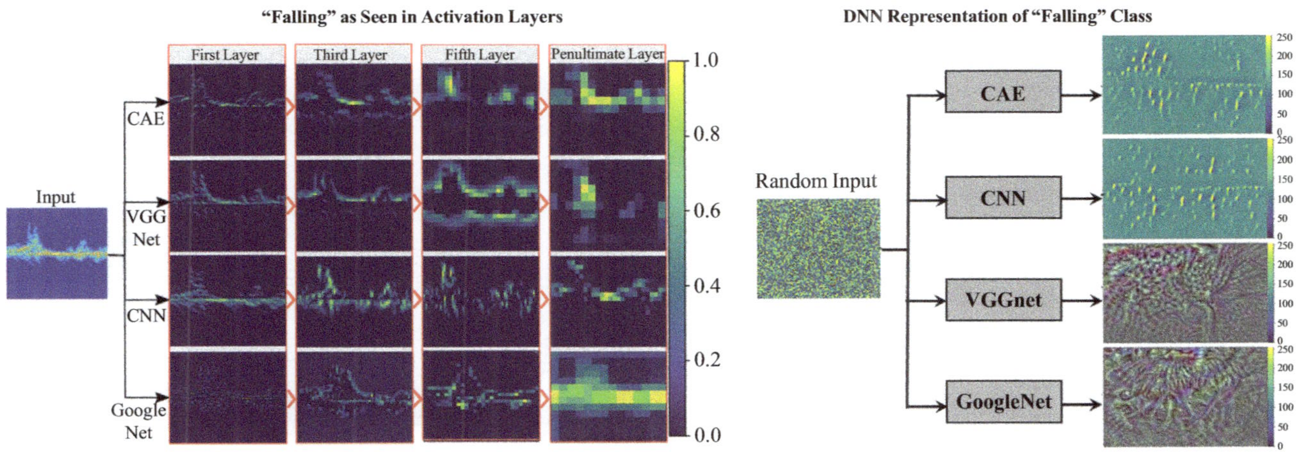

Figure 11.1 Visualization of "falling" in radar micro-Doppler data [18]

optical images [20,21] and hyperspectral images [22]. Various RNNs as well as long short-term memory (LSTM) RNNs have also been utilized on SAR time-series imagery for applications such as vegetation mapping [23] and agricultural crop classification [24,25]. One disadvantage of RNNs, however, is that they can be difficult to train because the number of training steps depends on the length of the time series, and, consequently, the patterns at the beginning of the time series are many layers away from the classification output. Moreover, RNNs can take a longer time to train than CNNs because the error must be backpropagated through each layer in turn [26].

As remote sensing data captures both spatial and temporal changes, several researchers have attempted to merge the benefits of spatial processing with convolutional neural networks (CNNs) with the temporal analysis afforded by RNNs. Examples include the use of a Convolutional Recurrent Neural Network (CRNN) for change detection in multispectral imagery [27] and temporal convolutional neural networks [26], which applies convolutions in the temporal dimension in order to automatically learn temporal (and spectral) features of multispectral data. Other ideas include processing the data as though they were videos using 3D CNNs [28] or exploiting the hierarchical nature of DNNs to map features across multiple timescales [29]. As an alternative to RNNs, a recent study [30] on agricultural crop mapping with SAR time-series data proposed instead using Markov random fields to capture temporal changes. This was merged with a CNN to both reduce speckle noise and extract spatial-contextual features to characterize texture and scattering mechanisms. Although a comprehensive comparison of the abovementioned techniques has not been conducted, a recent comparison of RNNs, CRNNs, and fully convolutional networks (FCNs) on SAR time-series data showed that FCNs outperformed both recurrent neural architectures [31,32].

11.1.6 How can the big data challenge presented by remote sensing data be addressed?

Every day, more and more remote sensing data is acquired from ground-based, airborne, and spaceborne assets from an increasing number of countries around the world. The need for data is driven by the plethora of applications of remote sensing data, ranging from environmental protection to precision agriculture, land-use and water resource management, geology, military applications, and more. According to IBM, 2.5 quintillion bytes of data are generated every day [33]. The staggering size of this number can be better understood by considering that "90% of the data in the world today has been created in the last 2 years alone [34]." This massive amount of data creates challenges in data management, loading, storage, and transmission as well as graphics processing unit (GPU)-based parallel processing and data analysis [35].

In regards to DNN design, research has focused on improving the convergence rates of large-scale DNNs via architectures such as residual networks [36], highway networks [37–40], and dense networks (DenseNets) [41]. Residual networks can mitigate training error by the inclusion of shortcut connections that enables an identity mapping and improves accuracy with depth. In SAR applications, residual networks have been used for a variety of tasks, including descalloping [42], urban change detection [43], and despeckling [44]. Highway networks have a cross-layer connectivity mechanism that increases training speed and convergence and can be categorized as a multipath-based CNN architecture. As examples, highway networks have been utilized in large-scale classification of SAR data [45] and for classification with limited labeled training data [46]. DenseNets concatenate outputs from previous layers, rather than summing as done in residual networks. Consequentially, DenseNets can help to mitigate the vanishing gradient problem, encourage feature reuse, and substantially reduce the number of parameters in the network. In SAR applications, DenseNets have been proposed for mitigating overfitting in polarimetric SAR (PolSAR) classification [47] and capturing small subtleties in SAR images, such as those observed in oil spill monitoring when seeking to differential actual oil films from those that are simply oil-like [48].

Another important research direction involves the application of compressed sensing [49] to reduce the number of required measurements when using radar systems. To give just a few examples, compressed sensing was used to reduce the number of frequency measurements required for image reconstruction in subsurface imaging with stepped frequency ground-penetrating radar (GPR) [50,51] and high-resolution diffraction tomographic GPR [52], reconstruction of full frequency band data when only a partial pulse is received from a stepped frequency GPR [53], and sparse spectral estimation in tomographic SAR [54]. In compressed sensing, a smaller number of random projections are typically used to construct the measurement matrix. Thus, one area of active research is the application of DNNs to learn optimal projections for improved image reconstruction [55–58]. The benefit of compressed sensing to data reduction is even more pronounced when applied to 3D SAR image reconstruction [59].

11.1.7 How can DL be used to leverage nontraditional data sources?

An emerging area of increasing interest that can benefit from the offerings of DL is the exploitation of nontraditional data sources, such as YouTube, Twitter, and Facebook, in remote sensing. There is a vast amount of information that is being shared over social media, and while some data is in the form of text, there are also quite a few cell phone or other images being shared. One significant advantage of social media data is that they can help compensate for the low temporal resolution of remote sensing data, providing imagery that could otherwise be unobtainable. As an example, consider the flooding and damage caused by Hurricane Harvey (a Category 4 hurricane that made landfall on Texas and Louisiana in August 2017). During a hurricane, it is difficult to collect airborne data via small aircraft or drone. Satellite data on the other hand may also not be available based on timing and revisit periods. Yet people on the ground and relief workers need immediate information to adequately respond to mitigate effects of such natural disasters. Social media provides a valuable means for communication and information regarding effected areas, damaged infrastructure, blocked roads, emergencies, and, in the case of Harvey, water depth, which can drive hydrological models to predict the potential extent and severity of flooding.

Social media images can also be used to augment and enrich SAR imagery. In a recent study [60,61], 1,766 social media images of the Berlin Cathedral were used to generate a 3D-point cloud, which was then projected to align with the perspective of TerraSAR-X high-resolution spotlight images. Tomographic SAR techniques were employed to generate a 3D point cloud from 102 TerraSAR-X images. Comparison of the two reconstructions (Figure 11.2) shows how the social media data can be used through comparison to increase the interpretability of the 3D TomoSAR reconstruction. Fusion of these two modalities can also lead to a 3D model that exploits the optical texture for scene understanding, while retaining the precise deformation parameters retrieved from SAR images.

11.1.8 How can DL be used in automotive autonomy?

Automotive autonomy is a nontraditional application in remote sensing. There are many challenges to be overcome in this area [62–65], and DL can play a significant role. Interferences from other radars, and even worse, jamming, are critical areas to be studied. Typical autonomy waveforms are frequency-modulated continuous wave, and in order to have high-range resolution, these radars use large bandwidths, typically in the 76–81-GHz range. This combination makes these radars susceptible to jamming and spoofing attacks [66]. In the future, vehicle radar systems are envisioned to also provide communications over the RF spectrum, potentially warning other vehicles and providing information to assist their systems with collision avoidance. Adaptive or reinforcement learning DL systems can potentially help in these scenarios. DL solutions can potentially be used to aid spectrum and time sharing between automotive radars and vehicle-to-vehicle communications.

Applications such as adaptive cruise control (acc) are well suited for a radar. In acc, the radar is used to identify vehicles and estimate their velocity and location, both of which are strengths of automotive radars. Here, DL can be utilized to (1) provide better detection capabilities, (2) help mitigate false alarms, such as could be caused by a manhole's large metallic cover, and (3) identify and classify objects (e.g., a car, a pedestrian, a cyclist), and even to augment tracking, traditionally performed via a

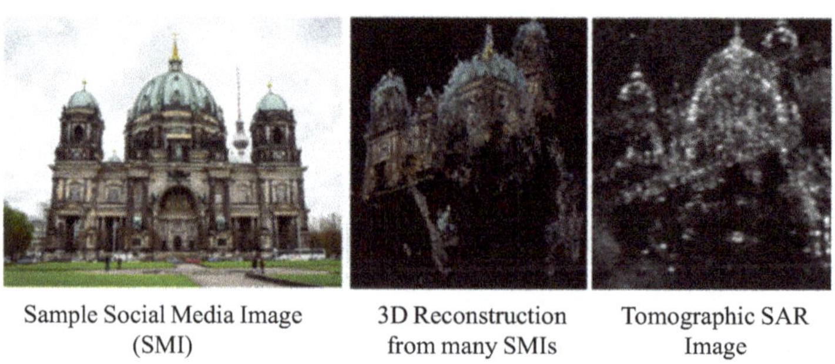

Figure 11.2 Comparison of 3D reconstruction of the Berlin Cathedral from social media images with that using tomographic SAR [60]

Kalman or particle filter. Clutter mitigation is always a challenge in radars, and it is not different in automotive radars [67]. As automotive radars' resolution improves, the data rates and information contained in the returns also increase. New DL networks can potentially leverage this information.

Radar by itself is not suitable as a sole sensor for automotive autonomy. DL solutions that can fuse radar, lidar, camera, and ultrasonic data will be needed [68–71].

11.2 Selected applications

Radar has been used in many remote sensing applications and provides a valuable source of information when optical images are ineffective, e.g., during the nighttime or under cloud cover. The phenomenological difference of radar versus optical imagery can also be exploited to ascertain additional information about a scene, such as depth or texture, based on electromagnetic scattering properties. Inggs and Lord [72] discuss many of the diverse applications of SAR imagery, including military intelligence gathering, reconnaissance, and weapon guidance, and civilian applications such as topographic mapping, geological, mining, oil spill monitoring, sea ice monitoring, oceanography, agricultural classification and assessment, and land-use monitoring. SAR imagery is useful in most of the aforementioned application areas. Some other areas where SAR is heavily utilized are oil spill detection [72], ocean wave studies [72], ice monitoring at high altitudes [72], change detection of urban areas [73], ecological studies [74], soil moisture [75–83], agriculture [84–88], digital elevation model (DEM) generation and analysis [89–93], fire detection and damage assessment [94,95], flood detection [96–101], and earthquake characterization [102–104]. SAR has even been used in tropical forest management [105], among others. There are also emerging applications in radar in automotive autonomy and human detection [106–113].

Interferometric SAR uses two or more SAR images to generate maps of surface deformation, and digital elevation models (DEM). Applications include, e.g., ice flow analysis, glacier studies, seismic analysis, change detection, [72], mapping urban extent [114], and DEM generation and analysis [89–93].

Signals of opportunity (SoOp) is an innovative technique that exploits reflected anthropogenic signals (multipath) for remote sensing of environment. In this scheme, the navigation or communication transmitters that provide free illumination and the receiver that captures the reflected signals can be considered as a multi-static radar system. This technique has been gaining attention in the remote sensing community since it only requires the development of small size, weight, and power–cost passive receivers, an analysis algorithm, and no onboard transmitter. In recent past, various SoOp studies have emerged to apply various machine learning approaches to space-borne data to estimate the geophysical parameters such as wind speed [115–118], soil moisture [119,120], wetland classification [121], and sea ice extent [122–125].

For a broader look into the role of radar in remote sensing, we would refer the readers to survey papers, such as that by Toth and Jóźków [126], which provides an in-depth look at airborne and satellite imaging systems, unmanned aerial systems

(UAS), and SAR satellite systems. In the next sections, we provide some examples of the way DL has been applied to several specific radar remote sensing case studies.

11.2.1 Land use and land cover (LULC) classification

Land use and land cover (LULC) classification is one of the most ubiquitous and important applications of remote sensing in general. Land cover processing seeks to determine the various land cover categories in an area, such as forest, urban, impermeable, buildings, and crops. Land-use analysis groups land areas in terms of agriculture, urban, etc. LULC estimation products are frequently used by city and state planners, developers, land surveyors, geospatial researchers and engineers, politicians, farmers, and researchers, to name just a few groups. LULC products are derived from a variety of remote sensing sources, including satellite and aerial imagery, multispectral or hyperspectral imagery, and SAR [72].

11.2.1.1 Land cover classification

As an example from agriculture, consider the classification of crops in Ukraine [127] using both optical data from the Landsat mission and radar data from the Sentinel-1 satellite mission. The proposed method is composed of a four-step process, including the use of self-organizing maps for preprocessing to remove atmospheric effects from the optical data and geospatial post-processing to combine results into a higher level LULC products.

The step of interest is the use of a CNN for LULC classification. This network is not only the first to use a CNN architecture but also is the first network to utilize radar data collected on different dates to allow for temporal considerations. The data used was created by combining eight images of the same region at different times. From the Sentinel data, 15 images were used with 2 bands each, and from the Landsat data 4 images were used with 6 bands each resulting in an image with 54 samples per pixel. These images were classified with two different CNN architectures: one-dimensional (1D) and two-dimensional (2D) CNNs. The 1D CNN takes in a vector input and classifies each pixel independently. Per pixel methods tend to produce speckled distortion [128] as disregarding spatial information decreases the classifier's robustness relative to noise. In contrast, the 2D CNN does compute local spatial correlations across a neighborhood of pixels provided in the input image. While this method is more robust to speckled distortion, it introduces the capability for small class regions such as roads to be misclassified.

These CNNs were applied to classify LULC for 11 categories: winter wheat, winter rapeseed, spring crops, maize, sugar beet, sunflower, soybean, forest, grassland, bare land, and water. These CNNs were compared to results from a traditional neural network composed of perceptrons and a random forest classifier. The CNN results slightly outperformed the other methods with an overall accuracy of 94.6% with the 2D CNN, 93.5% with the 1D CNN, 92.7% with the traditional neural network, and only 88.8% for the random forest classifier.

11.2.1.2 Land use classification

Next, consider the use of deep belief networks (DBNs) with PolSAR imagery to identify agricultural regions, lakes, rivers, open spaces, and residential areas around the Toronto area [129]. DBNs, introduced in [130], are neural networks in which each pair of forms a restricted Boltzmann machine (RBM). As the RBM is an unsupervised learning algorithm, the intermediate layers of DBNs attempt to optimize the quality of the features output by each layer rather than just attempting to maximize the accuracy of the output of the final layer as with gradient descent. However, they are more complicated to train as they require a multistage training process, where the network begins as one RBM and is trained as each additional RBM is appended until finally terminating with a softmax layer and a final training run. In addition, this method is not pixel based thus will not have the speckled results common in pixel-based approaches. The network used in [129] was tested with a variable number of RBM layer sets. Overfitting was observed with more than two RBM layer sets, thus the network used consisted of five layers: two RBMs and a softmax layer. The network operated on two C-band PolSAR images collected by RADARSAT-2. These images were segmented into ten classes using four methods: a support vector machine (SVM), stochastic expectation–maximization, a traditional neutral network, and the author's proposed DBN. The DBN achieved and overall accuracy for all classes of 81.74%. This was an improvement of 5% compared to the SVM, which outperformed the remaining two methods. Figure 11.3 shows the LULC output for all assessed methods compared to the collected SAR data and a Google Maps image of the same region. The classes in Figure 11.3 are water, golf course (golf), pasture, construction (Cons.), low-density residential, high-density residential, two unspecified crops (Crop1, Crop2), forest, and industrial (Ind.).

Alternatively, Geng *et al.* [131] explored SAR classification using deep convolutional autoencoders (DCAEs). In this study, pixel-wise classification was used to identify ground cover in SAR images; namely, rivers, forests, grassland, buildings, and roads. By using an autoencoder, the authors freed the neural network to select and learn its own feature representation and compensate for speckle noise. Unlike typical DCAE's, the proposed network has its early convolutional layers augmented with mean, variance, and 2D Gabor filters as extra convolutional units. To encourage scale transformation to overcome SAR speckle noise, average pooling is used after the convolutional layer. After the convolution and pooling, the network has stacked sparse autoencoder layers to perform dimensionality reduction. The output can then be classified. The classified image is then up-sampled to undo the downscaling of the pooling layer. Morphological smoothing is used to smooth and improve the up-sampled classification image. The trained network outperformed classical methods such as the SVM with a radial basis kernel.

11.2.2 Change detection

Briefly, change detection entails analyzing images of the same scene taken at different times to identify changes in the scene. It is critical in many civil and military applications making it an important function in remote sensing. Military analysts use

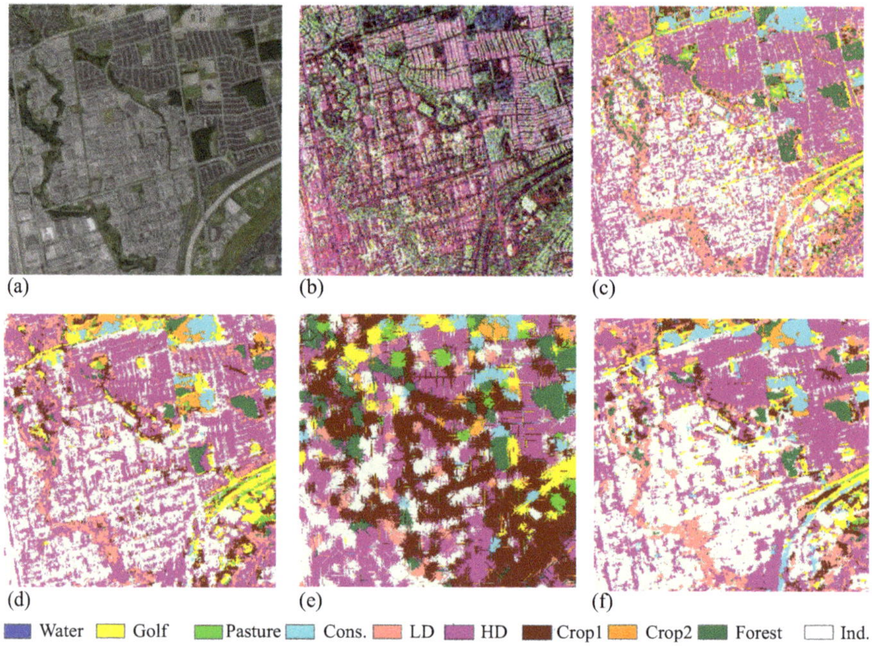

Figure 11.3 Comparison of Google Earth (a), PolSAR image (b), and LULC results from SVM (c), NN (d), SEM (e), and DBN (f) [129]

change detection to identify and quantify developments of airfields and other man-made structures and identify new fortifications. In the civil sphere, change detection is frequently applied when managing natural resources and in urban planning. With growing deployment of orbital SAR sensors, interest in change detection of SAR imagery has grown steadily. SAR images introduce particular complications, such as speckle noise and shadowing, that can confound more traditional change detection techniques.

As an example, consider the DNN developed by Gong *et al.* [132] for SAR change detection. This work pursues an unsupervised approach by first passing the two SAR images into a preclassifier. This preclassifier uses less accurate classical techniques to develop initial estimates for whether a pixel has changed or not between the two SAR images. Next, local neighborhoods in the SAR images with high probability of being correctly labeled are selected as training samples for the DNN. Using these samples, supervised training can then be performed. To initialize the DNN layer weights, the network weights are initialized by learning a stack of radial basis machines (RBM). This provides each layer in the network with a good starting point for backpropagation. To evaluate the proposed method, the first two identical networks, with one using the preclassifier labels and the other using the actual ground-truth labels, are trained. While the DNN trained with ground truths performed better, the classification accuracy of the DNN trained on unsupervised preclassification data performed only marginally worse, often within a few percentage points of error. This verified the efficacy of preclassification to generate labels when no ground truth is available. Next, the DNN itself was validated by comparing performance to classical thresholding and clustering techniques. For the three datasets evaluated, the proposed DNN outperformed both clustering and thresholding techniques, demonstrating the efficacy of the DNN method.

While it is most common for change detection algorithms to operate on images from the same modality, Liu *et al.* [133] proposed an unsupervised technique to change detection on heterogeneous images. Namely, it performs change detection between optical and SAR images taken at different dates. In tasks such as post-disaster evaluation, SAR imagery from before the disaster is often unavailable. However, the relative economy of optical imaging often means high-resolution optical images are available for a region pre-disaster. Post-disaster, SAR images can be readily captured and are advantageous compared to optical imaging due to its resilience to cloud cover, smoke, dust, and other optical contaminants. The proposed network is a symmetric network consisting of two convolutional networks with a coupling layer linking the feature spaces of the two convolutional networks. To accelerate training, the two convolutional networks can be formed from the encoders of pretrained autoencoders. In order to perform pixel-wise change detection, these convolutional networks are trained without any pooling layers. To train the coupling layers, an iterative scheme is used. A loss function consisting of the pixel-wise difference between the two convolutional networks' feature space is used. This difference is multiplied by the estimated probability that said pixel is unchanged. The network is then iteratively trained by alternating between updating network weights and updating the network output, the estimated probability of pixel change. This has the effect of encouraging the network to only use unchanged pixels for updating its weights while iteratively learning to identify such pixels. For evaluation of the network, it can be trained and tested with homogeneous images and compared to classical methods such as mean and log-ratio and wavelet fusion. When tested on homogenous SAR and optical datasets, the proposed network performed better than the traditional methods.

11.2.3 Ship detection

Ship detection is a civilian application of ATR algorithms, which are traditionally rooted in a military context. Utilization of SAR imagery in ship detection is important as monitoring of ship traffic and fishery monitoring can be performed over wide areas in almost all weather conditions and at any time of day [134]. Ships are typically present in SAR imagery as small colocated collections of bright coherent scatterers. Most of the traditional methods in SAR ship detection attempt to (1) model the clutter and detect ships via a specialized constant false alarm rate (CFAR) method, (2) detect differences in the more coherent ship returns versus the more noncoherent sea returns, (3) apply transforms such as wavelets to the SAR imagery, or (4) apply a specialized algorithm specifically designed for SAR imagery. Most of the traditional methods include specialized filter techniques and CFAR processors [135–141], custom feature extraction methods [134,142,143], and specialized image processing methods [144].

CNNs have been applied to SAR data for ship detection due to the spatial nature of the data. Zhan *et al.* [145] utilize a traditional CNN to detect ships and icebergs in small SAR images. In [146], region-based CNNs (R-CNNs) are used for ship detection from SAR data. R-CNNs divide the input data into a fixed number of candidate regions and then attempt to classify those regions. This allows for larger SAR image inputs as the network can properly classify data with multiple ships. Liu *et al.* [147] utilize a two-stage process to imitate an R-CNN for ship detection. First, traditional methods are used to preprocess SAR images to extract salinity and perform sea–land segmentation. Then a sliding window is applied over the region classified as sea, and window of data is input to a CNN. Figure 11.4 shows a map of ship detections in Tokyo Bay using this method.

Schwegmann *et al.* [148] utilize very deep highway networks to classify ships in SAR imagery. Highway networks utilize an additional transfer function to implement adaptive gating in hidden layers to minimize the effect of vanishing gradient. The resulting layer output is a weighted sum of the layer input and the output from an unmodified hidden layer. These highway layers thus allow for networks with hundreds of hidden layers. In addition, highway networks can train with smaller datasets. Four networks were trained with layer counts of 2, 20, 50, and 100. The dataset used was constructed from SAR imagery taken by the RADARSAT-2 and Sentinel-1 missions. The data used was from the South African exclusive economic zone. Each SAR image was divided into 21 by 21 pixel subimages that were classified as positive (ship), false positive (ship-like), and negative. False-positive images were defined as images classified as false positives by a CA-CFAR. The 100-layer network demonstrated the highest accuracy of 96.26%, which was only slightly better than the cubic SVM that had 95.53% accuracy. However, when disregarding the classification accuracy of false positives and negatives, it is observed that the classification accuracy for positives is 83.98% for the 50-layer network, but only 70.87% for the cubic SVM.

Figure 11.4 Ship detections in Tokyo Bay [147]

11.2.4 Geophysical parameter estimation

In remote sensing, geophysical parameter estimation refers to the process of extracting quantitative assessments of the properties of the surface or medium being perceived; e.g., soil moisture, surface roughness, ice thickness, crop height, or aboveground biomass. Estimating these parameters is difficult, and empirical or semiempirical models are often used.

11.2.4.1 Drought monitoring

Shen *et al.* [149] use DL to fuse remote sensing data with meteorological and soil data to perform drought monitoring in the Henan province of China. For the remote sensing data, radar-based precipitation measurements from the Tropical Rainfall Measuring Mission and temperature data from thermal sensors from the Moderate Resolution Imaging Spectroradiometer mission were used. Drought is related to many measurable factors: precipitation, temperature, soil moisture and type, etc. While the relationships between some of these parameters are currently defined, a comprehensive model relating all of these factors is not easily obtained. The authors attempt to use DL to construct this comprehensive model inputting all available parameters into their network. The deep learner used was a forward network with a modified training model. This process began by inputting the data into only the first layer and alloying sparsity constraints to perform unsupervised feature learning. Likewise subsequent layers were appended and trained until the final layer was added and the entire network was trained with the data labels. This training method increases the complexity of implementation but divides the deep learner's task into multiple but less difficult subproblems. The final network constructed was of 8 layers, with 500 neurons. The network outputs an estimated comprehensive meteorological drought index. For test data, the network had a root mean square error (RMSE) of 0.357 that corresponds to a consistency rate of 79.8%. In addition, the network's results were compared with the standard precipitation evaporation index to verify the results.

11.2.4.2 Precipitation nowcasting

Precipitation nowcasting involves making short-term predictions or forecasts of rainfall intensity based on current atmospheric conditions, radar, and other sensor measurements. Nowcasting requires higher resolution and accuracy as the information is used in determinations for immediate actions in areas such as air traffic control. While robust physical models for forecasting exist, computer vision is typically applied to radar products due to the computational complexity of the robust models. DL has been shown to outperform computer vision in most applications; however, nowcasting is a spatiotemporal problem and deep learners are typically spatial or temporal. In [150], Shi *et al.* accounted for this by combining the spatial capabilities convolutional layers with the temporal capabilities of RNNs to create a convolutional long short-term memory (ConvLSTM) network. Two types of data were used to assess the performance of the ConvLSTM. Sequences of 2D radar echo maps were used to assess its performance for precipitation nowcasting, and the MNIST dataset was used to assess the performance compared to fully connected RNNs. For the MNIST dataset, the ConvLSTM achieves a cross-entropy loss of 3,670.85, which is a reduction of 24% from the traditional LSTM network results. For precipitation nowcasting, the ConvLSTM is compared to a traditional LSTM network and computer vision techniques currently used for this problem. Because traditional LSTM networks do not consider spatial information, both the ConvLSTM and computer vision techniques outperformed it. The ConvLSTM had a mean square error in rainfall estimation of 1.420. The next best computer vision technique had a mean square error of 1.684. This work is continued in [151] where the trajectory gated recurrent unit is proposed to allow for locational invariance.

11.2.4.3 Snow depth estimation

In [152], a deep learner is used to estimate snow depth on sea ice from radiometry data. Data from the Advanced Microwave Scanning Radiometer 2 (AMSR2) mission and Soil Moisture and Ocean Salinity (SMOS) mission are used. The AMSR2 is preprocessed into three input values: two gradient ratios and one polarization ratios of the brightness temperature at selected bands and polarizations. Likewise a single polarization ratio is computed from the SMOS data. Two networks are proposed with one processing the three AMSR2 ratios, and the second processing all four ratios from the AMSR2 and SMOS data. These networks differ in input size as well as the number of layers. Both networks begin with a 15-neuron, fully connected layer with a sigmoid function followed by batch normalization. The first network then follows with three sets of 15-neuron, fully connected layers with rectified linear unit (ReLU) activation functions. The second network only has two of these layer sets. The final hidden layer in both networks consists of a 20-neuron, fully connected layer with a ReLU activation function. Finally, both networks output a single value for estimated snow depth using a single perceptron with a hyperbolic tangent activation function. These networks were compared with three empirically derived models for snow depth estimation from radiometry data. The neural networks outperformed all three methods with the four input network having the lowest RMSE of 0.040 m. The three input network had an RMSE of 0.053 m. The best empirical method had an RMSE of 0.07 m [152].

11.2.4.4 Ice layer tracking

In the news, there have been increasing numbers of reports that major ice sheets on the Earth, such as the Greenland ice sheet, are melting and contributing the global rise of sea level [153]. Radar ice sounding has been an invaluable tool for measuring ice thickness, and, hence, predicting the rise of sea levels in the future. Ice sounding is conducted through the application of SAR imaging with GPR. This results in images of a vertical slice of the ice, also known as the radar echogram, which can clearly show ice layers, and, at the very bottom, the ice bedrock, as illustrated in Figure 11.5. The difference in depth between the top layer (air–ice interface) and bottom layer (ice–rock interface) is the measurement of ice thickness.

Thus, ice thickness measurement relies upon accurately detecting and tracking the top and bottom layers of the radar echogram. Over the years, many techniques have been developed for edge detection, including filtering, thresholding, and statistical

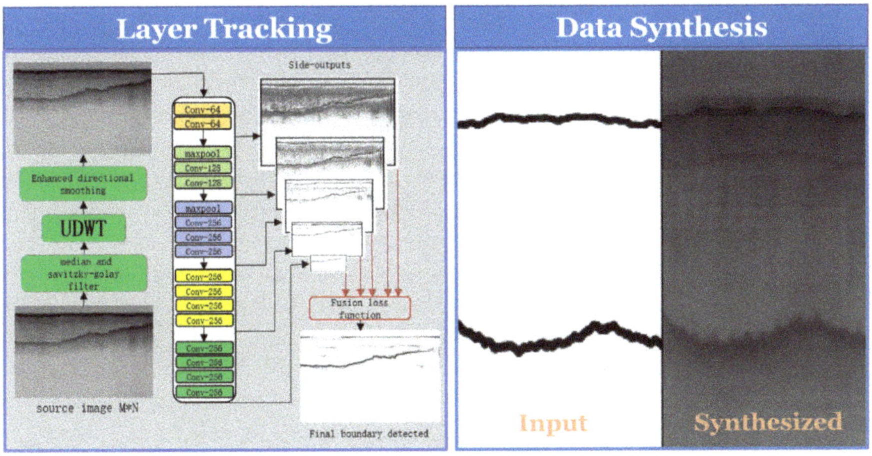

Figure 11.5 Example applications of deep learning to layer tracking [161] and radar ice sounding data synthesis [163]

analysis [154], active contours [155], the level-set technique [156], and graphical models [157], which have been applied to radar ice sounding. The capability of DNNs to learn higher dimensional representation has yielded advances in edge detection as well, which several DL-based approaches having been proposed [158–160] to realize hierarchical, multi-scale architectures. A multi-scale approach for layer tracking in radar echograms [161] has recently proposed, which merges wavelet transforms with a convolutional neural network. Wavelet transforms have been effective in speckle noise reduction and can be used to decompose and smooth the signal. A fused loss function is computed from features extracted at different layers of the CNN. Results show that the proposed deep hybrid net yields an improved F-score in comparison to conventional techniques when applied to hand-annotated datasets.

The internal layers of the echogram can also be tracked and reveal important scientific information about the structure and formation of ice. However, as with many real-world problems, the presence of noise limits the attainable performance, even when large amounts of data are available. Initial work [162] has shown that DNNs have been able to track a low percentage of internal contours in echogram data. However, recent work in generating synthetic echograms [163], shown in Figure 11.5, via CycleGANs [164], has yielded improvements in bedrock layer tracking that could be leveraged for data annotation and extension to internal layer mapping.

11.2.5 Radar aeroecology

Aeroecology is an interdisciplinary science focused on the study of airborne organisms with the aim of deepening understanding about the ecological functions of the aerosphere and the bio-organisms that move through it. In addition to having important applications to the understanding of animal migration and foraging movements, global pest and disease control, biodiversity and conservation issues, and monitoring of the effects of climate change, aeroecology has also been critical in ensuring the safety of military and civilian aircraft from bird strikes.

The use of radar in ornithology dates back to 1945, when the presence of birds was first observed in data collected from two Royal Air Force radar stations. In the early years, ornithological studies were mainly done using "radars of opportunity," i.e., radars that were actually designed for other applications, such as surveillance, tracking, weather, and marine radar, but exploitable for bird studies. Examples include the "Superfledermaus" military tracking radar, the airport surveillance radars (ASR)-4 and ASR-8 , and the WSR-57 weather surveillance radar network of nine radars positioned about the Gulf of Mexico. More recently, the WSR-88D Doppler weather surveillance radars deployed in Oklahoma and Florida and the Next Generation Weather Radar (NEXRAD) and Operational Program for the Exchange of Weather Radar Information (OPERA) network of weather radars in the United States and Europe, respectively, have all been utilized in studies of bird migration.

11.2.5.1 Birds in ground radar data

Over the past 20 years, efforts have been made to design radars specific for the detection and identification of birds and avian biological studies [165,166]. One of the first such "bird radars" is the Radar Observation of Bird Intensity (ROBIN) system developed by the Dutch Research Institute for Applied Science (TNO) in 1989. The ROBIN system was specifically designed for bird strike applications—an area of increasing concern as it has been reported that collisions have increased at least 50% over the past 10 years, especially near airports and military airbases. Other examples include the Swiss Bird Radar, designed for ornithological research, and the MERLIN Avian Radar System and Turkish KUSRADAR, designed for airport bird strike prevention [167].

Depending on whether the radar system includes a quadrature demodulator, two types of avian signatures may be extracted from radar returns: (1) the received power as a function of time and (2) the micro-Doppler signature. Examples of both are illustrated in Figure 11.6 for insects and a variety of bird species. Visually, it may be observed that different types of birds exhibit differences in wing flapping frequency as well as the duration of wing flapping—two features that are significant for identification [168]. A variety of machine learning approaches have been applied to identify bird species with radar, such as the extraction of features with wavelet transforms for classification with SVM [169]. A significant body of work involving the classification of bird micro-Doppler signatures is also done in the context of the detection and identification of UAS, where bird species are often grouped together into a single "bird" class, and distinguished from various types of UAS. DBNs [170] and CNNs [171] have been proposed for bird/drone identification.

11.2.5.2 Birds and insects in meteorological radar data

Long-range, intercontinental studies of bird migration can be conducted by exploiting existing networks of meteorological radar, which cover large expanses of territory.

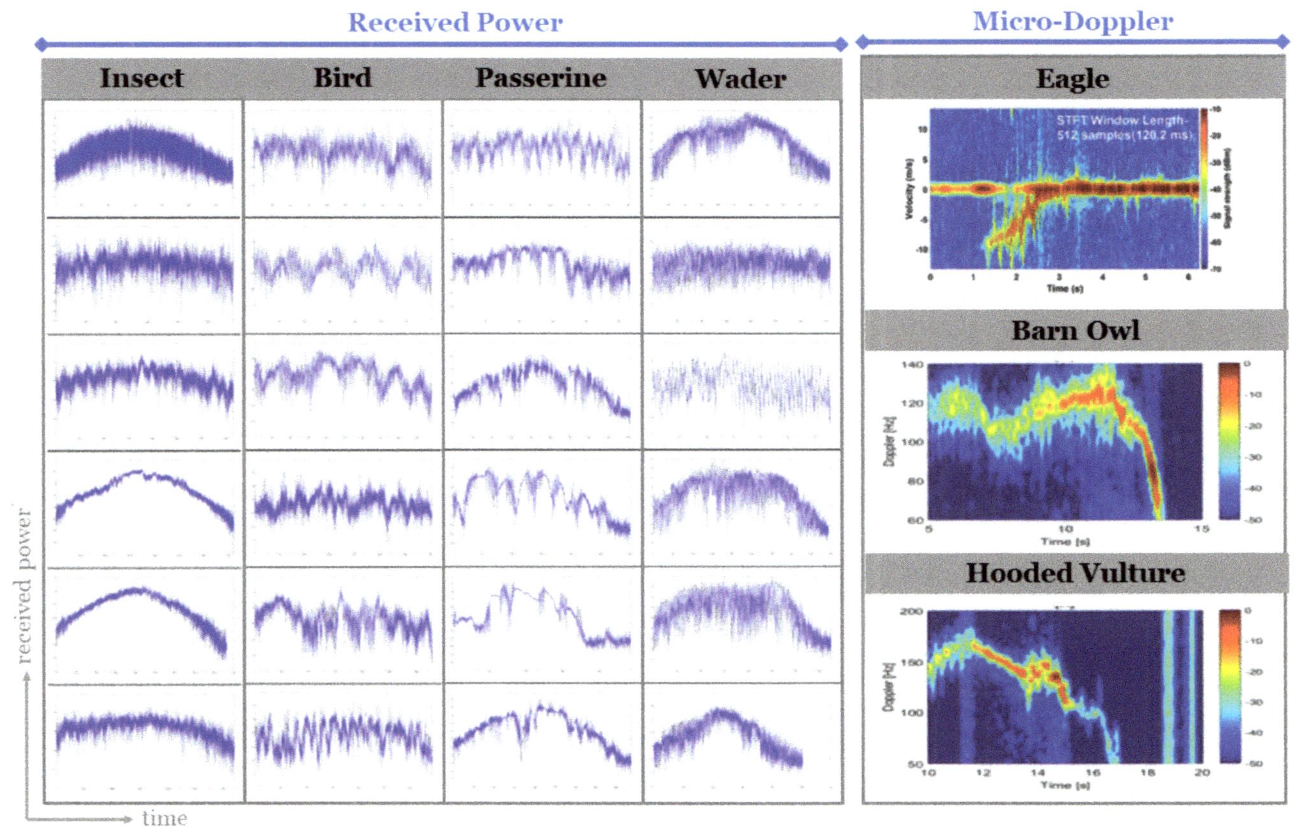

Figure 11.6 Example bird radar signatures [168,172,173]

NEXRAD encompasses most of North America, while OPERA has coverage over much of Europe. In addition to precipitation nowcasting, meteorological radars also provide a valuable resource for the analysis of cloud formations, weather-related phenomena across different atmospheric layers, and identification of different types of precipitation. Examples in which DL has been exploited include the identification of precipitation clouds [174], detection of hail [175,176], and hydrometeor classification [177].

Meteorological radar data, however, is often "contaminated" with returns from birds, bats, or insects, which are detrimental to weather radar data products, especially radial wind products key to accurate short-term weather prediction [165]. While weather radar algorithms strive to remove biological returns from data, these same returns provide an opportunity for studies of radar aeroecology. CNNs have been exploited to study a number of biological phenomena, including MistNET [178] for measuring historical bird migration in the United States, the tracking of animal migration patterns across weather radar data [179,180], and the detection of bird roosting using NEXRAD radar data [181]. Large-scale projects, such as ENRAM (European Network for the Radar surveillance of Animal Movement, 2013–17, www.enram.eu), have aimed at exploiting the vast amount of weather radar data for aeromycological studies. Currently, BioRADAR (www.biodarproject.org) in the United Kingdom aims at leveraging machine learning to monitor insect biodiversity using weather radar. In all these ongoing research endeavors, DL is providing the big data analysis capabilities to achieve large-scale results that would have otherwise been unobtainable.

11.3 Additional resources

There are several recently published survey papers relating to the application of DL to remote sensing (in general, but which also discuss radar), such as Zhu *et al.* [182], Ball *et al.* [183], Tsagkatakis *et al.* [184]. Given the rapid pace at which DL-related research is progressing, however, the best sources for information are remote sensing–related conference venues and journals. Many SAR remote sensing papers are published in IEEE Geoscience and Remote Sensing Letters, IEEE Transactions on Geoscience and Remote Sensing, the Canadian Journal of Remote Sensing, Electronics, Geophysical Research Letters, IEEE Access, IEEE Transactions on Aerospace and Electronic Systems, IEEE Journal of Selected Topics in Applied Earth Observations and Remote Sensing, IEEE Transactions on Antennas and Propagation, IET Radar, Sonar and Navigation, the International Journal of Remote Sensing, ISPRS Journal of Photogrammetry and Remote Sensing, Remote Sensing, Remote Sensing of Environment, Sensors, and many others. Some popular conferences for SAR publications are the IEEE Radar Conference (RadarCon), IEEE Geoscience and Remote Sensing Symposium (IGARSS), and SPIE Defence and Commercial Sensing (DCS) Symposium, among others.

11.4 Concluding remarks

There is much research that remains to be addressed in regards to the application of DL to remote sensing. DNNs have paved the way for significant advances in accuracy for a broad range of earth science applications. But, a DNN with millions of parameters is impossible for us to understand fully. Remote sensing analysts and researchers not only want a good answer: they want to know how it works. This need has driven research in explainable AI, or AI that explains its decisions.

Additionally, there is an urgent need to develop ways to deal with the limited amount of data available. Radar remote sensing data is hard and expensive to collect. Moreover, datasets, such as MSTAR, are saturated and not difficult enough for more modern DL approaches. Datasets with hard cases and large variety of data are needed, not just for training, but also for benchmarking and comparison. Moreover, data needs to be made publicly available, possibly targeted and specifically designed in the context of challenge problems.

The computer vision community, in general, does a great job at sharing data, as well as providing codes and their papers often compare five to ten other state-of-the-art systems, while many (not all) remote sensing and radar papers only compare one to two methods (possibly because implementing the comparison methods is very difficult without original codes). The radar community and remote sensing community could do a much better job of providing codes with new methods, allowing researchers to more easily replicate results.

When real data for training is limited, however, synthetic data generation may be able to compensate for the deficiency. Methods such as GANs or medium- or high-fidelity modeling can potentially fill the gap, as might transfer learning.

Models are powerful tools that embody the constraints imposed by physics. They should not be tossed aside to leave researchers reliant on DNNs for figuring everything out using massive amounts of data—data that is limited in remote sensing. Physics-aware, hybrid approaches to DL that integrate the benefits of data-driven learning with model-based knowledge can help us reap the benefits of both approaches. This will especially become significant when constructing models to collaboratively utilize disparate data sources, which have different phenomenology but are related through observation of the same scenes or events, usually along the same timeline. Novel DNN architectures can be devised that exploit the interrelations between multimodal, multi-temporal datasets for improved performance.

Related to this challenge is that current DNN architectures are not well suited to learn from radar remote sensing data because they are real valued. Yet, most radar data and imagery are inherently complex valued. Investigations into the design of complex-valued DNNs [185] is one area that holds the potential for immediate performance improvements, simply by better matching the network to the underlying structure of the input data.

DNNs have already broadened the realm of possibilities in remote sensing. As novel architectures evolve to address current challenges, it will be exciting to see applications advance beyond even what we may dream of today.

References

[1] Ross WD, Waxman AM, Streilein WW, *et al.* Multi-sensor 3D image fusion and interactive search. In: *Proceedings of the Third International Conference on Information Fusion*, Paris, France. vol. 1; 2000. p. TUC3/10–TUC3/17.

[2] Aksu R, Rahman MM, and Gurbuz SZ. 3D scene reconstruction from multi-sensor EO-SAR data. In: *Proceedings of the SPIE, Defense+Commercial Sensing Symposium*; 2020.

[3] Park J, Javier RJ, Moon T, *et al.* Micro-Doppler based classification of human aquatic activities via transfer learning of convolutional neural networks. In: *Sensors*; 2016;16(12).

[4] Pan Z, Yu W, Yi X, *et al.* Recent progress on generative adversarial networks (GANs): A survey. *IEEE Access.* 2019;7:36322–36333.

[5] Zhu S, Huang C, Su Y, *et al.* 3D Ground penetrating radar to detect tree roots and estimate root biomass in the field. *Remote Sensing.* 2014;6(6):5754–5773. Available from: http://dx.doi.org/10.3390/rs6065754.

[6] Zheng J, Teng X, Liu J, *et al.* Convolutional neural networks for water content classification and prediction with ground penetrating radar. *IEEE Access.* 2019;7:185385–185392.

[7] Guyon I and Elisseeff A. An introduction to variable and feature selection. *Journal of Machine Learning Research.* 2003;3(null):1157–1182.

[8] Camps-Valls G, Mooij J, and Scholkopf B. Remote sensing feature selection by kernel dependence measures. *IEEE Geoscience and Remote Sensing Letters.* 2010;7(3):587–591.

[9] Tekeli B, Gurbuz SZ, and Yuksel M. Information-theoretic feature selection for human micro-Doppler signature classification. *IEEE Transactions on Geoscience and Remote Sensing.* 2016;54(5):2749–2762.

[10] Hariharan S, Mandal D, Tirodkar S, *et al.* A novel phenology based feature subset selection technique using random forest for multitemporal PolSAR crop classification. *IEEE Journal of Selected Topics in Applied Earth Observations and Remote Sensing.* 2018;11(11):4244–4258.

[11] Saeys Y, Inza I, and Larrañaga P. A review of feature selection techniques in bioinformatics. *Bioinformatics.* 2007;23(19):2507–2517. Available from: https://doi.org/10.1093/bioinformatics/btm344.

[12] Zeiler MD and Fergus R. Visualizing and understanding convolutional networks. CoRR. 2013;abs/1311.2901. Available from: http://arxiv.org/abs/1311.2901.

[13] Erhan D, Bengio Y, Courville AC, *et al.* Visualizing Higher-Layer Features of a Deep Network. Also presented at the *ICML 2009 Workshop on Learning Feature Hierarchies*, Montreal, Canada; 2009.

[14] Simonyan K, Vedaldi A, and Zisserman A. Deep inside convolutional networks: Visualising image classification models and saliency maps. CoRR. 2013;abs/1312.6034.

[15] Murdoch WJ, Singh C, Kumbier K, *et al.* Definitions, methods, and applications in interpretable machine learning. *Proceedings of the National Academy of Sciences.* 2019;116(44):22071–22080.

[16] Doshi-Velez F and Kim B. Towards A Rigorous Science of Interpretable Machine Learning; 2017. arXiv1702.08608.

[17] Bach S, Binder A, Montavon G, *et al.* On Pixel-Wise Explanations for Non-Linear Classifier Decisions by Layer-Wise Relevance Propagation; 2015. Available from: http://dx.doi.org/10.17169/refubium-18639.

[18] Seyfioglu MS and Gurbuz SZ. Deep neural network initialization methods for micro-Doppler classification with low training sample support. *IEEE Geoscience and Remote Sensing Letters.* 2017;14(12):2462–2466.

[19] Reichstein M, Camps-Valls G, Stevens B, *et al.* Deep learning and process understanding for data-driven Earth system science. *Nature.* 2019;566: 195–204.

[20] Ienco D, Gaetano R, Dupaquier C, *et al.* Land cover classification via multitemporal spatial data by deep recurrent neural networks. *IEEE Geoscience and Remote Sensing Letters.* 2017;14(10):1685–1689.

[21] Sun Z, Di L, and Fang H. Using long short-term memory recurrent neural network in land cover classification on Landsat and Cropland data layer time series. *International Journal of Remote Sensing.* 2019;40(2): 593–614.

[22] Mou L, Ghamisi P, and Zhu XX. Deep recurrent neural networks for hyperspectral image classification. *IEEE Transactions on Geoscience and Remote Sensing.* 2017;55(7):3639–3655.

[23] Ho Tong Minh D, Ienco D, Gaetano R, *et al.* Deep recurrent neural networks for winter vegetation quality mapping via multitemporal SAR Sentinel-1. *IEEE Geoscience and Remote Sensing Letters.* 2018;15(3):464–468.

[24] Ndikumana E, Baghdadi N, Dominique C, *et al.* Deep recurrent neural network for agricultural classification using multitemporal SAR Sentinel-1 for Camargue, France. *Remote Sensing.* 2018;10:1217.

[25] Zhou Y, Luo J, Feng L, *et al.* Long-short-term-memory-based crop classification using high-resolution optical images and multi-temporal SAR data. *GIScience & Remote Sensing.* 2019;56(8):1170–1191.

[26] Pelletier C, Webb G, and Petitjean F. Temporal convolutional neural network for the classification of satellite image time series. *Remote Sensing.* 2019;11:523.

[27] Mou L, Bruzzone L, and Zhu XX. Learning spectral-spatial-temporal features via a recurrent convolutional neural network for change detection in multispectral imagery. *IEEE Transactions on Geoscience and Remote Sensing.* 2019;57(2):924–935.

[28] Ji S, Chi Z, Xu A, *et al.* 3D convolutional neural networks for crop classification with multi-temporal remote sensing images. *Remote Sensing.* 2018;10:75.

[29] Zhong L, Hu L, and Zhou H. Deep learning based multi-temporal crop classification. *Remote Sensing of Environment.* 2018;221:430–443.

[30] Danilla C, Persello C, Tolpekin V, *et al.* Classification of multitemporal SAR images using convolutional neural networks and Markov random fields. In: *2017 IEEE International Geoscience and Remote Sensing Symposium (IGARSS)*; 2017. p. 2231–2234.

[31] Chamorro Martinez J, Nigri Happ P, Castro J, *et al.* A Comparison Of Fully Convolutional And Recurrent Networks For Multi-Temporal Crop Recognition Using SAR Images. In *Proc. XIX Brazilian Symposium on Remote Sensing*, Santos, Brazil; 2019(19).

[32] Mullissa AG, Persello C, and Tolpekin V. Fully convolutional networks for multi-temporal SAR image classification. In: *IGARSS 2018 – 2018 IEEE International Geoscience and Remote Sensing Symposium*, Valencia, Spain; 2018. p. 6635–6638.

[33] Chi M, Plaza A, Benediktsson JA, *et al.* Big data for remote sensing: Challenges and opportunities. *Proceedings of the IEEE*. 2016;104(11):2207–2219.

[34] Laney D. 3D Data Management: Controlling Data Volume, Velocity, and Variety. META Group; 2001. Available from: http://blogs.gartner.com/doug-laney/files/2012/01/ad949-3D-Data-Management-Controlling-Data-Volume-Velocity-and-Variety.pdf.

[35] Ma Y, Wu H, Wang L, *et al.* Remote sensing big data computing: Challenges and opportunities. *Future Generation Computer Systems*. 2015;51:47–60. Special Section: A Note on New Trends in Data-Aware Scheduling and Resource Provisioning in Modern HPC Systems. Available from: http://www.sciencedirect.com/science/article/pii/S0167739X14002234.

[36] He K, Zhang X, Ren S, *et al.* Deep residual learning for image recognition. CoRR. 2015;abs/1512.03385. Available from: http://arxiv.org/abs/1512.03385.

[37] Srivastava RK, Greff K, and Schmidhuber J. Highway networks. CoRR. 2015;abs/1505.00387. Available from: http://arxiv.org/abs/1505.00387.

[38] Srivastava RK, Greff K, and Schmidhuber J. Training very deep networks. CoRR. 2015;abs/1507.06228. Available from: http://arxiv.org/abs/1507.06228.

[39] Greff K, Srivastava RK, and Schmidhuber J. Highway and residual networks learn unrolled iterative estimation. CoRR. 2016;abs/1612.07771. Available from: http://arxiv.org/abs/1612.07771.

[40] Zilly JG, Srivastava RK, Koutník J, *et al.* Recurrent highway networks. CoRR. 2016;abs/1607.03474. Available from: http://arxiv.org/abs/1607.03474.

[41] Huang G, Liu Z, and Weinberger KQ. Densely connected convolutional networks. CoRR. 2016;abs/1608.06993. Available from: http://arxiv.org/abs/1608.06993.

[42] Xu S, Qiu X, Wang C, *et al.* Desnet: Deep residual networks for descalloping of ScanSAR images. In: *IGARSS 2018 – 2018 IEEE International Geoscience and Remote Sensing Symposium*, Valencia, Spain; 2018. p. 8929–8932.

[43] Li L, Wang C, Zhang H, *et al.* Residual Unet for urban building change detection with Sentinel-1 SAR data. In: *IGARSS 2019 – 2019 IEEE International Geoscience and Remote Sensing Symposium*, Yokohama, Japan; 2019. p. 1498–1501.

[44] Gu F, Zhang H, Wang C, *et al.* Residual encoder-decoder network introduced for multisource SAR image despeckling. In: *2017 SAR in Big Data Era:*

Models, Methods and Applications (BIGSARDATA), Beijing, China; 2017. p. 1–5.
[45] Guo Y, Chen E, Li Z, *et al.* Convolutional highway unit network for large-scale classification with GF-3 dual-Pol SAR data. In: *IGARSS 2018 – 2018 IEEE International Geoscience and Remote Sensing Symposium*, Valencia, Spain; 2018. p. 2424–2427.
[46] Lin Z, Ji K, Kang M, *et al.* Deep convolutional highway unit network for SAR target classification with limited labeled training data. *IEEE Geoscience and Remote Sensing Letters*. 2017;14(7):1091–1095.
[47] Dong H, Zhang L, and Zou B. Densely connected convolutional neural network based polarimetric SAR image classification. In: *IGARSS 2019 – 2019 IEEE International Geoscience and Remote Sensing Symposium*, Yokohama, Japan; 2019. p. 3764–3767.
[48] Yaohua X and Xudong M. A SAR oil spill image recognition method based on DenseNet convolutional neural network. In: *2019 International Conference on Robots Intelligent System (ICRIS)*, Haikou, China; 2019. p. 78–81.
[49] Ender J. A brief review of compressive sensing applied to radar. *Proceedings International Radar Symposium.* 2013;1:3–16.
[50] Gurbuz AC, McClellan J, and Scott Jr WR. Compressive sensing for subsurface imaging using ground penetrating radar. *Signal Processing.* 2009 10;89:1959–1972.
[51] Gurbuz AC, McClellan JH, and Scott WR. A compressive sensing data acquisition and imaging method for stepped frequency GPRs. *IEEE Transactions on Signal Processing.* 2009;57(7):2640–2650.
[52] Qu L, Yin Y, Sun Y, *et al.* Diffraction tomographic ground-penetrating radar multibistatic imaging algorithm with compressive frequency measurements. *IEEE Geoscience and Remote Sensing Letters*. 2015;12(10):2011–2015.
[53] Xu D, Du L, Liu H, *et al.* Compressive sensing of stepped-frequency radar based on transfer learning. *IEEE Transactions on Signal Processing.* 2015;63(12):3076–3087.
[54] Zhu XX and Bamler R. Super-resolution power and robustness of compressive sensing for spectral estimation with application to spaceborne tomographic SAR. *IEEE Transactions on Geoscience and Remote Sensing.* 2012;50(1):247–258.
[55] Weiss Y, Chang HS, and Freeman WT. Learning Compressed Sensing. In *Proceedings of Allerton Conference on Communication, Control, and Computing*, Monticello, IL; 2007.
[56] Duarte-Carvajalino JM and Sapiro G. Learning to sense sparse signals: Simultaneous sensing matrix and sparsifying dictionary optimization. *IEEE Transactions on Image Processing.* 2009;18(7):1395–1408.
[57] Rafi RHM and Gurbuz AC. Learning to sense and reconstruct a class of signals. In: *2019 IEEE Radar Conference (RadarConf)*, Boston, MA; 2019. p. 1–5.
[58] Yao H, Dai F, Zhang D, *et al.* DR2-Net: Deep residual reconstruction network for image compressive sensing. *Neurocomputing.* 2019;359:483–493.

[59] Álvarez Y, Rodriguez-Vaqueiro Y, Gonzalez-Valdes B, *et al.* Three-dimensional compressed sensing-based millimeter-wave imaging. *IEEE Transactions on Antennas and Propagation.* 2015;63(12):5868–5873.

[60] Wang Y and Xiang Zhu X. Earth observation using SAR and social media images. In: *2017 IEEE Conference on Computer Vision and Pattern Recognition Workshops (CVPRW)*, Honolulu, HI; 2017. p. 1580–1588.

[61] Wang Y, Kang J, and Xiang Zhu X. Fusion of SAR Interferometry and Social Media Images for 4D Urban Modeling. In *2018 21st International Conference on Information Fusion*, Cambridge, UK; 2018. pp. 1601–1606.

[62] Patole SM, Torlak M, Wang D, *et al.* Automotive radars: A review of signal processing techniques. *IEEE Signal Processing Magazine.* 2017;34(2):22–35.

[63] Van Brummelen J, O'Brien M, Gruyer D, *et al.* Autonomous vehicle perception: The technology of today and tomorrow. *Transportation Research Part C: Emerging Technologies.* 2018;89:384–406.

[64] Bagloee SA, Tavana M, Asadi M, *et al.* Autonomous vehicles: Challenges, opportunities, and future implications for transportation policies. *Journal of Modern Transportation.* 2016;24(4):284–303.

[65] Hussain R and Zeadally S. Autonomous cars: Research results, issues, and future challenges. *IEEE Communications Surveys & Tutorials.* 2018;21(2): 1275–1313.

[66] Kapoor P, Vora A, and Kang KD. Detecting and mitigating spoofing attack against an automotive radar. In: *2018 IEEE 88th Vehicular Technology Conference (VTC-Fall).* IEEE Explore, Chicago, IL, USA; 2018. p. 1–6.

[67] Murad M, Bilik I, Friesen M, *et al.* Requirements for next generation automotive radars. In: *2013 IEEE Radar Conference (RadarCon13).* IEEE Explore, Ottawa, ON; 2013. p. 1–6.

[68] Gao H, Cheng B, Wang J, *et al.* Object classification using CNN-based fusion of vision and LIDAR in autonomous vehicle environment. *IEEE Transactions on Industrial Informatics.* 2018;14(9):4224–4231.

[69] Stiller C, León FP, and Kruse M. Information fusion for automotive applications—An overview. *Information Fusion.* 2011;12(4):244–252.

[70] Kocić J, Jovičić N, and Drndarević V. Sensors and sensor fusion in autonomous vehicles. In: *2018 26th Telecommunications Forum (TELFOR).* IEEE, Belgrade; 2018. p. 420–425.

[71] Giacalone JP, Bourgeois L, and Ancora A. Challenges in aggregation of heterogeneous sensors for Autonomous Driving Systems. In: *2019 IEEE Sensors Applications Symposium (SAS).* IEEE, Sophia Antipolis, France; 2019. p. 1–5.

[72] Inggs MR and Lord RT. Applications of Satellite Imaging Radar. South African Institute of Electrical Engineers, SAIEE, South Africa; 2000.

[73] Liao M, Jiang L, Lin H, *et al.* Urban change detection based on coherence and intensity characteristics of SAR imagery. *Photogrammetric Engineering & Remote Sensing.* 2008;74(8):999–1006.

[74] Kasischke ES, Melack JM, and Dobson MC. The use of imaging radars for ecological applications: A review. *Remote Sensing of Environment.* 1997;59(2):141–156.

[75] Bindlish R and Barros AP. Parameterization of vegetation backscatter in radar-based, soil moisture estimation. *Remote Sensing of Environment.* 2001;76(1):130–137.

[76] Colliander A, Jackson TJ, Bindlish R, *et al.* Validation of SMAP surface soil moisture products with core validation sites. *Remote Sensing of Environment.* 2017;191:215–231.

[77] Rahman MM, Moran MS, Thoma DP, *et al.* Mapping surface roughness and soil moisture using multi-angle radar imagery without ancillary data. *Remote Sensing of Environment.* 2008;112(2):391–402.

[78] Oh Y. Quantitative retrieval of soil moisture content and surface roughness from multipolarized radar observations of bare soil surfaces. *IEEE Transactions on Geoscience and Remote Sensing.* 2004;42(3):596–601.

[79] Oh Y, Sarabandi K, and Ulaby FT. An inversion algorithm for retrieving soil moisture and surface roughness from polarimetric radar observation. In: *Proceedings of IGARSS'94-1994 IEEE International Geoscience and Remote Sensing Symposium.* vol. 3. IEEE; 1994. p. 1582–1584.

[80] Kim Y and van Zyl JJ. A time-series approach to estimate soil moisture using polarimetric radar data. *IEEE Transactions on Geoscience and Remote Sensing.* 2009;47(8):2519–2527.

[81] Dubois PC, Van Zyl J, and Engman T. Measuring soil moisture with imaging radars. *IEEE Transactions on Geoscience and Remote Sensing.* 1995;33(4):915–926.

[82] Moran MS, Peters-Lidard CD, Watts JM, *et al.* Estimating soil moisture at the watershed scale with satellite-based radar and land surface models. *Canadian Journal of Remote Sensing.* 2004;30(5):805–826.

[83] Shi J, Wang J, Hsu AY, *et al.* Estimation of bare surface soil moisture and surface roughness parameter using L-band SAR image data. *IEEE Transactions on Geoscience and Remote Sensing.* 1997;35(5):1254–1266.

[84] McNairn H and Brisco B. The application of C-band polarimetric SAR for agriculture: A review. *Canadian Journal of Remote Sensing.* 2004;30(3): 525–542.

[85] Lopez-Sanchez JM and Ballester-Berman JD. Potentials of polarimetric SAR interferometry for agriculture monitoring. *Radio Science.* 2009;44(02): 1–20.

[86] Lopez-Sanchez JM, Hajnsek I, and Ballester-Berman JD. First demonstration of agriculture height retrieval with PolInSAR airborne data. *IEEE Geoscience and Remote Sensing Letters.* 2011;9(2):242–246.

[87] Skriver H, Dall J, Le Toan T, *et al.* Agriculture classification using POLSAR data. In: *Proceedings of the 2nd International Workshop POLINSAR 2005 (ESA SP-586).* Editor: H. Lacoste. ESRIN, Frascati, Italy. vol. 586; 2005. p. 32.

[88] Ferrazzoli P. SAR for agriculture: Advances, problems and prospects. In: *Proceedings of 3rd International Symposium Retrieval of Bio-and Geo-Physical Parameters from SAR Data for Land Applications*, Sheffield, UK. vol. 475; 2002. p. 47–56.

[89] Abdelfattah R and Nicolas JM. Topographic SAR interferometry formulation for high-precision DEM generation. *IEEE Transactions on Geoscience and Remote Sensing.* 2002;40(11):2415–2426.

[90] Shimada M. Ortho-rectification and slope correction of SAR data using DEM and its accuracy evaluation. *IEEE Journal of Selected Topics in Applied Earth Observations and Remote Sensing.* 2010;3(4):657–671.

[91] Rao Y, Venkataraman G, Rao K, *et al.* SAR interferometry for DEM generation and movement of Indian glaciers. In: IGARSS 2004. *2004 IEEE International Geoscience and Remote Sensing Symposium*, Anchorage, AK. vol. 2. IEEE; 2004. p. 1128–1131.

[92] Ferraiuolo G, Meglio F, Pascazio V, *et al.* DEM reconstruction accuracy in multichannel SAR interferometry. *IEEE Transactions on Geoscience and Remote Sensing.* 2008;47(1):191–201.

[93] Ferretti A, Prati C, and Rocca F. Multibaseline InSAR DEM reconstruction: The wavelet approach. *IEEE Transactions on Geoscience and Remote Sensing.* 1999;37(2):705–715.

[94] Siegert F and Hoffmann AA. The 1998 forest fires in East Kalimantan (Indonesia): A quantitative evaluation using high resolution, multitemporal ERS-2 SAR images and NOAA-AVHRR hotspot data. *Remote Sensing of Environment.* 2000;72(1):64–77.

[95] Ruecker G and Siegert F. Burn scar mapping and fire damage assessment using ERS-2 SAR images in East Kalimantan, Indonesia. *International Archives of Photogrammetry and Remote Sensing.* 2000;33(B7/3; PART 7):1286–1293.

[96] Nico G, Pappalepore M, Pasquariello G, *et al.* Comparison of SAR amplitude vs. coherence flood detection methods-a GIS application. *International Journal of Remote Sensing.* 2000;21(8):1619–1631.

[97] Schlaffer S, Matgen P, Hollaus M, *et al.* Flood detection from multi-temporal SAR data using harmonic analysis and change detection. *International Journal of Applied Earth Observation and Geoinformation.* 2015;38:15–24.

[98] Martinis S, Twele A, and Voigt S. Towards operational near real-time flood detection using a split-based automatic thresholding procedure on high resolution TerraSAR-X data. *Natural Hazards and Earth System Sciences.* 2009;9(2):303–314.

[99] Matgen P, Hostache R, Schumann G, *et al.* Towards an automated SAR-based flood monitoring system: Lessons learned from two case studies. *Physics and Chemistry of the Earth, Parts A/B/C.* 2011;36(7–8):241–252.

[100] Townsend P. Relationships between forest structure and the detection of flood inundation in forested wetlands using C-band SAR. *International Journal of Remote Sensing.* 2002;23(3):443–460.

[101] Martinis S, Kuenzer C, Wendleder A, *et al.* Comparing four operational SAR-based water and flood detection approaches. *International Journal of Remote Sensing.* 2015;36(13):3519–3543.

[102] Michel R, Avouac JP, and Taboury J. Measuring ground displacements from SAR amplitude images: Application to the Landers earthquake. *Geophysical Research Letters.* 1999;26(7):875–878.

[103] Pathier E, Fielding E, Wright T, *et al.* Displacement field and slip distribution of the 2005 Kashmir earthquake from SAR imagery. *Geophysical Research Letters*. 2006;33(20).
[104] Fujiwara S, Nishimura T, Murakami M, *et al.* 2.5-D surface deformation of M6. 1 earthquake near Mt Iwate detected by SAR interferometry. *Geophysical Research Letters*. 2000;27(14):2049–2052.
[105] van der Sanden JJ. *Radar Remote Sensing to Support Tropical Forest Management*. Wageningen Agricultural University; 1997.
[106] Angelov A, Robertson A, Murray-Smith R, *et al.* Practical classification of different moving targets using automotive radar and deep neural networks. *IET Radar, Sonar & Navigation*. 2018;12(10):1082–1089.
[107] Wheeler TA, Holder M, Winner H, *et al.* Deep stochastic radar models. In: *2017 IEEE Intelligent Vehicles Symposium (IV)*, Los Angeles, CA. IEEE; 2017. p. 47–53.
[108] Kim S, Lee S, Doo S, *et al.* Moving target classification in automotive radar systems using convolutional recurrent neural networks. In: *2018 26th European Signal Processing Conference (EUSIPCO)*. IEEE, Rome; 2018. p. 1482–1486.
[109] Seyfioğlu MS, Özbayoğlu AM, and Gürbüz SZ. Deep convolutional autoencoder for radar-based classification of similar aided and unaided human activities. *IEEE Transactions on Aerospace and Electronic Systems*. 2018;54(4):1709–1723.
[110] Patel K, Rambach K, Visentin T, *et al.* Deep learning-based object classification on automotive radar spectra. In: *2019 IEEE Radar Conference (RadarConf)*. IEEE, Boston, MA, USA; 2019. p. 1–6.
[111] Lombacher J, Hahn M, Dickmann J, *et al.* Potential of radar for static object classification using deep learning methods. In: *2016 IEEE MTT-S International Conference on Microwaves for Intelligent Mobility (ICMIM)*. IEEE, San Diego, CA; 2016. p. 1–4.
[112] Brodeski D, Bilik I, and Giryes R. Deep radar detector. In: *2019 IEEE Radar Conference (RadarConf)*. IEEE, Boston, MA; 2019. p. 1–6.
[113] Mun J, Kim H, and Lee J. A deep learning approach for automotive radar interference mitigation. In: *2018 IEEE 88th Vehicular Technology Conference (VTC-Fall)*. IEEE, Chicago, IL, USA; 2018. p. 1–5.
[114] Grey W and Luckman A. Mapping urban extent using satellite radar interferometry. *Photogrammetric Engineering & Remote Sensing*. 2003;69(9): 957–961.
[115] Kasantikul K, Yang D, Wang Q, *et al.* A novel wind speed estimation based on the integration of an artificial neural network and a particle filter using BeiDou GEO reflectometry. *Sensors*. 2018;18(10):3350.
[116] Liu Y, Collett I, and Morton YJ. Application of neural network to GNSS-R wind speed retrieval. *IEEE Transactions on Geoscience and Remote Sensing*. 2019;57(12):9756–9766.
[117] Reynolds J, Clarizia MP, and Santi E. Wind speed estimation from CYGNSS using artificial neural networks. *IEEE Journal of Selected Topics in Applied Earth Observations and Remote Sensing*. 2020;13:708–716.

[118] Wang F, Yang D, Zhang B, *et al.* Waveform-based spaceborne GNSS-R wind speed observation: Demonstration and analysis using UK TechDemoSat-1 data. *Advances in Space Research.* 2018;61(6):1573–1587.
[119] Eroglu O, Kurum M, Boyd D, *et al.* High spatio-temporal resolution CYGNSS soil moisture estimates using artificial neural networks. *Remote Sensing.* 2019;11(19):2272.
[120] Senyurek V, Lei F, Boyd D, *et al.* Machine learning-based CYGNSS soil moisture estimates over ISMN sites in CONUS. *Remote Sensing.* 2020;12(7):1168.
[121] Rodriguez-Alvarez N, Podest E, Jensen K, *et al.* Classifying inundation in a tropical wetlands complex with GNSS-R. *Remote Sensing.* 2019;11(9):1053.
[122] Yan Q and Huang W. Sea ice sensing from GNSS-R data using convolutional neural networks. *IEEE Geoscience and Remote Sensing Letters.* 2018;15(10):1510–1514.
[123] Yan Q, Huang W, and Moloney C. Neural networks based sea ice detection and concentration retrieval from GNSS-R delay-Doppler maps. *IEEE Journal of Selected Topics in Applied Earth Observations and Remote Sensing.* 2017;10(8):3789–3798.
[124] Yan Q and Huang W. Detecting sea ice from TechDemoSat-1 data using support vector machines with feature selection. *IEEE Journal of Selected Topics in Applied Earth Observations and Remote Sensing.* 2019;12(5):1409–1416.
[125] Yan Q and Huang W. Sea ice remote sensing using GNSS-R: A review. *Remote Sensing.* 2019;11(21):2565.
[126] Toth C and Jóźków G. Remote sensing platforms and sensors: A survey. *ISPRS Journal of Photogrammetry and Remote Sensing.* 2016;115:22–36.
[127] Kussul N, Lavreniuk M, Skakun S, *et al.* Deep learning classification of land cover and crop types using remote sensing data. *IEEE Geoscience and Remote Sensing Letters.* 2017;14(5):778–782.
[128] Niu X and Ban Y. An adaptive contextual SEM algorithm for urban land cover mapping using multitemporal high-resolution polarimetric SAR data. *IEEE Journal of Selected Topics in Applied Earth Observations and Remote Sensing.* 2012;5(4):1129–1139.
[129] Lv Q, Dou Y, Niu X, *et al.* Urban land use and land cover classification using remotely sensed SAR data through deep belief networks. *Journal of Sensors.* 2015;2015:1–10.
[130] Hinton GE, Osindero S, and Teh Y. A fast learning algorithm for deep belief nets. *Neural Computation.* 2006;18(7):1527–1554.
[131] Geng J, Fan J, Wang H, *et al.* High-resolution SAR image classification via deep convolutional autoencoders. *IEEE Geoscience and Remote Sensing Letters.* 2015;12(11):2351–2355.
[132] Gong M, Zhao J, Liu J, *et al.* Change detection in synthetic aperture radar images based on deep neural networks. *IEEE Transactions on Neural Networks and Learning Systems.* 2016;27(1):125–138. Available from: http://ieeexplore.ieee.org/document/7120131/.

[133] Liu J, Gong M, Qin K, *et al.* A deep convolutional coupling network for change detection based on heterogeneous optical and radar images. *IEEE Transactions on Neural Networks and Learning Systems.* 2018;29(3): 545–559.

[134] Gambardella A, Nunziata F, and Migliaccio M. A physical full-resolution SAR ship detection filter. *IEEE Geoscience and Remote Sensing Letters.* 2008;5(4):760–763.

[135] Marino A, Walker N, and Woodhouse I. Ship detection with RADARSAT-2 quad-pol SAR data using a notch filter based on perturbation analysis. In: *2010 IEEE International Geoscience and Remote Sensing Symposium. IEEE*, Honolulu, HI; 2010. p. 3704–3707.

[136] Stastny J, Hughes M, Garcia D, *et al.* A novel adaptive synthetic aperture radar ship detection system. In: OCEANS'11 MTS/IEEE KONA. IEEE, Waikoloa, HI; 2011. p. 1–7.

[137] Leng X, Ji K, Yang K, *et al.* A bilateral CFAR algorithm for ship detection in SAR images. *IEEE Geoscience and Remote Sensing Letters.* 2015;12(7):1536–1540.

[138] Xu F, Wang H, Song Q, *et al.* Intelligent ship recognition from synthetic aperture radar images. In: *IGARSS 2018-2018 IEEE International Geoscience and Remote Sensing Symposium.* IEEE, Valencia, Spain; 2018. p. 4387–4390.

[139] Gao G. A Parzen-window-kernel-based CFAR algorithm for ship detection in SAR images. *IEEE Geoscience and Remote Sensing Letters.* 2010;8(3):557–561.

[140] Ai J, Qi X, Yu W, *et al.* A new CFAR ship detection algorithm based on 2-D joint log-normal distribution in SAR images. *IEEE Geoscience and Remote Sensing Letters.* 2010;7(4):806–810.

[141] Wang C, Liao M, and Li X. Ship detection in SAR image based on the alpha-stable distribution. *Sensors.* 2008;8(8):4948–4960.

[142] Chen J, Chen Y, and Yang J. Ship detection using polarization cross-entropy. *IEEE Geoscience and Remote Sensing Letters.* 2009;6(4):723–727.

[143] Shirvany R, Chabert M, and Tourneret JY. Ship and oil-spill detection using the degree of polarization in linear and hybrid/compact dual-pol SAR. *IEEE Journal of Selected Topics in Applied Earth Observations and Remote Sensing.* 2012;5(3):885–892.

[144] An W, Xie C, and Yuan X. An improved iterative censoring scheme for CFAR ship detection with SAR imagery. *IEEE Transactions on Geoscience and Remote Sensing.* 2013;52(8):4585–4595.

[145] Zhan C, Zhang L, Zhong Z, *et al.* Deep learning approach in automatic iceberg – Ship detection with SAR remote sensing data. CoRR. 2018;abs/1812.07367. Available from: http://arxiv.org/abs/1812.07367.

[146] Li J, Qu C, and Shao J. Ship detection in SAR images based on an improved faster R-CNN. In: 2017 SAR in Big Data Era: Models, Methods and Applications (BIGSARDATA), IEEE, Beijing, China; 2017. p. 104–107.

[147] Liu Y, Zhang M, Xu P, *et al.* SAR ship detection using sea-land segmentation-based convolutional neural network. In: *2017 International Workshop on Remote Sensing With Intelligent Processing (RSIP)*; 2017. p. 1–4.

[148] Schwegmann CP, Kleynhans W, Salmon BP, *et al.* Very deep learning for ship discrimination in synthetic aperture radar imagery. In: *2016 IEEE International Geoscience and Remote Sensing Symposium (IGARSS)*; 2016. p. 104–107.

[149] Shen R, Huang A, Li B, *et al.* Construction of a drought monitoring model using deep learning based on multi-source remote sensing data. *International Journal of Applied Earth Observation and Geoinformation.* 2019;79:48–57.

[150] Shi X, Chen Z, Wang H, *et al.* Convolutional LSTM network: A machine learning approach for precipitation nowcasting. CoRR. 2015;abs/1506.04214. Available from: http://arxiv.org/abs/1506.04214.

[151] Shi X, Gao Z, Lausen L, *et al.* Deep learning for precipitation nowcasting: A benchmark and a new model. In: Guyon I, Luxburg UV, Bengio S, *et al.*, editors. Advances in Neural Information Processing Systems 30. Curran Associates, Inc.; 2017. p. 5617–5627. Available from: http://papers.nips.cc/paper/7145-deep-learning-for-precipitation-nowcasting-a-benchmark-and-a-new-model.pdf.

[152] Braakmann-Folgmann A and Donlon C. Estimating snow depth on arctic sea ice using satellite microwave radiometry and a neural network. *The Cryosphere Discussions.* 2019:1–26.

[153] Leahy S. Greenland's ice is melting four times faster than thought – What it means. In: National Geographic; 2019. Retrieved from https://www.nationalgeographic.com/environment/2019/01/greeland-ice-melting-four-times-faster-than-thought-raising-sea-level/.

[154] Ferro A and Bruzzone L. Analysis of radar sounder signals for the automatic detection and characterization of subsurface features. *IEEE Transactions on Geoscience and Remote Sensing.* 2012;50(11):4333–4348.

[155] Gifford CM, Finyom G, Jefferson M, *et al.* Automated polar ice thickness estimation from radar imagery. *IEEE Transactions on Image Processing.* 2010;19(9):2456–2469.

[156] Rahnemoonfar M, Fox GC, Yari M, *et al.* Automatic ice surface and bottom boundaries estimation in radar imagery based on level-set approach. *IEEE Transactions on Geoscience and Remote Sensing.* 2017;55(9):5115–5122.

[157] Crandall DJ, Fox GC, and Paden JD. Layer-finding in radar echograms using probabilistic graphical models. In: *Proceedings of the 21st International Conference on Pattern Recognition (ICPR2012)*, Tsukuba, Japan; 2012. p. 1530–1533.

[158] Xie S and Tu Z. Holistically-nested edge detection. CoRR. 2015;abs/1504.06375. Available from: http://arxiv.org/abs/1504.06375.

[159] El-Sayed M, Estaitia Y, and Khafagy M. Edge detection using convolutional neural network. *International Journal of Advanced Computer Science and Applications.* 2013;4:11–17.

[160] Liu Y, Cheng M, Hu X, *et al.* Richer convolutional features for edge detection. CoRR. 2016;abs/1612.02103. Available from: http://arxiv.org/abs/1612.02103.

[161] Kamangir H, Rahnemoonfar M, Dobbs D, *et al.* Deep hybrid wavelet network for ice boundary detection in radar imagery. In: *IGARSS 2018 – 2018 IEEE*

International Geoscience and Remote Sensing Symposium, Valencia, Spain; 2018. p. 3449–3452.

[162] Yari M, Rahnemoonfar M, Paden J, Oluwanisola, LK and Montgomery L. Smart Tracking of Internal Layers of Ice in Radar Data via Multi-Scale Learning 2019 IEEE International Conference on Big Data (Big Data), Los Angeles, CA, USA; 2019. p. 5462–5468.

[163] Rahnemoonfar M, Johnson J, and Paden J. AI radar sensor: Creating radar depth sounder images based on generative adversarial network. *Sensors* (Basel, Switzerland). 2019;19.

[164] Zhu J, Park T, Isola P, *et al.* Unpaired image-to-image translation using cycle-consistent adversarial networks. CoRR. 2017;abs/1703.10593. Available from: http://arxiv.org/abs/1703.10593.

[165] Gürbüz SZ, Reynolds DR, Koistinen J, *et al.* Exploring the skies: Technological challenges in radar aeroecology. In: *2015 IEEE Radar Conference (RadarCon)*, Washington, DC; 2015. p. 0817–0822.

[166] Hüppop O, Ciach M, Diehl R, *et al.* Perspectives and challenges for the use of radar in biological conservation. *Ecography.* 2019;42(5):912–930. Available from: https://onlinelibrary.wiley.com/doi/abs/10.1111/ecog.04063.

[167] Özcan AH, Baykut S, Şahinkaya DSA, *et al.* Micro-Doppler effect analysis of single bird and bird flock for linear FMCW radar. In: *2012 20th Signal Processing and Communications Applications Conference (SIU)*, IEEE, Mugla, Turkey; 2012. p. 1–4.

[168] Gurbuz SZ, Seyfioglu MS, and Liechti F. Identification of Biological Signals with Radar. EU ENRAM Short-Term Scientific Mission Technical Report, Swiss Ornithological Institute (Vogelwarte), Sempach, Switzerland; 2015.

[169] Zaugg S, Saporta G, Loon E, *et al.* Automatic identification of bird targets with radar via patterns produced by wing flapping. *Journal of the Royal Society, Interface/the Royal Society.* 2008;5:1041–1053.

[170] Mendis GJ, Randeny T, Wei J, and Madanayake, A. Deep learning based Doppler radar for micro UAS detection and classification. In: *MILCOM 2016 – 2016 IEEE Military Communications Conference*, Baltimore, MD. 2016; p. 924–929.

[171] Rahman S and Robertson DA. Classification of drones and birds using convolutional neural networks applied to radar micro-Doppler spectrogram images. *IET Radar, Sonar Navigation.* 2020;14(5):653–661.

[172] Ritchie M, Fioranelli F, Griffiths H, *et al.* Monostatic and bistatic radar measurements of birds and micro-drone. In: *2016 IEEE Radar Conference (RadarConf)*, Philadelphia, PA; 2016. p. 1–5.

[173] Rahman S and Robertson D. Radar micro-Doppler signatures of drones and birds at K-band and W-band. *Scientific Reports.* 2018;8.

[174] Wang H, Shao N, and Ran Y. Identification of precipitation-clouds based on the dual-polarization Doppler weather radar echoes using deep-learning method. *IEEE Access.* 2019;7:12822–12831.

[175] Pullman M, Gurung I, Maskey M, *et al.* Applying deep learning to hail detection: A case study. *IEEE Transactions on Geoscience and Remote Sensing.* 2019;57(12):10218–10225.

[176] Wang P, Lv W, Wang C, *et al.* Hail storms recognition based on convolutional neural network. In: *2018 13th World Congress on Intelligent Control and Automation (WCICA)*, IEEE, Changsha, China; 2018. p. 1703–1708.
[177] Lu Y and Kumar J. Convolutional neural networks for hydrometeor classification using dual polarization Doppler radars. In: *2019 International Conference on Data Mining Workshops (ICDMW)*, IEEE, Beijing, China; 2019. p. 288–295.
[178] Lin TY, Winner K, Bernstein G, *et al.* MistNet: Measuring historical bird migration in the US using archived weather radar data and convolutional neural networks. *Methods in Ecology and Evolution*. 2019.
[179] Hu CH, Li S, Wang R, *et al.* Extracting animal migration pattern from weather radar observation based on deep convolutional neural networks. *The Journal of Engineering*. 2019;2019(4):6541–6545. Available from: https://digital-library.theiet.org/content/journals/10.1049/joe.2019.0041.
[180] Cui K, Hu C, Wang R, *et al.* Deep-learning-based extraction of the animal migration patterns from weather radar images. *Science China Information Sciences*. 2020;63:140304.
[181] Chilson C, Avery K, Mcgovern A, *et al.* Automated detection of bird roosts using NEXRAD radar data and Convolutional neural networks. *Remote Sensing in Ecology and Conservation*. 2018.
[182] Zhu XX, Tuia D, Mou L, *et al.* Deep learning in remote sensing: A comprehensive review and list of resources. *IEEE Geoscience and Remote Sensing Magazine*. 2017;5(4):8–36.
[183] Ball JE, Anderson DT, and Chan CS. Comprehensive survey of deep learning in remote sensing: Theories, tools, and challenges for the community. *Journal of Applied Remote Sensing*. 2017;11(4):042609-1–042609-54.
[184] Tsagkatakis G, Aidini A, Fotiadou K, *et al.* Survey of deep-learning approaches for remote sensing observation enhancement. *Sensors*; 2019.
[185] Trabelsi C, Bilaniuk O, Serdyuk D, *et al.* Deep complex networks. CoRR. 2017;abs/1705.09792. Available from: http://arxiv.org/abs/1705.09792.

Epilogue: looking toward the future

Sevgi Zubeyde Gurbuz[1]

This book has introduced technical background, fundamental challenges, and applications of deep learning to radar applications. As such, we hope that it proves to be a text that is accessible to all researchers interested in this field, whether that person be a radar engineer, computer scientist, mathematician, scientist who uses radar remote sensing data, data analyst, hardware designer, or student seeking a resource to get started. Recent progress has shown that deep learning can enable significant advancements in RF sensing. But to truly unleash its potential, there are many fundamental challenges that must be addressed.

Currently, most approaches convert the RF data in real 2D or 3D representations primarily to enable exploitation of existing deep neural network (DNN) architectures that require such representations at their input. This choice is not driven by the optimization of some performance requirement but by a pragmatic and practical need for an expedient solution. RF sensors provide an extremely rich, temporally variant, complex, and high dimensional source of data, which current DNN architectures only partially exploit. Future DNN designs should be matched to optimize extraction of information from the data, else performance will be inherently bounded by the narrow prism with which current architectures are bounded. Innovations to advance complex neural networks, sequential/temporal analysis, joint domain processing, dimension reduction, sparse signal processing, compressed sensing, and RF signal understanding are needed as part of a larger view on computational intelligence.

Using human intellect as a model, the more fundamental question is how can we use deep learning as part of a greater scheme to model various layers of cognitive radar processing? While deep learning leverages massive amounts of data to learn underlying representations, how can we use domain knowledge, physics, simulation, and modeling to reduce complexity and increase robustness by teaching DNNs these constraints? The challenge of training under low sample support is just a symptom of a greater challenge of how prior knowledge can be exploited, and how new knowledge should be generated, stored, and updated. At the core of this aim lies a need to advance implementations of memory and temporal analysis via recurrent neural networks, long short-term memory, and reinforcement learning. A single network may not be enough to implement desired functionality: a more complex, dynamic system with multiple layers of low-level and high-level processing will be needed. Surely, deep learning will lie at the heart of next-generation, cognitive radar design.

[1]Department of Electrical and Computer Engineering, The University of Alabama, Tuscaloosa, AL, USA

Equally important is ensuring the robustness of radar systems and analysis that rely on DNNs. Radar systems must be able to maintain performance under adverse conditions, which are caused not just by noise and clutter but by open threats, such as jamming, and (perhaps unintentional) sources of interference, such as communications signals, and missing data. In learning from the data, we must be alert to the possibility that DNNs can be misled by manipulated data and other forms of attack. Ensuring the robustness of DNNs is a topic that requires more attention to secure the reliability of future designs.

Thus, a related practical need is that for challenging real-world datasets upon which proposed DNN designs for RF sensing can be benchmarked and compared. In part because historically RF data has been costly to acquire, may be classified or sensitive in nature, or restricted as corporate proprietary data, there are not very many publicly available datasets. The advent of software-defined radios and low cost, embedded RF sensors design for health and automotive applications has greatly increased the availability of RF data, while the culture of sharing such measured datasets is starting to change. This is evidenced by some researchers deciding to publish their datasets as companions to papers or theses. As more data becomes available, domain adaptation techniques for exploiting such "datasets of opportunity" as well as techniques for generating and adapting synthetic data will become more important.

We have also started to see DNNs applied in more and more fundamental aspects of radar signal processing, including target detection, image reconstruction, denoising, waveform design, and the proposal of DNNs for end-to-end data analysis and signal classification. The past few years have seen a tremendous increase in research activities, as special sessions, panel discussions, and workshops on machine learning, deep learning, and cognitive radar have become a regular occurrence in radar conferences. We look forward to our community growing as new ideas fusing advancements in artificial intelligence with radar signal processing and system design are brought forth. Hopefully, this book can provide a first stepping stone for many more researchers to be intrigued by the potential of deep learning for RF sensing and motivate them to contribute to the field.

Index

www.ingramcontent.com/pod-product-compliance
Lightning Source LLC
LaVergne TN
LVHW051048210926
848486LV00001B/48
9781785618529